Servitization and Physical Asset Management

Servitization and Physical Asset Management

INTERNATIONAL®

Warrendale, Pennsylvania, USA

400 Commonwealth Drive
Warrendale, PA 15096-0001 USA
E-mail: CustomerService@sae.org
Phone: 877-606-7323 (inside USA and Canada)
724-776-4970 (outside USA)

Library of Congress Catalog Number 2018948882
SAE Order Number R-479
http://dx.doi.org/10.4271/R-479

ISBN-Print 978-0-7680-9486-2
ISBN-PDF 978-0-7680-9573-9
ISBN-ePUB 978-0-7680-9488-6
ISBN-PRC 978-0-7680-9489-3
ISBN-HTML 978-0-7680-9487-9

To purchase bulk quantities, please contact: SAE Customer Service

E-mail: CustomerService@sae.org
Phone: 877-606-7323 (inside USA and Canada)
724-776-4970 (outside USA)

Visit the SAE International Bookstore at books.sae.org

contents

CHAPTER 4

Analysis 87

CHAPTER 5

Observability 99

CHAPTER 6

Time Series Analysis 109

preface

Since the mid-1980s (and before that, if you count involvement in the thermodynamic simulation and analysis of Rolls-Royce plc's large civil aircraft engines), the author has been involved in the creation and development of a number of novel engineering and business methodologies that underpin physical asset management.

This experience began in the Performance Office in Rolls-Royce plc, where initial work on extensions to the Kalman Filter as applied to aeroengine module performance and sensor bias analysis was carried out. This, together with work on applying the Kalman Filter to time series analysis, inspired the creation and development of the Rolls-Royce COMPASS™ condition monitoring system, as well as leading to the award of a PhD from Cranfield University.

Involvement in the engineering and business aspects of railway asset management during the author's employment at Data Systems and Solutions Ltd. (now part of Rolls-Royce plc) led to an opportunity to move to Bombardier Transportation, where the author led the engineering development of the rail data visualization and analysis capabilities behind the Bombardier ORBITA™ railway asset management system.

The author then moved to Intelligent Energy, based in Loughborough, where a team was set up to develop asset management capabilities for proton exchange membrane fuel cells applied to distributed power, automotive, and consumer electronics products, as well as gathering and analyzing data to support the company's research and development efforts.

The author has also applied several of the analysis and visualization techniques developed for processing aeroengine data to manage his own health, bringing his weight and blood pressure under control and increasing his general fitness and well-being.

Exposure to a good deal of academic research in this area and some of the asset management capabilities of several energy, utility, and vehicle companies at a number of seminars and conferences held over the last two decades has expanded the author's experience beyond that gained by paid employment. This knowledge has also been enhanced by a large library of books, presentations, and papers on asset management and related subjects collected over nearly five decades.

This diverse set of experiences (described in *Provost 2012*, a presentation given to several audiences over the last few years) has resulted in what is regarded by many people as a unique and wide-ranging perspective on physical asset management theory and practice, which the author has attempted in this book to distil into an informed view of useful ideas and references that can contribute to physical asset management success. The author's own material is, by necessity, heavily augmented by references to the work of many experts in this field.

It is impossible for any single individual to fully comprehend, let alone write up, all the issues surrounding asset management. *The Goal*, Eli Goldratt's best-selling novel on the theory of constraints (*Goldratt and Cox 2013*), shows how Alex Rogo, the hero of the story, is guided rather than instructed by Jonah, his mentor. Jonah says to him:

> *Alex, if I simply told you what to do, ultimately you would fail. You have to gain the understanding for yourself in order to make [things] work.*

This is definitely true when applied to physical asset management. Readers should think of this book as a guide to physical asset management knowledge and techniques, rather than a list of recipes and formulae.

Good luck, and enjoy the book!

Michael Provost
Bramcote, Nottinghamshire, UK
December 2018

acknowledgments

The author could not have done this alone; many people, too numerous to mention, at Rolls-Royce plc, Data Systems and Solutions Ltd., Cranfield University, Bombardier Transportation, and Intelligent Energy Ltd. have helped him over the last four decades.

The author has been very privileged over the years to meet a wide cross section of current and potential customers, suppliers, and competitors of Rolls-Royce, Bombardier Transportation, and Intelligent Energy during his working life, as well as other external organizations that interact with these companies. The list is a long and varied one: airlines, airframers, representatives from military and government organizations, technical committees, train builders and operators, automotive manufacturers, consultants, academics, journalists, trade show attendees, etc. from all over the world and at all levels, from the most powerful minister, chairman, and chief engineer to the most humble (but hugely important...) mechanic at the "sharp end" of the operation. The author thanks each and every one of them for providing him with new perspectives and challenges and forcing him to think beyond the narrow confines of thermodynamic and mathematical analysis.

There are a large number of very able experts in the field of asset management across the world that have created an incredible body of knowledge on all aspects, from wide-ranging business strategy to the details of effective lubrication. References to many of them appear throughout the book: thanks go to them for what they have contributed. Readers are urged to refer what they have to say directly, rather than relying on the author's summaries.

Thanks are also due to those who graciously gave permission to use their figures and other material in the book, as well as to the authors of the references and websites who between them have created a very large body of knowledge and understanding that has been and continues to be a huge source of inspiration to all. The author hopes that his efforts have done them justice.

Finally, Sheila, Nicholas, and Simon have put up with a husband and father too often distracted during evenings and weekends by the demands of creating and developing asset management ideas and techniques for planes, trains, clean energy, and human health. Thanks and sorry...

The author offers his gratitude to them and others too numerous to mention who have helped him over the years. Any errors and misunderstandings that have crept into this work are his, not theirs.

A Note on the References

References are listed in alphabetical order for each chapter.

The reader is recommended to use Google (www.google.com) to search for these references. References to site URLs have deliberately not been provided, since "link rot" means such information goes out of date quickly. Google finds the downloadable file directly (usually on several websites) or a website where the hardcopy or electronic book can be purchased: costs can vary, so it is wise to look at several of the choices that Google provides. The book by Tara Calishain and others (*Calishain et al. 2003*) is a useful reference guide to the full capabilities of Google.

FIGURE 1 U.S. states and their abbreviations (*Miller 2009*; reproduced with permission).

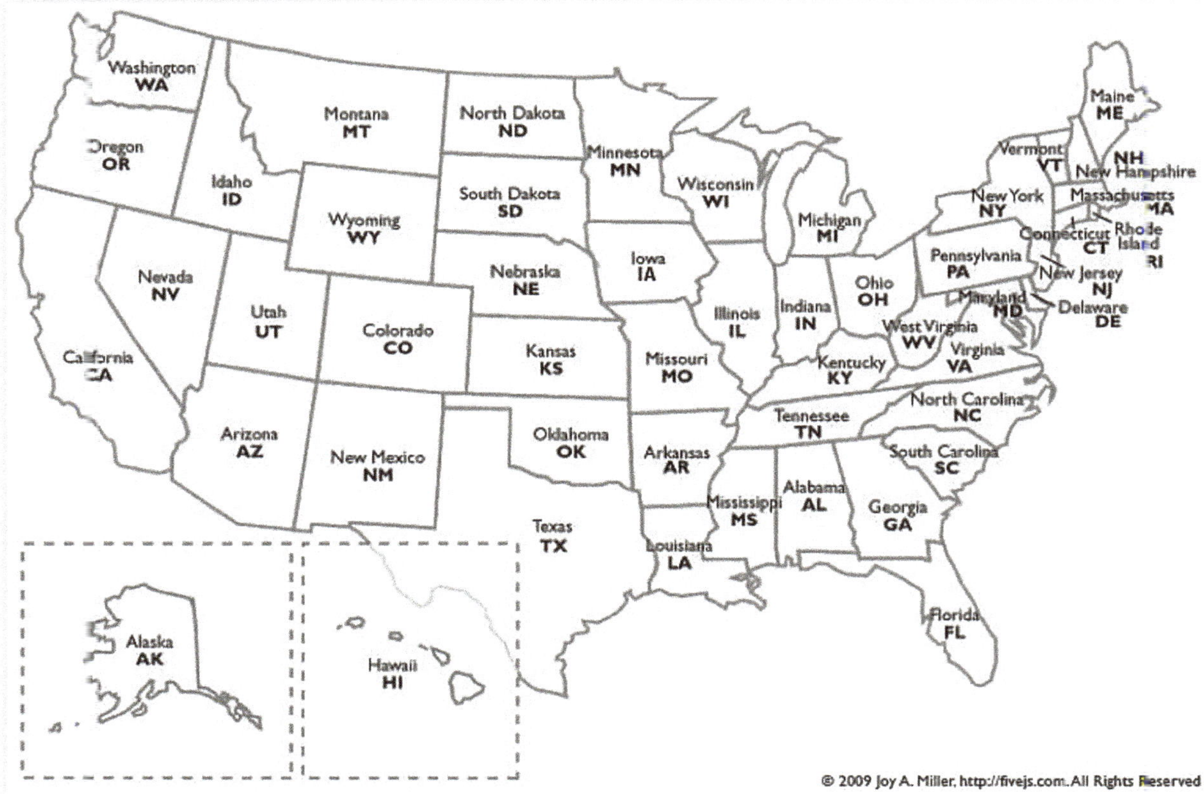

The year shown is the year of publication. If that information is not available, then the year of retrieval is stated. All the references quoted contain very useful information. However, books and articles that the author has found particularly helpful or inspiring are indicated with an asterisk (*).

Many of the references have been published in the United States. For the benefit of international readers, Figure 1 (taken from Joy Miller's e-book (*Miller 2009*)) gives the abbreviations used for the states in which these sources are based.

A Note on the Websites

Websites are listed in alphabetical order for each chapter.

Since web URLs are subject to change for technical or commercial reasons, using a search engine may produce more reliable results than using the website addresses quoted here.

A Note on the Additional Material

Additional references and websites that the author has been made aware of since the main body of the book was written are shown for convenience for each chapter, in alphabetical order.

list of acronyms

This list also includes a number of acronyms that readers may find elsewhere in the literature. The website www.acronymfinder.com is a good source of acronym meanings.

AAAF	Association Aéronautique et Astronautique de France
AAD	Advanced Anomaly Detection
AAKR	Auto Associative Kernel Regression
AC	Alternating Current
ACMS	Aircraft Condition Monitoring System
ADVISOR®	ADvanced VehIcle SimulatOR
AEAT	AEA Technology (formerly Atomic Energy Authority)
AI	Artificial Intelligence
AIAA	American Institute of Aeronautics and Astronautics
AISM	Association Internationale de Signalisation Maritime
AM	Asset Management
AMMJ	Asset Management & Maintenance Journal
APM	Asset Performance Management
ASD	AeroSpace and Defence Industries Association of Europe
ASEE	American Society for Engineering Education
ASME	American Society of Mechanical Engineers
AURA	Advanced Uncertain Reasoning Architecture
BB	Broadband (vibration)
BI	Business Intelligence
BINDT	British Institute of Non-Destructive Testing
BM	Breakdown (reactive/run-to-failure) Maintenance
BRIC	Brazil, Russia, India, China
BSI	British Standards Institution
CAA	Civil Aviation Authority
CAD	Computer-Aided Design
CALCE	Center for Advanced Life Cycle Engineering
CBM	Condition-Based Maintenance
CBR	Case-Based Reasoning
CDS	Controls and Data Services
CFA	Contracting For Availability
CFAR	Constant False Alarm Rate
CHP	Combined Heat and Power
CM	Condition Monitoring

CM	Corrective Maintenance
CMMS	Computerized Maintenance Management System
CMO	Collaborative Market Orientation
COMPASS™	COndition Monitoring and Performance Analysis Software System
CRISP-DM	CRoss-Industry Standard Process for Data Mining
CRM	Customer Relationship Management
DAME	Distributed Aircraft Maintenance Environment
DC	Direct Current
DCF	Discounted Cash Flow
DES	Discrete Event Simulation
DFA	Design For Availability
DFM	Design For Maintainability
DFR	Design For Reliability
DGLR	Deutsche Gesellschaft für Luft- und Raumfahrt
DNV	Det Norsk Veritas
DS&S	Data Systems and Solutions
EAM	Enterprise Asset Management
EASA	European Aviation Safety Agency
EHM	Engine Health Monitoring
EIU	Economist Intelligence Unit
EPSRC	Engineering and Physical Sciences Research Council
ERP	Enterprise Resource Planning
ESP	Engine Simulation Program
ESRC	Economic and Social Research Council
ETSU	Energy Technology Support Unit
EV	Electric Vehicle
FAA	Federal Aviation Administration
FMEA	Failure Modes and Effects Analysis
FMECA	Failure Modes, Effects, and Criticality Analysis
FORTRAN	FORmula TRANslation
FRACAS	Failure Reporting, Analysis, and Corrective Action System
GAMBICA	Group of Association of Manufacturers of British Instruments, Control and Automation
GDP	Gross Domestic Product
GE	General Electric
GFMAM	Global Forum on Maintenance & Asset Management
GMT	Greenwich Mean Time
GPC	Geometric Process Control
HAT	Hub of All Things
HBR	Harvard Business Review
HOMER	Hybrid Optimization Model for Electric Renewables
HUMS	Health and Usage Monitoring System

IAC	Integral Asset Care
IALA	International Association of Marine Aids to Navigation and Lighthouse Authorities
IAM	Institute of Asset Management
iAPM	intelligent Asset Performance Management
ICE	Institution of Civil Engineers
ICT	Information and Communication Technologies
IEC	International Electrotechnical Commission
IEE	Institution of Electrical Engineers (now Institution of Engineering and Technology)
IEEE	Institute of Electrical and Electronics Engineers
IET	Institution of Engineering and Technology (formerly Institution of Electrical Engineers)
IIoT	Industrial Internet of Things
IMechE	Institution of Mechanical Engineers
IoT	Internet of Things
ISEAM	International Society of Engineering Asset Management
ISO	International Standards Organisation
IT	Information Technology
IVHM	Integrated Vehicle Health Management
IVHM	Integrated Vehicle Health Monitoring
KBE	Knowledge-Based Engineering
KEEL®	Knowledge Enhanced Electronic Logic®
KM	Knowledge Management
KNN	K-Nearest Neighbors
KPI	Key Performance Indicator
LCC	Life-Cycle Cost
LCE	Life-Cycle Engineering
LNG	Liquid (or Liquefied) Natural Gas
LOESS	LOcalized regrESSion
M2M	Machine to Machine
MACRO	MAintenance Cost/Risk Optimization
MCR	Maintenance Criticality Reassessment
MIMOSA	Machinery Information Management Open System Alliance
MOOC	Massive Open Online Course
MPS	Maintenance Planning System
MRO	Maintenance, Repair, and Overhaul
MSET	Multivariate State Estimation Technique
MTBF	Mean Time Between Failures
NASA	National Aeronautics and Space Administration
NFF	No Fault Found
NN	Neural Network (or Net)

NPV	Net Present Value
NREL	National Renewable Energy Laboratory
OBC	Outcome-Based Contracting
OEM	Original Equipment Manufacturer
OLAP	Online Analytical Processing
OM	Opportunity-based Maintenance
OMDEC	Optimal Maintenance DECisions
OPC®	Open Productivity and Connectivity
ORBITA™	*(Not an acronym: name deliberately chosen in order not to stand for anything)*
P&W	Pratt & Whitney
PAM	Predictive Asset Management
PAS	Publicly Available Specification
PBL	Performance-Based Logistics
PBTH™	Power By The Hour™
PCA	Principal Component Analysis
PCN	Process Chain Network
PdM	Predictive Maintenance
PEM	Proton Exchange Membrane
PHM	Prognostics and Health Management
PHM	Prognostic Health Monitoring
PLM	Product Life-cycle Management
PM	Preventive Maintenance
PPCL	Process Plant Computing Ltd
PPM	Preventive and Predictive Maintenance
PSS	Product-Service System
PV	Photovoltaic
RAeS	Royal Aeronautical Society
RBD	Reliability Block Diagram
RCA	Root Cause Analysis
RCM	Reliability-Centered Maintenance
RCO	Reliability-Centered Operations
RFID	Radio-Frequency IDentification
RM	Routine Maintenance
ROI	Return On Investment
RR	Rolls-Royce
RSSB	Rail Safety and Standards Board
RUL	Remaining Useful Life
SaaS	Software as a Service
SAE	Society of Automotive Engineers
SALVO	Strategic Assets Lifecycle Value Optimization
SAMP	Strategic Asset Management Plan

SCADA	Supervisory Control And Data Acquisition
SDE	Signal Data Explorer
SFC	Specific Fuel Consumption
SHI	System Health Indicator
SIG	Special Interest Group
SM	Scheduled Maintenance
SMRP	Society for Maintenance and Reliability Professionals
SOM	Self-Organizing Map
STRAPP	trusted digital Spaces through Timely Reliable And Personalized Provenance
SVM	Support Vector Machine
TES	Through-life Engineering Services
TPM	Total Productive Maintenance
TWPL	The Woodhouse Partnership Ltd
UTC	United Technologies Corporation
UTC	Universal Time Coordinated
USDOE	United States Department of Energy
USEPA	United States Environmental Protection Agency
VDM®	Value-Driven Maintenance®
VHM	Vibration Health Monitoring
VRLA	Valve-Regulated Lead-Acid (battery)
WAVE	*(Not an acronym: name chosen to represent the fluid flow basis of the calculations)*
WiLCO®	Whole Life Cost Optimization
WMG	Warwick Manufacturing Group

References

* denotes source that the author found particularly helpful and/or inspiring.

Calishain, T., Dornfest, R., and Adams, D., *Google Pocket Guide* (Sebastopol, CA: O'Reilly and Associates Inc, 2003). Pg. xvii

Goldratt, E. and Cox, J., *The Goal: A Process of Ongoing Improvement*, 3rd ed. (Farnham, UK: Gower Publishing Ltd., 2013). Pg. xv

Miller, J., *Learn the States and Postal Abbreviations* (Aubrey, TX: Five J's Homeschool, 2009). Pgs. xviii, xviii

* Provost, M., "Planes, Trains and Clean Energy: A Life of Simulation, Analysis, Monitoring and Asset Management," Unpublished presentation, Intelligent Energy Ltd., Loughborough, UK, 2012. Pg. xv

Websites

Acronym Finder Pg. xix	www.acronymfinder.com
Google Pg. xvii	www.google.com

CHAPTER 1

Introduction

Servitization and Physical Asset Management has been written to provide a high-level overview of servitization and physical asset management, giving the reader a structured source of guidance and reference information on the management of physical assets for people at all levels in industry:

- Senior executives considering the expansion of their businesses into servitization and the provision of asset management services for the products they design and manufacture
- Middle management wishing to know what needs to be done to look after the assets they are responsible for and who to approach for help
- "Hands-on" engineers looking for contacts and advice on detailed tools and techniques

Readers should always bear in mind the aim of good asset management: that everything, including business relationships between suppliers, customers, and users as well as asset operation, should benefit.

A "Map" of the Book

Figure 1.1 shows a "map" of the topics the book covers and how they interconnect.

The author has managed, over the last four decades, to collect a large number of books, papers, and presentations on the myriad of topics that fall under the physical

FIGURE 1.1 Map of the book.

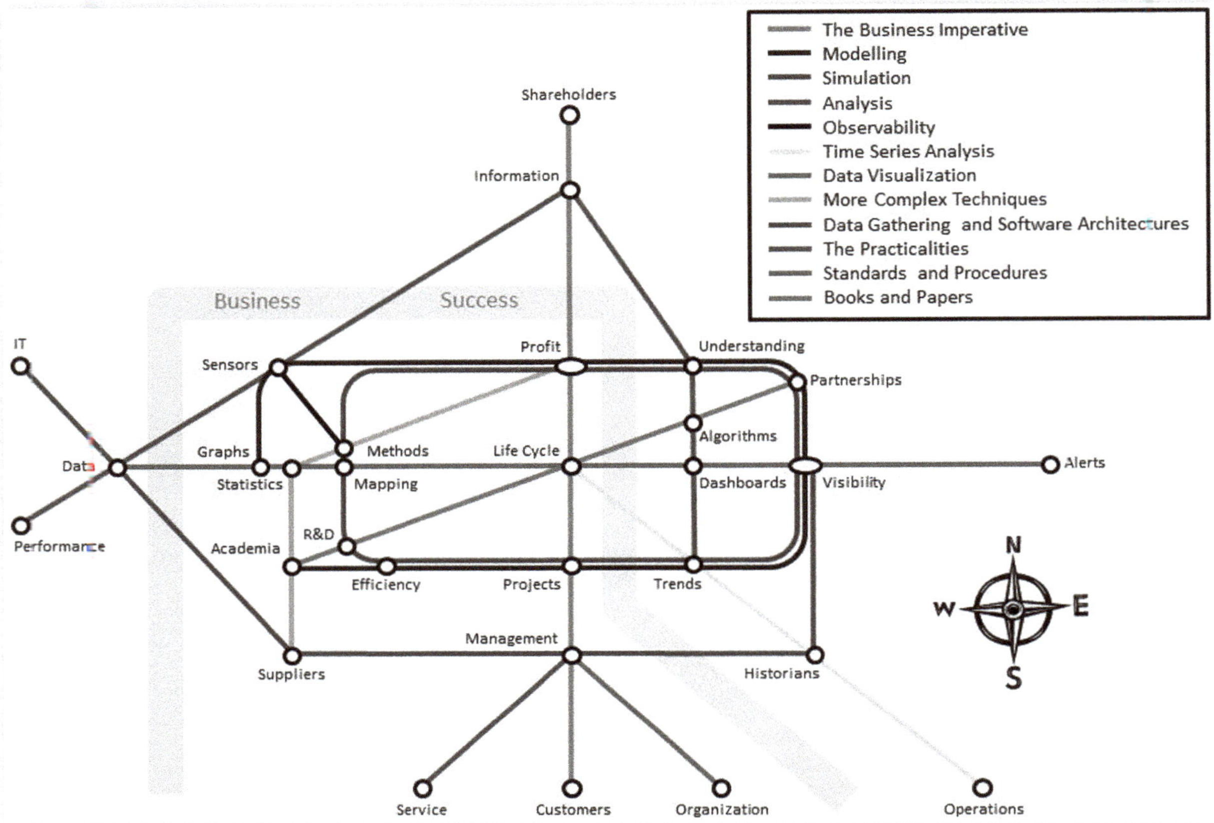

asset management umbrella. This book is an attempt to curate and bring together this material into a coherent view of how all the various ideas link together.

A book of this nature cannot hope to repeat the work of experts and go into every detail on all aspects of physical asset management, so readers are encouraged to seek out and read the relevant references when they require in-depth coverage of particular topics.

What Are Assets?

Assets, in their broadest sense, are items used by businesses to create value and generate profits. In order to deliver business benefit, they must be available for use at the right time and place, and must function and operate correctly. They must function and operate on demand, since lack of functional delivery at the right time and place could have costly repercussions. The time, effort, and material resources needed to keep them functioning and operating correctly must be minimized. They are also items of value, on the balance sheet(s) of one or more businesses, since their owners might not be their operators.

Why Worry about Asset Management?

Assets need to be looked after; they are not always "fit and forget" business fixtures. They cost time and effort to create and consume materials during manufacture. Also, they may have to last for a long time, since replacement may be technically and/or financially difficult or even impossible. Making rational business decisions based on information about the assets used by a business can yield significant financial benefits (increased revenues, reduced costs, increased shareholder value from more efficient and longer-lived assets, etc.). Simply put, proper asset management can mean the difference between business success and business failure.

What Do We Need to Know about Our Assets?

Understanding what has happened, is happening, and may be about to happen to assets used by a business is crucial. Questions that need to be asked and answered include:

- Are they working as required?
- How are they being treated?
- What are the identities of their component parts and how are they behaving?
- What has gone wrong and when do the faults/failures need to be fixed?
- What is about to go wrong and can/should faults be fixed before failure?
- Where are the assets?
- When and where do they need to be available, for operation or maintenance?
- Are substitutes needed, because of functional or operational failures?
- Do the assets need maintaining or replacing?

How Do We Find Out What Is Happening?

Accurate information about the functional and operational performance of the assets used in a business enables better business planning and results in fewer unpleasant surprises. Current technologies allow data to be gathered as often as needed to understand many of the above issues. Sensors that can measure a wide variety of parameters are cheap, accurate, and reliable, because they are usually fitted as part of control systems. Communication is cheap and reliable, with high bandwidths across any distance. Processing large quantities of raw data to produce information a business can use is becoming easier and cheaper. The processed outputs are also becoming easier to understand.

What Is the Business Benefit?

Since most businesses want the *functionality* an asset provides, rather than the asset itself (what it *does*, not what it *is*), asset management represents a significant business opportunity for asset manufacturers, asset users, and third parties. The guarantee of functionality is usually worth as much as or more than the asset itself. Those with the knowledge and mind-set required to manage assets effectively gain considerable competitive advantage. Asset management removes or contains significant business risks, such as business development, cost, and balance sheet value risks, as well as regulatory, competition, revenue, and customer satisfaction risks.

Who Is in the Best Position to Carry Out Asset Management?

Looking across the stakeholders, it can be seen that:

- Customers and end users have a deep but narrow knowledge of a particular operation, but lack knowledge about performance of similar assets used by others.
- Asset owners have an interest in maintaining "cradle-to-grave" asset performance beyond the time horizons of many users, in order to retain and maximize asset value. However, the necessary engineering and operational skills are not usually part of their core businesses.
- Suppliers have a very deep knowledge of asset subsystem capabilities and performance, but lack the necessary knowledge of asset operational contexts.
- IT/communications vendors have a very deep knowledge of software and communication technology capabilities, but lack core asset engineering and operational expertise.
- Manufacturers and system integrators have relationships with all the key stakeholders, giving them a good overview of "cradle-to-grave" asset performance and operation. However, they lack detailed knowledge of some of the drivers of asset performance and operation.

Everyone mentioned above has a role to play. However, manufacturers and system integrators, working in partnership with other stakeholders, are in an ideal position to coordinate asset management because of the engineering knowledge they have of the assets they created and the broad overview they may have of the operation of these assets by their customer base. Asset management offers an excellent opportunity for every stakeholder, but particularly equipment manufacturers and system integrators, to leverage their knowledge of the assets in order to generate maximum business benefit for everyone. It is a big mistake for any one stakeholder to try to "shut out" the others; this breeds an atmosphere of mistrust that hampers information flow; encourages duplication of hardware, software, and business effort; and diminishes the value that can be created from asset management.

Figure 1.2 tabulates some asset management perspectives and techniques, as a foretaste of the book's contents and the questions it sets out to answer.

FIGURE 1.2 Asset management perspectives and techniques (*Provost 2012*).

Time period	Asset questions	Information sources	Business questions	Information outputs
Short term	• What's it doing? • What's it costing/earning? • Is it failing/broken?	• Health Monitoring • Maintenance Reports • Operational Monitoring	• How do we fix it? • How much are we owed?	• Work Orders • Operational advice • Billing
Medium term	• What could it do? • What could happen to it?	• Reliability Analysis • Failure Analysis • Failure Reporting • Fault Trees	• How should we prepare for and solve problems?	• Maintenance Plans • Supply Chain Management
Long term	• What do we want it to do? • How do we make money from it?	• Simulation • Forecasting • Life Cycle Costing	• How do we set up the business, from cradle to grave?	• Business Plans • Logistics Plans • Enterprise Plans

What Is Servitization?

Professor Andy Neely of the Cambridge Service Alliance has produced a succinct description of servitization in his blog (*Neely 2013*: http://andyneely.blogspot.co.uk). It is a transformation involving firms (often manufacturing firms) developing the capabilities that they need to provide services and solutions that supplement their traditional product offerings. Firms that have fully embraced servitization recognize that they are selling *solutions* that deliver *value* and *outcomes* that customers are seeking, not *products*. They see the products that they make and sell as platforms that they can use to deliver services. Servitization (a term that first appeared in the late 1980s) is discussed more fully in Chapter 2.

A summary of the business benefits of servitization (also referred to as service infusion: product-service system (PSS) is another related term) is given in Figure 1.3.

Summary of the Book

The business imperatives of servitization and asset management are addressed: the value that lies in what an asset *does*, not what it *is*, is emphasized, and a willingness for customers to pay more for the former than the latter is discussed. Tools and techniques for assessing asset life-cycle costs and the need for asset monitoring are introduced, emphasizing the decision-maker's requirement for information rather than data. Some of the excellent work done by academics and consultants on servitization is introduced and the successes that companies have achieved from embracing this concept are also summarized.

The power that the use of modelling and simulation capabilities gives the producers and users of assets is discussed. References are given to sources of detailed information for both general modelling and simulation capabilities and modelling of particular asset types. Analysis methods that can be used to determine asset health and the component performance changes that drive overall asset performance shifts are reviewed. A technique developed for optimizing asset sensor suites is presented, and methods

FIGURE 1.3 Benefits of servitization (www.everythingworkswonderfully.com, based on information from *Aston Business School 2013* and *Baines and Lightfoot 2013*).

You've designed and developed your products. What next?

- How do you ensure that they will work the way you want them to?
- How do you find out how they actually work in the field?
- How do you delight your customers and build solid relationships with them?
- How do you avoid becoming just another commodity producer, competing purely on price?
- How do you turn your efforts into long-term business value?

Many firms now realize that they need to sell **solutions** to customer problems, rather than just **products**: this change of mind-set is called **Servitization**. It requires you to move from a 'make it and sell it' mentality to one that considers what your customers' needs really are and how to help your customers operate and manage the products they buy from you over the long term to achieve their business objectives.

As Rolls-Royce, MAN Trucks, Caterpillar, Xerox, Alstom, GE and many others have discovered, the benefits from making this transition are genuinely transformative:-

- Enhanced revenue - increases of between **2x** and **4x** have been reported;
- Improved margins - increases of between **3x** and **10x** have been reported;
- Significant and sustainable business growth - increases of **5%-10%/year** have been reported;
- Increased market share - can your competitors match you or catch you?
- Enhanced customer satisfaction - you are selling them what they **really want**;
- More repeat business - your customers like you, and keep coming back for more;
- More predicable income streams - from revenue 'lumpiness' to recession-beating annuity;
- Improved business reputation - you are seen to consistently **deliver benefits** to stakeholders.

Servitization increases business growth by up to 10% whilst reducing customers' costs by over 25%, according to interviews with 33 key executives by the Aston University Business School. **All companies, from smallest to largest, can reap the benefits of Servitization.**

Adoption of servitization can be inhibited by a lack of awareness: the transition from pure manufacturing requires nurturing by people skilled in its implementation. How do you go about this? A great deal has been written about Servitization and Physical Asset Management, but much of it is unstructured and written in language that is hard to understand. However, help is at hand...

for analyzing time series data gathered from assets are reviewed. The importance of providing compelling visualizations of data and information is discussed, before an overview of more complex analysis techniques is presented, augmented by references to more detailed works on a wide range of ideas, tools, and techniques. The main requirements of data gathering and software architectures are also reviewed.

The practicalities of asset reliability, failure mode analysis, alarms, maintenance, management of maintenance and spare parts, availability, and condition monitoring are each briefly covered by referring the reader to the extensive literature that is available on these important subjects. Standards (particularly PAS 55 and ISO 55000) that have been created recently to guide and advise asset management practitioners are presented.

An appendix presents the economics behind a simplified servitization opportunity, while another appendix goes into more detail on most of the analysis methods described in the book using a fictitious power system for a small Pacific island as an example.

Finally, some important asset management lessons learned by the author over the years are summarized.

There are many valid perspectives to servitization and asset management: it is impossible to be overly proscriptive or describe every detail of these complex and wide-ranging ideas in a book such as this. The books, papers, and articles that are referred to throughout this book have all been written by experts in their individual fields and should be consulted if the reader requires more information. This book aims to open the reader's eyes to what is possible, how important servitization and asset management are, and where to look for advice.

References

* denotes sources that the author found particularly helpful and/or inspiring.

Aston Business School, "Servitization Impact Study: How UK Based Manufacturing Organisations Are Transforming Themselves to Compete through Advanced Services," Aston Business School, Aston University, Birmingham, UK, 2013. Pg. 6

* Baines, T. and Lightfoot, H., *Made to Serve* (Chichester, UK: John Wiley & Sons Ltd, 2013). Pg. 6

Neely, A., "What Is Servitization?," Cambridge Service Alliance, Cambridge University, Cambridge, UK, 2013, http://andyneely.blogspot.co.uk, November 2013. Pg. 5

* Provost, M., "Planes, Trains and Clean Energy: A Life of Simulation, Analysis, Monitoring and Asset Management," Unpublished presentation, Intelligent Energy Ltd, Loughborough, UK, 2012. Pg. 5

Websites

Andy Neely's Blog Pg. 05 http://andyneely.blogspot.co.uk

Everything Works Wonderfully Pg. 06 www.everythingworkswonderfully.com

CHAPTER 2

The Business Imperative

Introduction

Many manufacturing industries today have a real problem: even if their products are very sophisticated and technically capable, they face intense pressure from low-cost competitors who can manufacture and sell virtually identical equipment at prices they cannot match.

The way of avoiding this "race to the bottom" is illustrated by referring to Figure 2.1.

Up to now, Original Equipment Manufacturers (OEMs) have sold products to their customers in what is essentially a single major transaction, represented by the product focus area in the bottom-left corner of Figure 2.1. Many OEMs also sell spare parts, which moves them toward the right of the figure and increases their interactions with their customers in what is traditionally known as the aftermarket (moving upward in the figure). The aftermarket is usually seen as a high-margin activity because manufacturers are generally monopoly suppliers of spare parts that are unique to the products they make, and can exploit this position in an attempt to make up for lower margins on initial equipment sales. In many industries, often with encouragement from customers, third parties have been quick to enter this market, seizing opportunities to make and supply spare parts and offer other services in competition with the OEMs, undercutting them because they do not carry the high overheads associated with the design, development, and manufacture of the original product.

Enlightened manufacturers have seen the potential of increasing their margins and interactions with customers by moving toward the right and upward in Figure 2.1. Rather than selling what their products *are*, they sell what they *do*. The contrast

FIGURE 2.1 Sources of value in equipment sales.

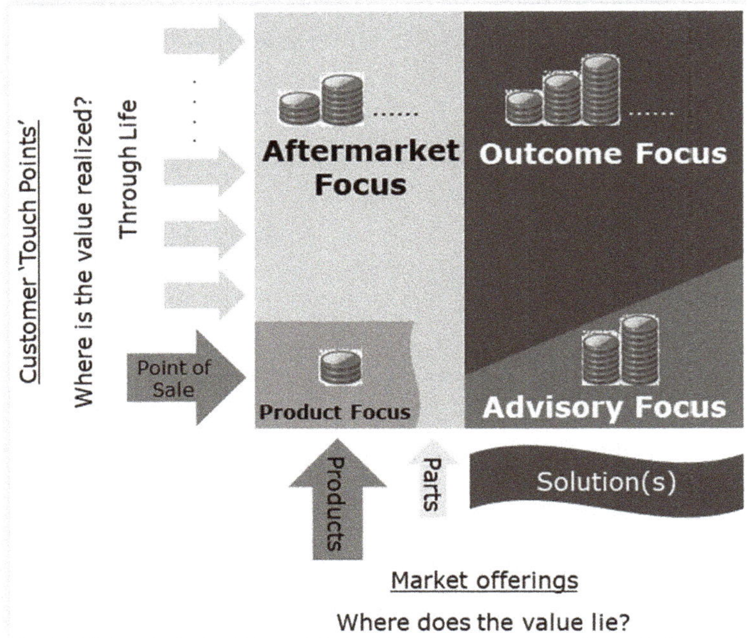

FIGURE 2.2 Typical production and consumption model (*Baines 2014b*: reproduced with permission).

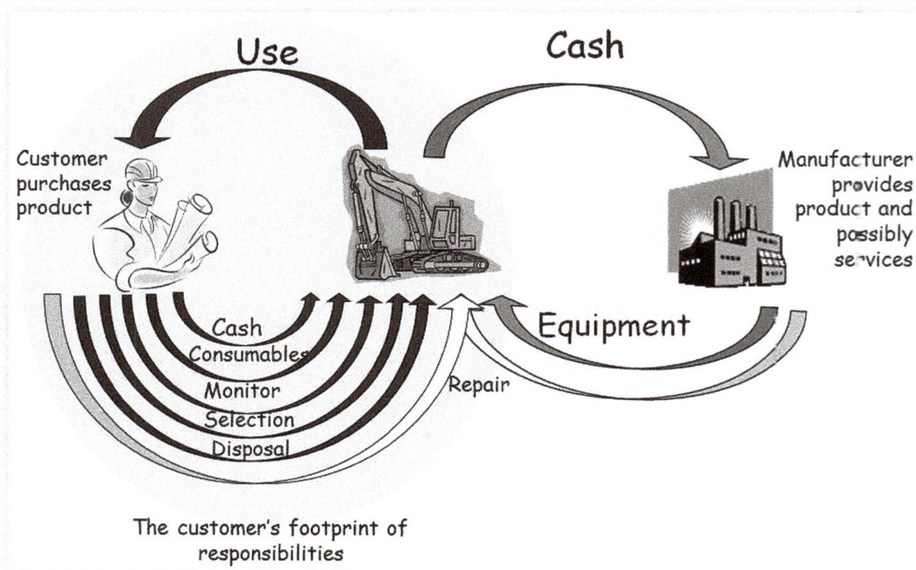

between the two approaches is well illustrated in Figures 2.2 and 2.3, using excavation equipment as an example.

Figure 2.2 shows that, in the traditional way of doing business with an OEM, the customer carries much of the burden for ensuring the final delivery of what he or she really wants, which in this case is a hole in the ground. Figure 2.3 shows the balance of responsibilities between customer and OEM moved toward the OEM, who can apply

FIGURE 2.3 A product-service system (*Baines 2014b*: reproduced with permission).

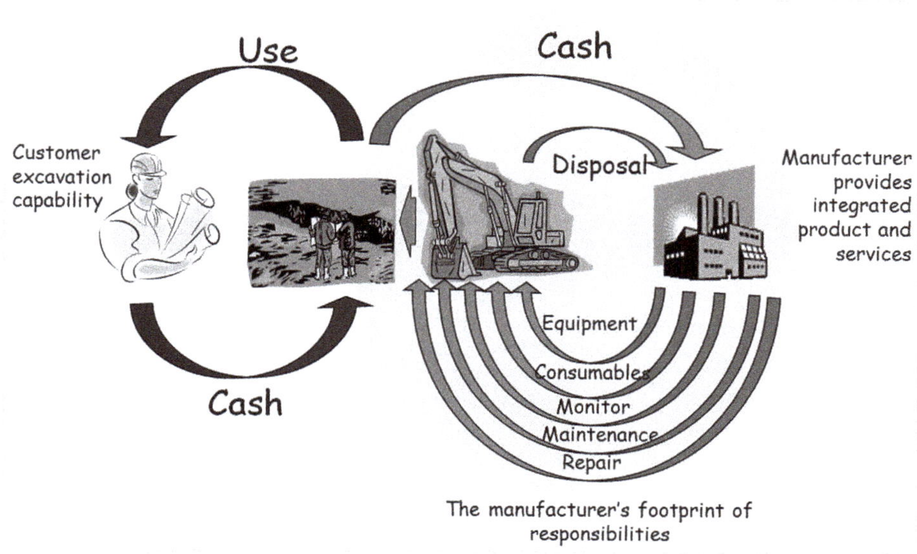

the expertise gained in designing, developing, and producing the equipment to looking after it throughout its life and ensuring that the customer gets what is really wanted: an *outcome*, not a piece of equipment. Both parties stand to gain from this arrangement, known as a Product-Service System (PSS). The OEM can add value by applying product knowledge gained during its design, development, and manufacture, while the customer can focus on his or her markets without the overheads of equipment ownership and maintenance. The customer sees the financial benefit of replacing large investments in capital equipment with operational expenditures that are usually more in phase with revenue generation, while the OEM receives a more predictable and even flow of payments that can support ongoing investments in new products and which, over time, can add up to more than the payments that would have been received from product sales alone.

Figure 2.4 shows how this concept (also known as servitization or service infusion) develops, from the provision of basic services focused around product delivery (the current situation for many OEMs) through product in-service support to advanced risk- and revenue-sharing arrangements that truly augment what the OEM's customer is setting out to deliver to his or her customer.

The transition an OEM must make from a product focus to a servitization focus can be (and usually is) an extremely difficult one. Virtually every business process is affected. Managers and employees, usually with vested interests in getting products delivered quickly and at the lowest cost, struggle with the need to change to a more long-term, customer-centric mind-set and the change of emphasis toward asset management rather than asset production. The servitization process can take many years, but the rewards, in terms of customer and investor perceptions, financial results, and even company survival, are enormous. Change has to be driven from the very top of the company, otherwise failure is almost inevitable. Figure 2.5 gives an indication of the range of the skills required to embrace asset management. Many of these will be new or not considered core activities by many OEMs: some of these can safely be outsourced, while others need to be kept in-house to secure maximum competitive advantage.

FIGURE 2.4 Types of service a manufacturer can offer (*Baines 2014b*: reproduced with permission).

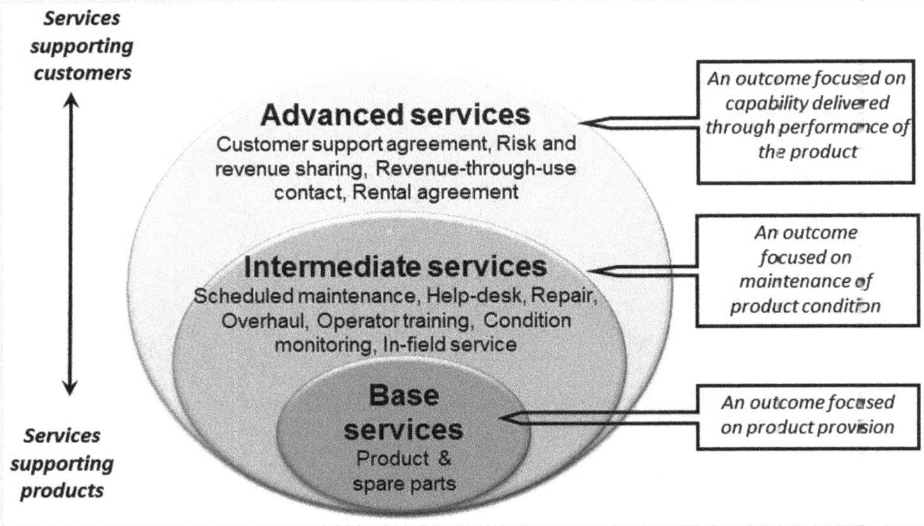

A presentation by the author (*Provost 2015*) describes the issues facing many UK businesses today following the rise of globalization and intensifying competition from emerging economies. It describes how UK manufacturers can and should be moving toward servitization using new thought processes and modern skills and capabilities to address the real needs of their markets. Appendix A summarizes the simplified servitization opportunity included in that presentation.

Companies that Have Embraced Servitization

Tanya Powley's article (*Powley 2015*) gives a high-level overview of the moves by UK manufacturers toward servitization and its impact on their businesses.

Rolls-Royce plc has been one of the major thought leaders in servitization since the early 1990s, when a strategy was devised to move Rolls-Royce's aeroengine business from a commodity supplier of aircraft engines and spare parts to a supplier of services, based on the notion that their customers want *thrust*, not *jet engines*. The Rolls-Royce Power-by-the-Hour™ concept (originally invented by Bristol Siddeley Engines) was trademarked over half a century ago (*Rolls-Royce 2012*), and the term has now entered the language. Partly inspired by Rolls-Royce's earlier development of the COMPASS™ condition monitoring system (*Provost 1989*), Rolls-Royce has developed and marketed the TotalCare™, CorporateCare™, and Mission Ready Management Solutions™ long-term support packages for airlines, business jet owners, and air forces, respectively, plus equivalents for their energy and marine customers. Rolls-Royce set up a subsidiary company (Data Systems & Solutions Ltd. (DS&S), now R^2 Data Labs) which has developed and managed much of the information technology infrastructure that supports Rolls-Royce's aftermarket businesses (*Data Systems & Solutions 2005*,

FIGURE 2.5 Scope of activities and skills needed for servitization and asset management (Intelligent Energy Ltd.: reproduced with permission).

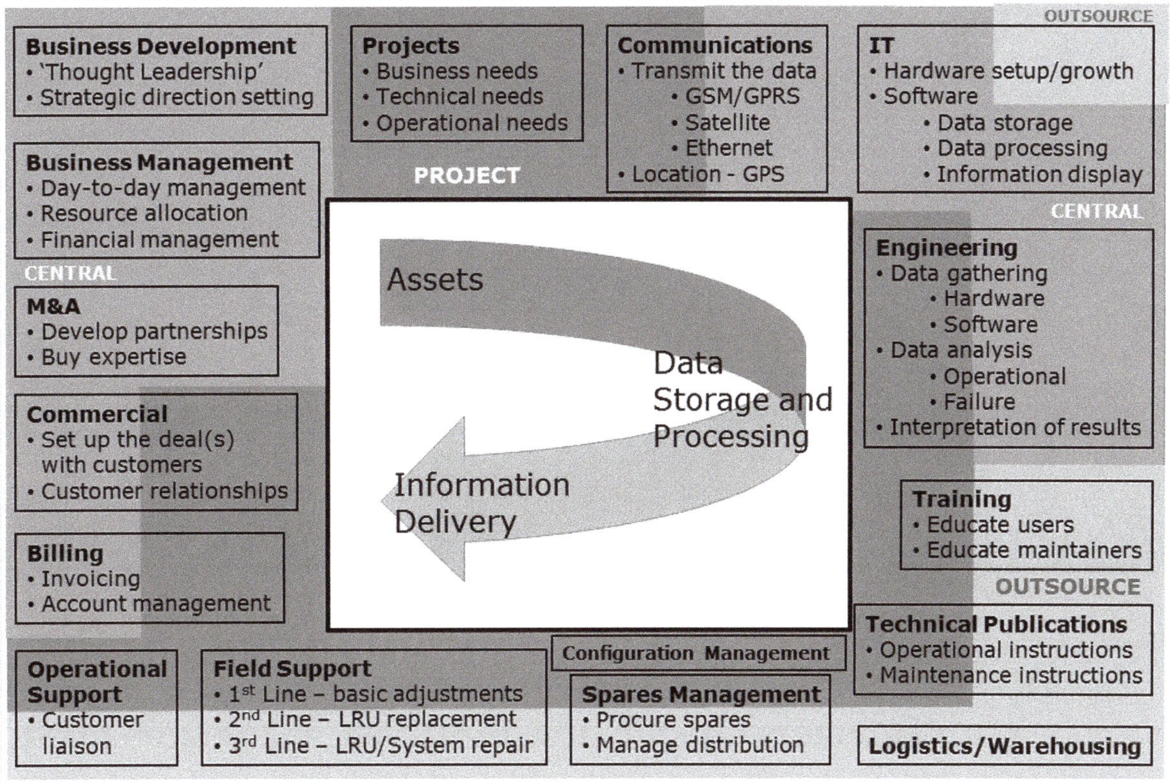

FIGURE 2.6 A tale of two industries (*Economist 2011*: reproduced with permission from David Simonds).

Morrison 2007). There are now many articles in the public domain that describe Rolls-Royce plc's approach, both from a technical standpoint (*Cook 2004*; *Robinson 2006*; *Jennions 2008*; *Waters 2009*; *Harrison 2012*; *King 2015*; *Harrison 2015*) and business perspective (*Economist 2009a*, *2009b*, *2011*). The presentation by Terry Hegarty (*Hegarty 2014*) summarizes the servitization journey that Rolls-Royce plc undertook from the defense aerospace perspective, while Kris Oldland's interview with David Gordon (*Oldland and Gordon 2014*) talks about the concept of "disruption-based availability," the next evolution of Rolls-Royce's service offerings. *The Times* newspaper has credited Rolls-Royce plc's servitization strategy with enabling the company to prosper during the recent recession (*Lee 2012*). Brochures describing Rolls-Royce plc's servitization market offerings can be downloaded from www.rolls-royce.com, while service case studies across a range of markets can be downloaded from www.controlsdata.com. Figure 2.6 shows a cartoon indicating the success Rolls-Royce plc has achieved with servitization.

Bombardier Transportation (www.bombardier.com) has begun to move toward servitization (despite the implication of Figure 2.6), partly inspired by some joint work carried out by their Derby-based Services Division personnel and DS&S in 2004-2005 that resulted in the creation of the ORBITA™ railway condition monitoring system. Bombardier ORBITA™ was launched in 2006 (*Bombardier 2006*, *Professional Engineering 2006*) and is fully described in two patent applications (*Forrest et al. 2008*, *2010*) and several presentations and publications (*Provost 2008*, *2009*, *2010*). The impact of its use has been widely reported (*Rackley 2006*; *European Rail Outlook 2007*; *Grantham 2007*; *Rail Engineer 2009*; *NVable 2009*; *Sawyer 2009*), including its contribution toward the smooth running of the railways during the London 2012 Olympics (*Bombardier 2013*). More recent developments of Bombardier Transportation's railway asset management capability that now uses more wayside equipment are discussed in a presentation by Simon Ellis (*Ellis 2015*). A UK Rail Safety and Standards Board (RSSB) report shows Bombardier ORBITA™ to be one of the leading systems of its type in the UK rail industry (*RSSB 2009*). Richard Clayton, seconded to Bombardier Transportation as part of his studies toward an engineering doctorate at Loughborough University in 2007-2011, describes the complexities of applying servitization principles to the privatized UK rail industry, with its mix of short-term and long-term incentives applied to the multitude of participants (*Clayton et al. 2009*): these and other issues have meant that Bombardier Transportation has made slow progress toward servitization. The books edited by Ken Cordner (*Cordner 2013, 2014*) discuss current UK rail vehicle fleet maintenance initiatives by Bombardier Transportation and others and provide succinct descriptions of the whole UK railway industry.

Boeing has introduced their GoldCare service for their aircraft, incorporating airplane health management (*Read 2006*). The latest brochures describing Boeing's fleet-wide servitization offerings can be downloaded from www.boeing.com: a more detailed technical description of them is also available (*Keller 2006*).

Des Evans, the CEO of MAN Region North (www.man.eu), has given several presentations of a system that makes use of sensors originally fitted to monitor the engine and powertrain of trucks to provide information on driver behavior (*Evans 2008, 2015a, 2015b*). With the active support of drivers, operators of truck fleets have seen substantial reductions in fuel consumption and brake, tire, and clutch wear, as well as reduced numbers of accidents and associated insurance costs, increased truck uptime, and improved regulatory compliance. The system has enabled MAN Trucks to become one of the United Kingdom's leading heavy goods vehicle suppliers. The papers by Aston Business School (*Aston Business School 2015a*) and Ali Bigdeli and Eleanor Musson (*Bigdeli and Musson 2015*) explore the adoption and implementation of advanced services in the United Kingdom's road transport industry, giving a picture of the current advanced services offerings available and the barriers and enablers for adoption.

Many of the world's leading automotive companies have realized that servitization is an important element in their quest for increased market share and business growth. BMW, Ford, General Motors, Daimler-Benz, and others are all expanding their service offerings to their customers to help them improve their driving experiences and control their costs. Steven Holland of General Motors has produced a paper describing their technical and business approaches (*Holland 2008*: www.gm.com) while BMW's website (www.bmw.com) has an entire section describing their ConnectedDrive service, which goes way beyond the usual car ownership model. Daimler-Benz is expanding its Car2Go service (www.car2go.com), where users rent Smart cars by the minute (though the service has been withdrawn in the United Kingdom), and General Motors has invested in a car-sharing service, hoping that drivers sharing cars will be more likely to buy one. Many people are now buying into the idea that ownership is not always the most efficient way of using things (*The Week 2014*). There are now a substantial number of companies offering vehicle tracking and fleet management services, including Quartix (www.quartix.net), Simplytrak (www.simplytrak.co.uk), Fleetmatics (www.fleetmatics.co.uk), FleetStar (www.fleetstar-online.com), NavStar (www.navstar.co.uk), and Microlise (www.microlise.com). A report by Secured by Design Ltd. on electric vehicle (EV) telematics (*Secured by Design 2010* is a flyer advertising this) concludes that there is little doubt that telematics will play a vital role in addressing many of the concerns that consumers and vehicle manufacturers have toward EV technologies, identifies the key opportunities and challenges that the automotive industry will face in deploying EV support services, and includes a table summarizing the main telematics plans of the main vehicle manufacturers. The presentation by Scott McCormick (*McCormick 2015*) of the Connected Vehicle Trade Association (www.connectedvehicle.org) discusses the connected vehicle ecosystem and the data and system security and privacy issues that affect connected vehicles, as well as the problems of adding connected vehicle technologies to existing fleets. The above is only a small sample of the servitization initiatives that many automotive companies are embarking on, as they realize that their customers want what their products *do*, not what they *are*. Chris Senior's presentation from Snap-on (*Senior 2014: www.snapon.com*) shows how suppliers to major OEMs are seizing the potential of servitization, driven mainly by the increasing software content of vehicles.

Essential Energy India Pvt. Ltd. has been set up to manage power production for mobile phone base stations in India (*Essential Energy 2014*), recognizing the fact that mobile phone operators do not regard the provision of electricity to drive their base stations as a core competence and are willing to outsource this activity to others. Intelligent Energy Ltd. (the parent of Essential Energy: www.intelligent-energy.com) sells clean, quiet, and efficient power as a service in its automotive, distributed power and consumer electronics sectors. The paper by John Cullen (*Cullen 2015*) looks at the implementation of servitization in the mining equipment industry and the issues that were tackled during that transition, while Kris Oldland's interview with Alex Bill (*Oldland and Bill 2015*) talks about Alstom's approach to servitization.

The panel discussion chaired by Kris Oldland (*Oldland et al. 2015*) looks at some of the issues of applying servitization to smaller companies; the need for behavior changes as well as software, hardware, and process changes; the data ownership questions; and the customer "pull" toward servitization. Services such as Airbnb (www.airbnb.com), Uber (www.uber.com), and Lyft (www.lyft.com) have led the way in personalizing servitization, as younger generations demand access, not ownership. Retailers, such as the U.S. DIY store Home Depot (www.homedepot.com), are renting out tools to compete with online tool sharers (*The Week 2014*).

Problems and Issues with Making Services a Separate Business Unit

There are three serious problems with making services a separate business unit with its own profit and loss accountability, as opposed to integrating manufacturing and services together into whole-of-life focused projects:

- *Conflicts of interest*. Projects focused on manufacturing and selling assets will want to maximize the number of assets sold and minimize their cost. The need for data gathering and transmission hardware to support the services division adds to the manufacturing division's costs while not necessarily adding to its revenues (particularly if competitors offer the same functionality more cheaply without data gathering). This can lead to refusal to fit data gathering hardware (sometimes aided and abetted by customers; see the next point) or fitting of minimum specification hardware that severely curtails potential services division offerings. Retrofit of proper data gathering and transmittal hardware after delivery to the customer is very costly and potentially opens the door to competitors, since customers may open up "upgrades" (and the service offerings that follow) to competitive tender.
- *Conflicts between customer offerings*. Projects focused on manufacturing and selling assets will approach customers with the lowest-price offering, telling them that their products are intrinsically reliable (so monitoring is not required) and wanting to sell extra assets as spares to cover customer availability requirements.

Customers therefore get two messages: one from the manufacturing division of the business effectively rubbishing service offerings in an effort to lower costs and clinch the deal, and another from the services division which appears to contradict what the manufacturing division has said. Customers can then become confused and angry, and end up playing the manufacturing and services divisions off against each other to the detriment of the whole business.

- *Financial conflicts*. The manufacturing division sees the costs, but the services division sees the revenues and profits. Unless some method is put in place to share equitably the whole life risks and rewards between the two divisions, there will be perpetual conflict and continued selling of suboptimal solutions.

The services division of a company needs to work together with the manufacturing division to produce the best result for the customer, under the umbrella of customer-facing projects that present a united front to customers, see the whole picture and integrate the offerings from each division in a way that benefits the whole company.

The Importance of Human Factors

Bernd Geropp, a very charismatic business leadership, strategy, and marketing coach (see www.more-leadership.com), makes the case in his presentation (*Geropp 2014*) that the ideas presented in this book will only make real headway in an organization if the interests of all participants are fully considered. Human factors are, therefore, central to making progress: readers will need to develop many "soft" skills in order to make the enormous changes that are needed to make servitization succeed. The articles by John Kotter (*Kotter 2007*) and John Kotter and Leonard Schlesinger (*Kotter and Schlesinger 2008*) discuss the stages of change management and the pitfalls unique to each stage, as well as tailoring change strategies round the resistance likely to be encountered.

Everyone knows people with poor relationship skills and everyone has, at times, needed a little friendly advice that would help them deal with friends and colleagues better. David Fraser (www.drdavidfraser.com) has written an excellent book (*Fraser 2010*) on how to improve abilities to get on with people. One book that has helped several people the author has mentored is by Stephen Bayley and Roger Mavity (*Bayley and Mavity 2007*), which makes many excellent points about getting ideas across and making the correct impression. Jo Owen's book (*Owen 2010*) looks at the fine art of developing influence and authority. Seth Godin's famous and inspiring book (*Godin 2010*) looks at what makes people indispensable, while the article by John Zenger and others (*Zenger et al. 2011*) looks at ways to become indispensable by adding complimentary skills to existing character and career strengths. The article by Phillip Slater (*Slater 2009*) looks at the attitudes and behaviors required to make a difference and demonstrate leadership, while the book by Adam Galinsky and Maurice Schweitzer (*Galinsky and Schweitzer 2015*) argues that the right balance must be struck between competition and cooperation to succeed in work and life. The article by Garrison Wynn (*Wynn 2014*) makes the point that being good at what you do these days is not enough: people have to like to like you as well as what you do. Chris Baréz-Brown's book (*Baréz-Brown 2014*) shows that liking your work is also good for you. The inspirational book by Chris Hadfield (*Hadfield 2013*) on the behavioral approaches necessary to achieve personal and professional ambitions and succeed in life is well worth reading.

Readers may wish to use the servitization opportunities presented in this book to redirect or reinvent their careers. The book by Tim Clark and others describing how to approach this (*Clark et al. 2012*) includes a template and many helpful examples; there is also an associated website (http://businessmodelyou.com). The article by Stephen Thomas (*Thomas 2009*) argues that there are experts at many levels inside every organization that could deliver additional value by using their knowledge and experience to act as consultants (see also Peter Block's book: *Block 2011*) without any additional cost. Finally, Robert Heller and Tim Hindle have provided a very useful compendium of general management skills (*Heller and Hindle 1998*) while Gordon MacKenzie has written an amusing book (*Mackenzie 1996*) on thwsssswe way that corporate processes and behaviors can destroy creativity and innovation.

Strategy

The books by Michael Porter (*Porter 1990*) and Gary Hamel and Coimbatore Prahalad (*Hamel and Prahalad 1994*) are classic business strategy texts that, while not mentioning servitization directly, touch on the concept as part of overall strategy formulation. The book and associated articles on Blue Ocean Strategy by W. Chan Kim and Renée Mauborgne (*Chan et al. 2004, 2005a, 2005b*; www.blueoceanstrategy.com) talk about the need for companies to truly differentiate their offerings: servitization is one way that they can do this effectively. Dmitrij Kabukin's thesis (*Kabukin 2014*) reviews the Blue Ocean Strategy approach in some detail. Avinash Dixit and Barry Nalebuff (*Dixit and Nalebuff 1991*) have written a very good introduction to strategic thinking which needs to be at the heart of any servitization strategy. The book by David Besanko and others (*Besanko et al. 2010*) extends these ideas with many examples based on economic theory.

Creating and explaining a servitization strategy needs brainstorming with all the potential stakeholders in an organization. Alexander Osterwalder and Yves Pigneur (*Osterwalder and Pigneur 2010*) have produced a book describing how to generate business models which includes a very helpful template and set of techniques (many of which are available on a companion website, www.businessmodelgeneration.com) while Dave Gray and others (*Gray et al. 2010*) introduce a number of more unusual brainstorming ideas to help engage participants and encourage different thought patterns. Mind mapping is a very useful tool which large numbers of people are very familiar with. Tony Buzan's book (*Buzan 2005*) is one of many that describe the technique, and there are a number of free and paid-for mind mapping software packages available. Jeff Conklin's book (*Conklin 2006*) describes a software package called Dialogue Mapping, which offers a structured approach to describing and documenting new concepts that may be difficult to understand and document. The software is available from the CogNexus Institute (http://cognexus.org/cognexus_institute). The excellent and thought-provoking book by Mark Payne (*Payne 2014*: www.fahrenheit-212.com) points out that innovations must satisfy both customer needs and business imperatives in order to succeed; a focus on only one aspect merely invites failure, so a combination of the "magic" element of idea generation with the "money" aspect of business need and commercial reality is critical to producing concrete innovation outcomes.

Ian Jennions has edited a series of useful and important books on Integrated Vehicle Health Management (IVHM) in which several experts discuss basic principles, business cases, technology, implementation, and lessons learned (*Jennions 2011, 2012, 2013, 2014*). Ian and others also discussed the application of IVHM principles in aerospace in more detail in an SAE International webcast (*Jennions et al. 2013*).

The books by Craig Kirkwood, John Morecroft, and Kim Warren (*Kirkwood 1998*; *Morecroft 2007*; *Warren 2008*) discuss methods for modelling the various feedback mechanisms that characterize business dynamics. Kim Warren's website (http://strategydynamics.com) contains a link to the Sysdea strategy modelling package (http://sysdea.com). Business Smart International (www.business-smart.com) contains a number of example aftermarket business simulations (including jet engines, razor blades, and inkjet printers) as well as providing the ability to create bespoke models.

The presentation by Padmakumar Easwapillai (*Easwapillai 2013*) discusses the information and communication technologies (ICT) behind servitization, showing the ICT contributions to a product support process map, the use of machine-to-machine (M2M) technologies to connect assets to business systems, and the impacts of mobility and big data.

The article by Dennis Belanger (*Belanger 2010*) explores what organizations need to do to make maintenance and reliability ideas work and create the ultimate reliable enterprise, while the article by Ricky Smith (*Smith 2009*) looks at the preparations companies can make to improve their operations in order to develop competitive advantage and benefit from economic upturns. The article by Rejeesh Gopalan and Sajit Kumar (*Gopalan and Kumar 2015*) discusses the synergies between servitization and asset management, the evolving asset management model in the servitization context, and parallel advances in technology that support servitization.

Finally, a paper by the author (*Provost 2004*) provides a high-level overview of the processes, issues, and value which can be gained from monitoring high-value assets. The books by Thomas Davenport and others (*Davenport and Harris 2007*; *Davenport et al. 2010*) look at the importance of analytics for both internal decision-making and creating a competitive edge, and the article by Phil Simon (*Simon 2014*) discusses the approaches companies need to make to use and manage data.

Financial Evaluation and Modelling

The cornerstone of the economic evaluation of servitization is the calculation of Net Present Value (NPV), also known as Discounted Cash Flow (DCF). This method, involving forecasting cash flows into the future before expressing them in current monetary values by taking into account current interest rates (adjusted for risk to determine the true cost of capital), is fully described in many textbooks and finance courses and will not be expanded on here. The book by Richard Brealey and Stewart Myers (*Brealey and Myers 1988*) is a classic, covering most aspects of financing, including NPV and why it is the best method of evaluating investments, financial risk and its measurement, financing decisions, and financial options. Robert Higgins' book (*Higgins 2009*) also discusses much of the same material. Kate Moran's book (*Moran 1995*) is an excellent introduction to the basic principles of investment appraisal, while Frank Crundwell's book (*Crundwell 2008*) has been written for engineers and scientists who need a working knowledge of NPV and the economic evaluation and funding of projects. The concept of shareholder value (based on NPV) and how it can drive strategic thinking is well explained in the book by Tom Copeland and others (*Copeland et al. 1996*). Other methods that go beyond NPV are mentioned in several other works referred to throughout this book. However, the reader is more likely to convince others successfully by applying this commonly understood method than using any others, particularly in initial discussions.

Servitization is often about creating options for an organization's future growth. The book by Avinash Dixit and Robert Pindyck (*Dixit and Pindyck 1994*) provides a very rigorous discussion of this complex subject. The papers by Armen Papazian

(*Papazian 2012, 2014*) discuss the idea of adding other important nonfinancial considerations, such as environmental, social, and reputational benefits, as extra "dimensions" to the time dimension that NPV and DCF are based on in order to give a more complete view of the true value of a project to an enterprise; space exploration is used as an example. Many servitization initiatives have succeeded because of this approach. The Finoptek financial value optimization tool under development (www.finoptek.com) is a cloud-based project analysis toolset that includes time-space, risk and ratio analyses, econometrics, charting, statistics, and reporting in one platform.

The book by Bill Hodges (*Hodges 1996*) provides an understanding of the main principles and practices of physical asset management, providing guidance to middle managers who have to suggest, promote, or control developments in their firm or organization but also giving senior managers a guide to promoting more economically aware staff.

Malcom Secrett's book (*Secrett 1993*) is an excellent introduction to the practicalities of using spreadsheets for budgeting and forecasting, while Michel Schlosser's book (*Schlosser 1992*) covers the theory and practice of financial modelling in more detail, and the books by Graham Friend and others (*Friend and Zehle 2004; Tennent and Friend 2005*) offer detailed guidance for business planning and modelling. These books should enable the reader to avoid many of the pitfalls of using spreadsheets for financial modelling. Spreadsheets can be very powerful tools for recording and communicating assumptions and carrying out "what-if" studies, as well as for calculation. However, they need very careful planning, programming, and checking to avoid serious (and embarrassing!) errors that could strongly affect the conclusions reached. Many organizations offer pre-prepared spreadsheet templates for business planning and modelling. The author has found these useful for guidance, but not a complete solution because there are always important aspects of the dynamics of any business that are not included in more generalized pre-packaged business modelling software, and it is sometimes very difficult to understand someone else's spreadsheet if they are not around to explain it.

There are many add-ins to the Microsoft Excel® spreadsheet program that expand on its native capabilities for financial analysis. @RISK® and the DecisionTools Suite from Palisade Corp (www.palisade.com) and Crystal Ball (www.oracle.com/us/products/applications/crystalball) add the ability to do Monte Carlo analysis. The book by Robert Clemen and Terence Reilly (*Clemen and Reilly 2014*) teaches the fundamental ideas of decision analysis and includes a student copy of the Palisade Corp. DecisionTools Suite.

The Value of In-Service Monitoring of Assets

When considering the value to be gained from monitoring an asset during its operation in service, the following questions should be asked:

- For reducing service failures and unplanned maintenance, in order to achieve material, labor, and service failure penalty savings:
  - Can failures be detected?
  - Can failures be diagnosed?
  - Can failures be prevented?

- For managing planned maintenance, in order to achieve material and labor savings:
  - Can maintenance actions and inspections be eliminated?
  - Can the periods between maintenance actions be extended?
  - Can a move to use-based maintenance be made?
  - Can a move to condition-based maintenance be made?
- For operations management:
  - Can the availability of the equipment be increased?
  - Can how and where the equipment is being used in service be determined?
  - Can the business impact of equipment use (revenues generated and costs incurred) be seen?
- What can be done now, with what sensors are fitted already, for control or other reasons?
- What could be done, if changes to the sensor fit were made?
- What new information could be generated that could benefit the customer?

To answer these questions, NPV methods can be used when inputs to the calculations are known with a reasonable degree of accuracy (supplemented by "what-if" or Monte Carlo analyses to understand the effects of uncertainties) while decision trees (which enable the consequences of many possible outcomes of an event to be added up, taking into account their probabilities) can be used when financial outcomes are principally governed by the probabilities of occurrence of significant events. The books by Sam Savage (*Savage 2003*) and Robert Clemen and Terence Reilly (*Clemen and Reilly 2014*) explain decision trees well, while a course by Craig Kirkwood (*Kirkwood 2002*) goes into more detail, and an example of their use in the petroleum industry is given in an article by Torben Riis (*Riis 1999*).

The following example shows how a decision tree can be used to determine the value that could be gained from monitoring an asset. A project manager has a choice: either monitor a system (cost **$X**) or do not monitor a system (cost **$0**). The system has a chance of failing of **α%**: if it fails, the resulting costs are **$Y** (if no monitoring is in place), while if it does not fail, the resulting costs are **$0**. If the system is monitored, there is a chance of picking up the problem of **β%**: if the problem is picked up and **$Z** is saved, the resulting costs are **$(Y-Z)**, while if the problem is not picked up, the resulting costs are **$Y** (the same as no monitoring). Figure 2.7 shows the resulting decision tree and how it is manipulated to produce the final result.

Stage 1 shows all the possible outcomes and costs and their probabilities of occurrence. The tree is then "folded" from right to left (Stages 2 and 3) with the costs at each node being the sum of each probability of occurrence multiplied by its cost, until the final result is reached (Stage 4). In this case, the cost of monitoring an asset (**$X**) must not exceed the probability of failure (**α%**) multiplied by the probability of picking up the failure (**β%**) (together creating a probability of detection) multiplied by the savings made (**$Z**). A more complex example, involving monitoring an asset to see if it requires refueling (rather than either refueling it to a schedule or only refueling it when the fuel runs out, which would stop it functioning and cause problems for the user), is shown in Figure 2.8.

The resulting formula is quite complex but can be programmed into a spreadsheet (*Provost 2013a*) in order to carry out "what-if" studies to justify the case for monitoring.

CHAPTER 2

FIGURE 2.7 Value of monitoring—simple example.

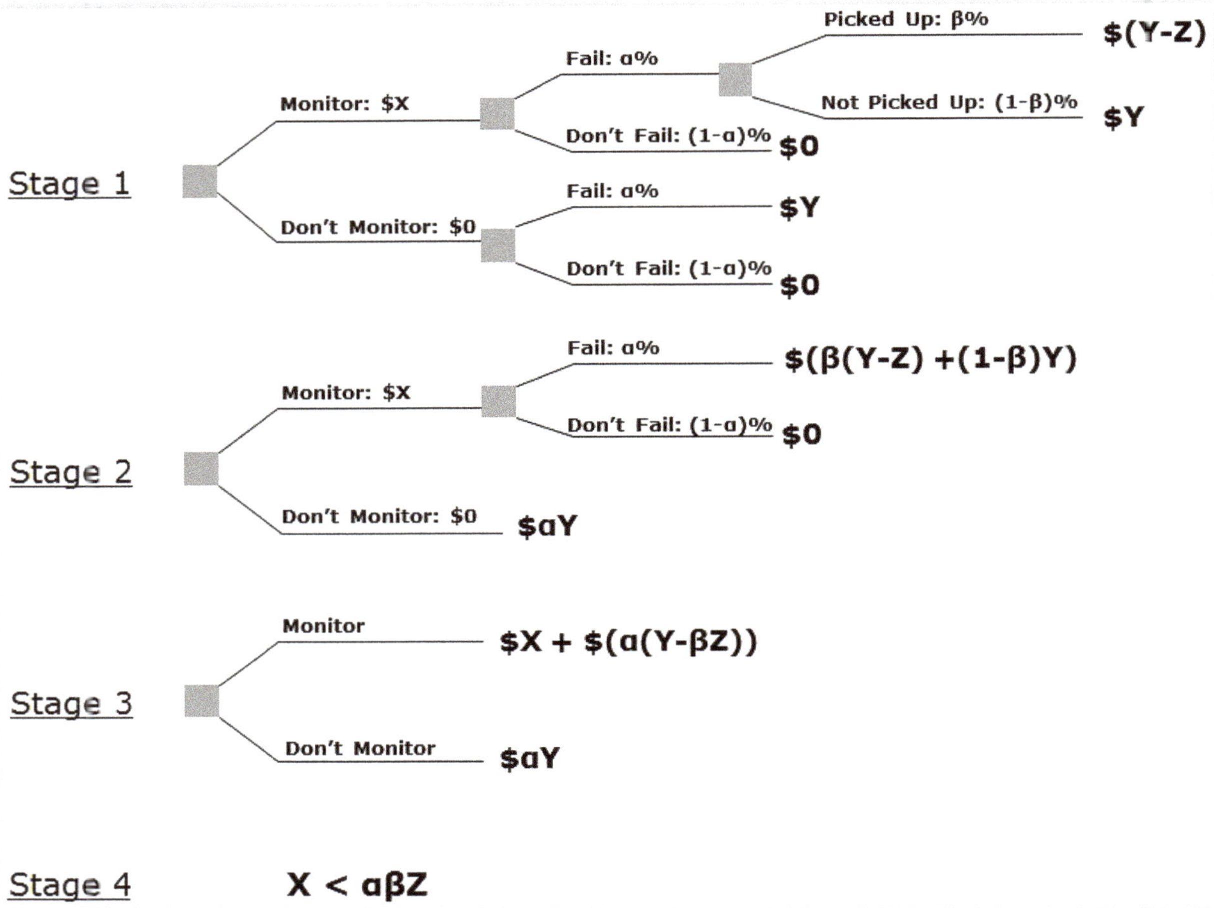

Another unpublished presentation by the author (*Provost 2013b*) discusses the use of Markov chains to justify the in-service monitoring of assets.

As a general rule, the monitoring of assets should be driven purely by business need. Given that current technologies allow almost anything to be monitored, it is easy to fall into the trap of gathering enormous quantities of data without thinking through what it will be used for in the hope that its mere existence will automatically provide solutions to business and technical issues. It is much better to determine the decisions that will be made as a result of monitoring an asset and the analysis requirements that support those decisions before specifying the sensors and associated on-board and off-board IT and communication systems that will feed the analysis.

The articles by José Guilherme Pinheiro Côrtes (*Côrtes 2011, 2014*) discuss the need for economics to recognize the realities of asset deterioration, the value added by maintenance, and the difficulties of providing both cash forecasts and discount rates for input into traditional finance tools, which ignore other costs such as loss of business reputation due to unscheduled downtime, delivery of poor quality goods, etc., that maintenance mitigates. Finally, the article by Michael Cook and Michael Muiter (*Cook and Muiter 2011*) discusses how to calculate return on investment (ROI) for a predictive maintenance program.

FIGURE 2.8 Value of monitoring—refueling example (Intelligent Energy Ltd.: reproduced with permission).

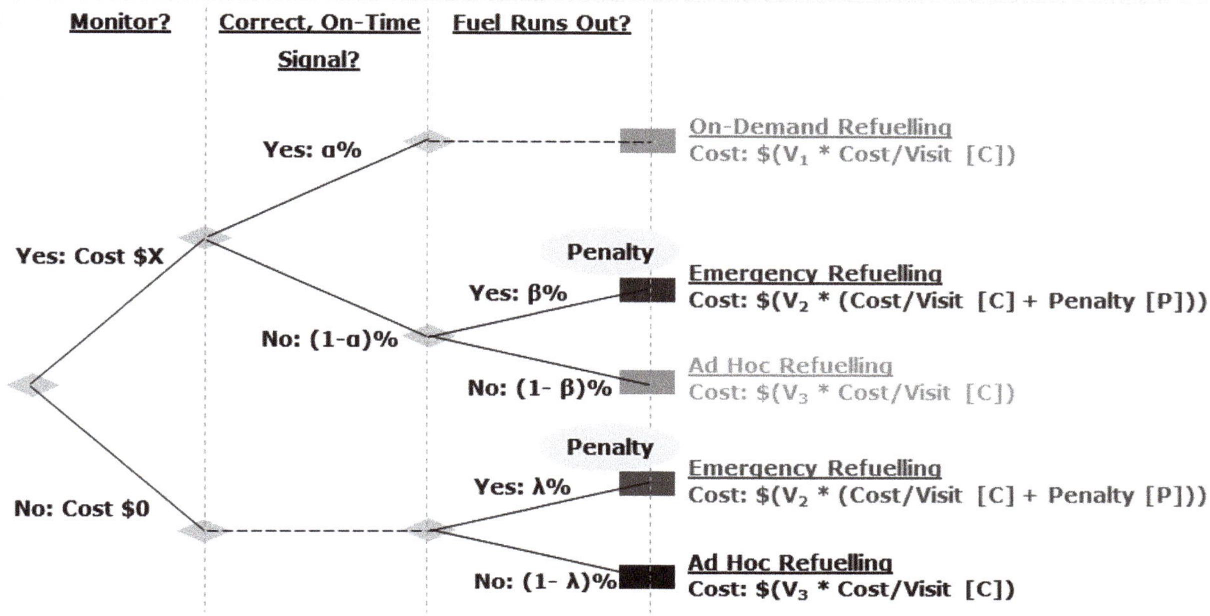

Life Cycle Costing

One of the most important aspects of servitization and asset management is Life Cycle Costing (LCC), in which all the costs associated with the purchase, use, and disposal of an asset are considered, rather than just the initial purchase price. By using NPV, these current and estimated future costs can be added together, different options can then be compared, and the one that minimizes the total costs (discounted over time) can then be chosen. The paper by the UK Treasury (*HM Treasury 1992*) introduces LCC and provides a simple example, while the paper by Paul Barringer (*Barringer 2003*) and John Farr's book (*Farr 2011*) go through the process in some detail. The tutorial by Paul Barringer and David Weber (*Barringer and Weber 1996*) and the good practices guide by Paul Barringer (*Barringer 1998*) are also good introductions to the subject, while the Paul Barringer and Associates Inc. website (www.barringer1.com) contains useful LCC resources. The National Aeronautics and Space Administration (NASA) has produced a detailed guide to their cost estimation processes (*NASA 2015*), while the executive summary and book by Europump and others (*Europump/Hydraulic Institute/US DoE 2000, Europump/Hydraulic Institute 2001*) provides a comprehensive approach and guidance for pump LCC.

A presentation by Walter Tomczykowski (*Tomczykowski 2011*) looks at the impact of reliability on life cycle costs and shows how decisions taken early in a project's lifetime can have far-reaching effects later on. The article by Ramesh Gulati (*Gulati 2013*) discusses Design For Reliability (DFR), an emerging discipline that refers to the process of designing reliability, maintainability, safety, etc., into products to improve availability, lower sustainment costs, and maximize asset utilization during the life of the asset. The presentation by Alec Erskine (*Erskine 2012*) looks at the use of operations research theories in LCC, while the presentation by Paul Sankey and Craig Gaughan (*Sankey and Gaughan 2011*) discusses a commercial software package (WiLCO®) for investment planning over the lifetime of assets that enables the user to balance operational and capital spend, costs and risks, and explore efficiency improvements. The article by Ian Miller (*Miller 2014*) discusses the element of risk in whole life cost estimating, an essential part of financial modelling in public finance initiatives and public private partnerships. The book by Richard de Neufville and Stefan Scholtes (*de Neufville and Scholtes 2011*) focuses on the challenge of producing best value in large, long-lasting projects, presenting tools such as Monte Carlo analysis and financial options theory that can be used to design assets with the potential for further expansion built in (such as designing a car park with foundations strong enough for further floors to be added in the future). Finally, the book by Ricardo Valerdi (*Valerdi 2008*) describes a model that can be used for quantifying the engineering effort in complex systems.

The Strategic Assets Lifecycle Value Optimization (SALVO: www.salvoproject.org) and MACRO (www.macroproject.org) projects take the concept much further. The SALVO project was a multi-sector collaborative program that researched, developed, and defined best practices in asset management decision-making: it addressed the problems encountered in the management of aging assets, decision-making in the face of risk and data uncertainty, and how to quantify "intangibles" and optimize whole life cycle value. The MACRO project was a 5-year, \$2.5m research and development (R&D) program exploring and developing best practice in risk-based industrial decision-making, in which leading European companies from many sectors collaborated to find and share the best methods for setting asset management strategies, particularly when hard data is limited or unavailable. The project has yielded a book of best practice guidelines, a comprehensive suite of process training courses and decision support tools (marketed by Decision Support Tools Ltd.: www.decisionsupporttools.com), and a broad range of case studies.

The excellent book by John Woodhouse (*Woodhouse 2014*) discusses the SALVO project in detail, explaining the whole field-proven process of making optimal asset management decisions and explaining and illustrating how to develop clear business justifications for what and when to spend in asset procurement, operation, maintenance, inspection, modifications, and renewals. John has also written an article (*Woodhouse 2015*) showing how the three different asset life cycle stages represent very different decision-making environments and offer different opportunities to influence the whole life cycle value. Both the papers and the presentation by Alex Thompson (*Thompson 2011, 2012a, 2013*) give a good overview of the SALVO project, while the presentation by Andy Hunt (*Hunt 2011*) goes into more detail. The article by Bill Reekie and others (*Reekie et al. 2015*) describes how the SALVO approach has been used in Scottish Water.

The two papers by Atai Winkler (*Winkler 2015a, 2015b*: www.pamanalytics.com) describe the PAM Analytics Predictive Asset Management (PAM) system, which models asset performance using survival analysis (also known as time to event analysis), the proportional hazards model, and discrete event simulation to model asset failure at individual asset (equipment) level for short-term operational maintenance planning and long-term strategic economic planning.

Peter Sandborn and others at the Center for Advanced Life Cycle Engineering (CALCE: www.calce.umd.edu) Prognostics and Health Management Group at the University of Maryland have produced a life cycle cost toolset that calculates the potential benefits of prognostics and health management (*Sandborn 2005, 2010, 2011; Sandborn and Feldman 2008*). They have also produced a number of papers that describe methodologies, models, and tools that address the design, manufacture, analysis, and management of electronic systems. The paper by Shungfeng Cheng and others (*Cheng et al. 2010*) discusses sensor systems for prognostic health management (PHM) and the emerging trends in sensor technologies. The papers by Kiri Feldman and others (*Feldman et al. 2008, 2009*) discuss the calculation of ROI for PHM activities, presenting studies that used a stochastic discrete event simulation model of the maintenance of a Boeing 737 multifunctional display to determine the potential ROI gains from using a precursor to failure PHM approach instead of relying on unscheduled maintenance. The paper by Estelle Scanff and others (*Scanff et al. 2007*) discusses using a similar model to predict the life cycle cost impact associated with the application of PHM to helicopter avionics. The papers by Gilbert Haddad and others (*Haddad et al. 2011a, 2011b*) provide optimization models based on real options and stochastic dynamic programming that maximize availability of offshore wind farms and commercial aircraft with prognostic capabilities. The papers by Taoufik Jazouli and Peter Sandborn (*Jazouli and Sandborn 2010, 2011*) present the application of PHM within a design for availability (DFA) approach that uses an availability requirement to predict the required logistics, design, and operation parameters with and without the application of PHM methods, using LCC analysis to quantify trade-offs of using PHM methods versus more traditional maintenance approaches. The paper by Peter Sandborn and Chris Wilkinson (*Sandborn and Wilkinson 2007*) presents a decision model that addresses how PHM results can best be interpreted to provide value to the system maintainer, including a methodology for determining an optimal safety margin and prognostic distance for various PHM approaches.

Maintainability is the degree to which a product can be maintained or repaired easily, economically, and efficiently. Design for maintainability (DFM) encompasses the measures taken to reduce the time and other resources spent in keeping a product performing well; it benefits the end user by reducing the total ownership costs through less downtime, lower maintenance costs, less inventory, fewer tools, and improved safety. Although it may increase the costs to manufacture a product, DFM can also increase market share and extend the product life cycle. The features of DFM products are obvious, yet they can easily be overlooked unless DFM is a deliberate consideration in the product development process. DFM should be considered early in the product design process when the product concept is flexible and change costs are low.

Process Chain Network Analysis

Scott Sampson's Process Chain Network (PCN) analysis framework helps visualize business processes, networks, and managerial issues. He has produced a paper and presentation (*Sampson 2011a, 2011b*: http://services.byu.edu/wp) that explain the PCN framework, which is a powerful approach to service innovation based on exploring process configuration alternatives. Innovation can be introduced into process chains by repositioning steps or sets of steps across the regions of a process domain or across the entities of a PCN. PCN diagrams allow the depiction of complex processes, identification of value propositions and cost drivers, and consideration of strategic process alternatives. Practitioners and researchers can use PCN analysis to visualize, analyze, and solve service management and configuration problems. Another presentation (*Sampson 2014*) reviews PCN analysis and discusses service performance measurement, concluding that this can be difficult to carry out because of various human factors.

Books and Papers

The book by Tim Baines and Howard Lightfoot (*Baines and Lightfoot 2013*) is rapidly becoming the classic text on servitization. It gives a complete review of the subject, detailing the main business issues (context, new processes, and implications) as well as covering service organization and delivery, facilities, Information Technology (IT), communications, and human factors. The authors have been able to use material from many of the companies that have led the development of servitization.

The excellent book by Valarie Zeithaml and others (*Zeithaml et al. 2014*) outlines the challenges of launching a service and solutions business within product-oriented organizations, providing a framework (called service infusion) that describes the different types of services and solutions that such companies can offer.

The book edited by Gunter Lay (*Lay 2014*) provides a comprehensive collection of sectoral studies of servitization in manufacturing, providing detailed analyses of manufacturing sectors that elucidate the options and barriers to servitization for each sector. The book also presents research that investigates the necessity to adapt various processes and departments inside manufacturing companies to servitized business models.

The book edited by Irene Ng and others (*Ng et al. 2011*) covers various aspects of service in complex engineering systems, with perspectives from engineering, management, design, operations research, strategy, marketing, and operations management that are relevant to different disciplines, organization functions, and geographic locations. Irene Ng's books (*Ng 2009, 2013, 2014a*) look at the latest concepts in pricing and revenue management for services, the fundamentals of value and the markets that are created from it, the differences between access and ownership that are driving new digital marketplaces, and the spin-offs that result. Many of her ideas apply to servitization as well.

The book by Lino Cinquini and others (*Cinquini et al. 2013*) looks at some servitization issues in more detail.

A publication by Raconteur (*Baines et al. 2014*) provides a high-level overview of product life cycle management.

The article by Lothar Schuh and Martin Schiefer (*Schuh and Schiefer 2012*) discusses how advanced technologies, new business processes, and modern workflow management now allows the delivery of more value-added services to customers, with modern IT systems opening up a variety of innovative business opportunities.

There are many "intangibles" and other hard-to-quantify variables and issues in servitization. Douglas Hubbard's book (*Hubbard 2014*) describes a number of thought processes and tools that give anyone the ability to produce estimates for such quantities that are good enough to enable progress to continue: there is an excellent discussion of some of the issues surrounding "intangibles" in a presentation by the same author (*Hubbard 2007*). A companion website (www.hubbardresearch.com) offers supporting materials, including a four-page laminated summary of the methods.

The book by Mark Stickdorn and Jacob Schneider (*Stickdorn and Schneider 2014*) describes and illustrates the emerging field of service design, drawing on the experience of 23 international authors from the global service design community as well as numerous other contributors.

Consultant Papers

The number of general articles on servitization written by consultants and appearing in the media is huge; only a selection of them can be mentioned here.

The paper by Barclays Corporate (*Barclays 2011*) previews a report examining how UK manufacturing competes globally and the steps the industry has taken to move up the value chain. Their survey shows that while the UK manufacturing sector has a sound base of companies with a service offering, the United Kingdom still lags behind other developed nations (and increasingly, emerging economies) and that servitization offers a real opportunity for UK manufacturing to move up the value chain, while ignoring servitization could affect global competitiveness.

The presentation by Nick Frank of Noventum IT Management Consulting (*Frank 2014*: www.noventum.com) gives the results of a survey which concludes that successful companies do not see service as a stand-alone revenue opportunity, but as a business solution to overcoming the overall growth challenges of a company and that advanced services are a key differentiator in driving company-wide revenue growth. Another article by Nick (*Frank 2015b*: www.serviceinindustry.com) discusses the reasons why service thinking is the pathway to profitable growth, while a third article (*Frank 2015a*) predicts that, in the future, assets will no longer be bought but only sold as part of a service. The UK Government's vision for UK manufacturing is discussed in two papers (*Foresight 2013a, 2013b*) which mention the need for new business models that include servitization.

The article by Morris Cohen and others (*Cohen et al. 2006*: http://hbr.org) concludes that firms must identify the products they want to support, design a portfolio of service products, use multiple business models, determine after-sales organizational structures, create an after-sales supply chain, and monitor performance in order to be successful.

The articles by Philipp Angehrn and others (*Angehrn et al. 2013*; *Roland Berger 2014*) conclude that OEMs need to move away from simply selling parts and services and move toward providing performance and asset management, shifting their focus from products to outcomes. Unless they adapt their service portfolios, they risk seeing their installed base taken over by lower-cost third-party service providers or by competitors who are better than them at generating revenue on the back of data analytics.

The paper by Arthur D Little (*Arthur D Little 2005*: www.arthurdlittle.com) surveys the development of service innovation in manufacturing, looking at factors that influence growth. This theme is also examined in the article by Jeffrey Glueck and others (*Glueck et al. 2007*) which also quotes a number of examples of service excellence.

The paper by the Aberdeen Group (*Aberdeen Group 2005*: www.aberdeen.com) looks at the strategies that best-in-class companies employ to succeed with servitization, including regarding aftermarket service as a top-line business opportunity, involving stakeholders early and often in the business transformation process, adopting an enterprise-wide perspective, bringing field service and parts logistics under one operational umbrella, addressing process deficiencies, and defining requirements and success criteria clearly before deploying technology.

The report published by Oxford Economics on manufacturing transformation (*Oxford Economics 2013*: www.oxfordeconomics.com) concludes that a confluence of external market pressures, new technologies, and new competition compels manufacturers to transform their business processes. Their survey data and interviews reveal that effective, lasting transformation requires a change in strategy and planning, an intense focus on creating service-based value, and an embrace of technology-driven innovation above and beyond traditional R&D.

The paper by PTC Inc. (*PTC 2015*: www.ptc.com) surveyed 370 service executives around the globe and validated the existence of a service continuum: a predictable model of service transformation that companies follow as they evolve toward advanced services offerings to improve revenue, profit margin, and customer value. Other papers by them (*PTC 2013a, 2013b*) analyze the market shifts that are driving manufacturing transformation and discuss how transforming service is key to driving profitable growth and competitive advantage.

The article by Parmar and others (*Parmar et al. 2014*) looks at how companies can use data to deliver new business value, while Richard Lucas' article (*Lucas 2013*) looks at how the increasing availability of data is impacting manufacturing, as data from in-service products affects new designs and enables new service offerings. Augmenting products to generate data, digitizing assets, combining data within and across industries, trading data via open platforms, and codifying a distinctive service capability are seen as very important.

The Capgemini Consulting/MIT Sloan Management School paper by George Westerman and others (*Westerman et al. 2011*: www.capgemini.com) presents the findings from a global study of how 157 executives in 50 large traditional companies are managing and benefiting from digital transformation, describing the elements of successful digital transformation and showing how to assess a firm's digital maturity. The research shows that, although large traditional firms are truly different from digital entrants, many are starting to transform their businesses successfully through digital technology.

The report by David Hart and others (*Hart et al. 2013*) gives a wide-ranging overview of the trends in servitization in Europe in 2013/2014.

The presentation by Daniela Buschak of the Fraunhofer Institute in Germany (*Buschak 2014*: www.fraunhofer.de) looks at the benefits to customers and service providers in the German machine tool industry of various after-sales services, while the presentation by Christian Lerch and others (*Lerch et al. 2014*) reviews the impact of servitization across a number of German industrial sectors, finding wide variations between different industries.

The article by David Houghton and Garry Lea (*Houghton and Lea 2009*) describes the approach used to deliver a high quality and effective contracting for availability (CFA) service.

Randy Thompson's presentation (*Thompson 2012b*) breaks down the move to a successful smart-services evaluation into ten well-explained steps.

Finally, mention should be made of Mary Meeker's widely read and well-respected annual surveys of internet trends (*Meeker 2014, 2015*: www.kpcb.com) which do not discuss servitization and asset management explicitly but do cover important hardware, software, and societal developments that will have an impact on these subjects.

Cambridge Service Alliance

Servitization has also attracted the attention of academia. The Cambridge Service Alliance, part of the University of Cambridge (www.cambridgeservicealliance.org), is one of the most active academic groups in the United Kingdom that is studying all aspects of servitization. The list of papers on the website and elsewhere is extensive. A summary of those the author has found is given below.

The presentation by Nigel Slack (*Slack 2005*) looks at why servitization is happening from the customer and supplier viewpoints, how servitization is happening as firms move from merely using product knowledge to capturing some of the value from end-use, and the risks involved from failing to meet market expectations and/or properly leveraging operational capabilities.

The presentation by Andy Neely (*Neely 2009*) surveys global trends in servitization, while the paper by Andy Neely and others (*Neely et al. 2011*) explores the extent of servitization in different countries.

The paper by Ivanka Visnjic and Bart Van Looy (*Visnjic and Van Looy 2011*) examines the relationship between servitization and performance in terms of the growth dynamics between products and services and the profitability of servitization. The findings not only suggest the importance of carefully considering business models in terms of complementarities, substitution effects, and their net impact but also suggest pathways to sustainable growth for servitizing manufacturing firms.

A working paper by Andy Neely (*Neely 2012*) explores the potential that different forms of service offer in addressing the grand challenges that society faces. The paper identifies five different forms of PSS (ranging from attempts to vertically integrate to services that completely replace products) and explores how each of these five forms might help address the challenges society faces.

The presentation by Andy Neely and Ornella Benedettini (*Neely and Benedettini 2010*) looks at the economic, strategic, and environmental rationales and the mind-set, timescale, and business model challenges of servitization.

The paper by Ivanka Visnjic and others (*Visnjic et al. 2013*) shows why service innovation is different. Conventional product-oriented thinking about the innovation process does not automatically translate into a services context, and firms that prepare properly before engaging in relational service provision (including hiring managers with service delivery experience or investing in a new stand-alone business unit to take on the services element) are more likely to be successful.

The paper by Ivanka Visnjic Kastalli and Bart Van Looy (*Visnjic Kastalli and Van Looy 2013*) analyzes the "servitization paradox" (in which servitization implementation leads to a potential performance decline) by disentangling the value creation and value appropriation processes of 44 national subsidiaries of a global manufacturing firm turned product-service provider in the 2001-2007 period, finding that the firm under study was able to successfully transcend the inherent substitution of products by services and to enact complementary sales dynamics between the two activities. Moreover, labor-intensive services such as maintenance, which imply higher levels of customer proximity, further enhance product sales.

An executive briefing by the group (*Cambridge Service Alliance 2013a*) identifies four key areas (effective decision-making, organizational changes, data capture, and predictive analytics) that must be adopted or utilized more effectively to improve asset management practice.

The papers by Ornella Benedettini and Andy Neely (*Benedettini and Neely 2012a, 2012b*) study the effects of service complexity by looking at the differences between simple and complex services.

The working paper by Taija Turunen and Andy Neely (*Turunen and Neely 2012*) identifies the structural changes that are needed when a manufacturer seeks to increase its service provision, illustrating how different organizational tensions emerge during the shift to services and how the service teams self-organize in response to these tensions.

The working paper by Ivanka Visnjic and Bart Van Looy (*Visnjic and Van Looy 2013*) shows that success in setting up a service business within a manufacturing firm is due to three operational capabilities: the skill set to extend the relationship with a broad client base, the capability to develop sophisticated service offerings for selected clients, and the ability to offer all the services efficiently.

Ivanka Visnjic and Andy Neely studied how twelve complex service providers changed their business models (*Visnjic and Neely 2012a, 2012b*). It concludes that success depends on a clear understanding of business risk, innovation, understanding and expansion of business offerings, and full customer engagement.

The presentation by Guang-Jie Ren and Mike Gregory (*Ren and Gregory 2009*) looks at the evolution of product support services, from break-and-fix to value-added service contracts and consultancy-led solutions that meet customer objectives.

Jingchen Hou and Andy Neely present a useful table showing the barriers to servitization in their paper (*Hou and Neely 2013*) together with a comprehensive series of references.

The working paper by Ornella Benedettini and others (*Benedettini et al. 2013*) looks at the differences between successful and unsuccessful servitized companies, concluding that successful firms tend to be older, larger, and more diversified than unsuccessful ones and appear to offer less variety of service types.

The working paper by Ivanka Visnjic and others (*Visnjic et al. 2012*) warns of the mixed impact of services on the financial performance for manufacturing firms. It identifies the negative effects of increasing service "breadth" (measured in number of services offered) and points out that increasing service "depth" (measured in completeness/sophistication of service offering) results in higher profit margins and an increase in market value.

The paper by Michael Barrett and others (*Barrett et al. 2011*) examines the challenges and opportunities that firms are likely to face as they move toward collaborative innovation with outside organizations, and how to address the readiness, trust, and governance issues.

The executive briefing by Alexandra Brintrup and others (*Brintrup et al. 2012*) imagines a complex supply chain that could organize and manage itself with comparatively little human intervention. The model the team developed incorporates software agents representing individual components and component communities on the demand side, as well as the suppliers. In addition, there are software agents that are tasked with searching for suppliers and resolving competition for resources through auctions. The self-serving asset agent platform has the potential to deliver reduced complexity, reduced time to service, less risk of system failure, and better decision-making.

The paper by Rachel Cuthbert and others (*Cuthbert et al. 2012*) examines whether, for a proposed contract based around complex equipment, current information systems are capable of providing information at an acceptable quality. It concludes that the control, ownership, and use of information differ across contract types as the operation and responsibility boundaries change.

The working paper by Jingchen Hou and Andy Neely (*Hou and Neely 2014*) explores the effects of social capital (supplier-customer relationships) on risks taken in outcome-based contracts by suppliers, exploring the risks of outcome-based contracts, construction and development of social capital, and effects of social capital on risks taken by suppliers.

The paper by Peter Fielder and others (*Fielder et al. 2014*) draws together the journey that BAE Systems has been and continues to travel along as it moves to delivery of a more service-oriented portfolio of products and the research that Cambridge University has been pursuing to better understand how accountabilities are managed for through-life service.

The paper by Veronica Martinez and Trevor Turner (*Martinez and Turner 2014*) presents a case study tracing the value proposition shifts of a single firm over 40 years, discussing the firm's strategic decisions, market adaptation, and influencing factors triggering the shifts to new offerings. A value proposition framework is introduced for organizations to diagnose the design and delivery of service value propositions that could track the endurance and adaptability of those propositions in the market over the long term.

The paper by John Mills and others (*Mills et al. 2012a*) highlights the challenges faced in translating client aspirations and fears into a complex service support contract and the potential benefits of understanding the client's full requirements, even though they may be unaffordable or too difficult to contract. The paper asserts that stakeholders must understand their mutual requirements fully to help generate a relationship where even the uncontracted service requirements are understood and respected: without that understanding service improvement will be difficult to achieve. Another paper by the same team (*Mills et al. 2012b*) describes the process of enterprise imaging, a way of providing a picture of a multi-organizational enterprise that provides products and/or service outputs.

The paper by Andy Neely (*Neely 2013*) presents data on the range and extent of servitization globally, contrasting levels in Germany with France, the United Kingdom, the United States, and the BRIC countries (Brazil, Russia, India, and China) and exploring the strategic, economic technological and environmental reasons driving firms to embrace servitization.

The paper by Sophie Tersago and Ivanka Visnjic (*Tersago and Visnjic 2011*) focuses on the recent business model innovation of healthcare providers in Belgium and discusses their drivers, characteristics, and the effect they have on these firms and the sector as a whole.

The paper by Chander Velu and Philip Stiles (*Velu and Stiles 2013*) examines the process of strategic decision-making when adopting a new business model that can disrupt an existing one. It offers a framework to help firms manage the decision-making and cannibalization processes when new and existing business models need to be run in parallel.

The paper by Chander Velu and others (*Velu et al. 2013*) introduces the concept of collaborative market orientation (CMO) which is defined as a set of capabilities that are jointly built, maintained, and exercised by members within an ecosystem. The authors highlight three such CMO capabilities: collaborative intelligence generation, collaborative intelligence dissemination, and collaborative responsiveness. By drawing upon literature to identify its constituent routines, the authors provide actionable steps for organizations to build CMO capabilities.

The paper by Anna Viljakainen and others (*Viljakainen et al. 2013*) tackles the issue of industrial transition into value-and-service-based business and offers a managerial tool to show how customer value is turned into profitable business. It suggests a new business model construct based on the service-dominant approach that analyzes customers as value co-creators, not as targets for selling to.

The executive briefing by Ivanka Visnjic Kastalli and Andy Neely (*Visnjic Kastalli and Neely 2013*) studies the business ecosystems of cities to see what lessons can be learned about the ecosystems that corporations inhabit and understand how business ecosystems unleash value.

The paper by Claire Weiller and Andy Neely (*Weiller and Neely 2013*) presents innovative business models that are being developed in four countries to support the commercialization of Electric Vehicles (EVs). The findings emphasize the importance of interindustry partnerships in new value chain configurations and an ecosystem view of value creation and capture.

The paper by Philip Woodall and others (*Woodall et al. 2013*) presents an approach that can help an organization to develop a suitable data quality assessment technique that leverages best practices.

The working paper by Ornella Benedettini and others (*Benedettini et al. 2014*) questions the general statement that turning to a service-oriented business model is a means for manufacturing companies to increase their chances to survive and prosper.

It suggests that the impact of a service orientation on a firm's likelihood of survival depends on the presence of certain preconditions within the firm and indicates characteristics that firms should try to protect when they expand into services if they do not want to incur greater risk of failure.

The working paper by Ruth Bolton and others (*Bolton et al. 2014*) argues that service organizations have focused too much on their core service performance and too little on designing the customer journey that enhances the entire customer experience. It points out that firms need to take a holistic view in order to deeply understand all customer-firm interactions at all "touch points" and should focus on the small details that make big differences to customers.

The working paper by Markus Eurich and Michael Burtscher (*Eurich and Burtscher 2014*) studies the situation in which customers are dependent on a single manufacturer or supplier for a specific service and cannot move to another vendor without substantial costs or inconvenience. This "lock-in" effect, typically considered by a business to be favorable and desirable, can also have negative consequences for the business, particularly when the customer becomes dissatisfied with the service.

The paper by Paulo Gaiardelli and others (*Gaiardelli et al. 2014*) develops a comprehensive model for classifying traditional and green product-service offerings, thus combining business and green offerings in a single model. It also describes the model-building process and its practical application in a case study.

The presentations by Andy Neely look at the role of big data and analytics in shifting a business model toward service provision (*Neely 2014a*) and provide insights into the servitization of manufacturing (*Neely 2014b*) and the shift to services (*Neely 2014c*), showing the value proposition, value delivery, and accountability spread that exist within a service provision ecosystem. Another article by Andy Neely (*Neely 2015*) shows that servitization is beginning to have a global impact.

The working paper by Mohamed Zaki and Andy Neely (*Zaki and Neely 2014*) reports on a study which provides a series of implications that may be particularly helpful to companies considering big data for their businesses.

The working paper by Jonathan Trevor and Peter Williamson (*Trevor and Williamson 2014*) discusses the benefits of moving toward an organization that looks more like a network of partnerships (a business ecosystem) recreated inside the company. This means viewing an organization not as a machine but as a set of species with different capabilities and knowledge and designing structures and incentives that encourage connections between them that can be constantly reconfigured.

The executive briefing by Florian Urmetzer and others (*Urmetzer et al. 2014*) gives an overview of the key considerations in asset management design, finding that few organizations have implemented asset management systems that are designed according to a specific methodology or set of design principles and that most firms do not view asset management as a strategic exercise but instead deal with it in an *ad hoc*, reactive way, responding to market conditions and the current trading and operational environment. The briefing includes several key elements that should be considered when designing and applying an effective asset management system in a service provision ecosystem.

The working paper by Josh Brownlow and others (*Brownlow et al. 2015*) presents an integrated framework that could help stimulate an organization to become data driven by enabling it to construct its own data-driven business model, centered around desired business outcomes, organization dynamics, resources, skills, and the business sector within which it sits, in coordination with the six fundamental questions for a data-driven business.

The working paper by Chara Makri and Andy Neely (*Makri and Neely 2015*) shows how, while the shift to services can significantly enhance the manufacturer's competiveness and sales revenues, the long-term nature of the resulting contracts and the multiple suppliers and partners involved expose manufacturers to an increased number of operational, performance, and financial risks. As a result, manufacturers need to develop a whole new set of capabilities, adapt their organizational structures in order to provide services to their customers in a successful way, and understand fully their responsibilities stemming from the services they provide.

The working paper by Ornella Benedettini and others (*Benedettini et al. 2015*) describes the initial steps of a study that examines 138 companies from the aerospace and defense industry to provide evidence on how manufacturing companies configure and orchestrate service-relevant capabilities in practice.

The executive briefing by Veit Dinges and others (*Dinges et al. 2015*) provides an invaluable guide to the technologies that are likely to play a pivotal role in the future of servitization. In doing so it offers integrated product-service providers some insights into how they can maintain or gain competitive advantage in their markets.

The working paper by Veronica Martinez and others (*Martinez et al. 2015*) explores the steps and practices regarding management of the shift to services in the animal health industry.

Finally, the KT-Box handbook, presenting a range of diagnostic and management tools to support engineering service operations, is also available from the Cambridge Service Alliance (*Cambridge Service Alliance 2013b*, www.cambridgeservicealliance.org/kt-box-learn.html).

Aston Business School

Another very strong academic group is led by Professor Tim Baines at Aston Business School (www.aston-servitization.com). Once again, the list of papers on the website and elsewhere is extensive. A summary of those the author has found is given below.

The proceedings of the 2013, 2014, and 2015 Spring Servitization Conferences held at Aston Business School (*Aston Business School 2013a, 2014, 2015c*) are a source of information on many topics to do with servitization. Kris Oldland's article (*Oldland 2015*) summarizes several of the key points from the 2015 Aston Spring Servitization Conference.

The paper from Aston Business School (*Aston Business School 2013b*) looks at the impact of servitization on UK manufacturing, describing the new and alternative services sector that is evolving which is rooted in the technological competences of manufacturing and showing strong growth opportunities, with a number of companies exploiting this strategy to create new and resilient revenue streams. Tim Baines' presentation (*Baines 2012*) describes the servitization process as one that uses advanced services based on intimate product knowledge to deliver high value to the customer, with customer outcomes, backed up by long-term risk and reward contracts, as deliverables.

An infographic describing servitization has been produced by the Aston Business School (*Aston Business School 2015d*); an executive education program is also available (*Aston Business School 2015b*).

The presentations and article by Tim Baines (*Baines 2013a, 2013c, 2014b, 2014c, 2015a, 2015b*) concisely summarize the main issues and impacts of servitization, which are also covered in four articles in *The Manufacturer* (*Baines 2013b, 2014a, 2014e, 2014f*) and an article in Raconteur (*Baines 2014d*).

CHAPTER 2

The paper by Tim Baines and others (*Baines et al. 2007*) discusses the concept of a Product-Service System (PSS), an integrated combination of products and services embracing a service-led competitive strategy, environmental sustainability, and the basis for differentiation from competitors who simply offer lower-priced products. The paper defines the PSS concept, reports on its origin and features, gives examples of applications along with potential benefits and barriers to adoption, summarizes available tools and methodologies, and identifies future research challenges.

The paper by Tim Baines and others (*Baines et al. 2008a*) reports the state of the art of servitization by presenting a review of literature currently available on the topic.

The paper by Tim Baines and others (*Baines et al. 2008b*) presents a framework that will help manufacturing firms to configure their internal production and support operations to enable effective and efficient delivery of products and their closely associated services.

The paper by Tim Baines and others (*Baines et al. 2010*) presents a survey that explores the extent, motivations, challenges, and successes of servitization within the business-to-business sector. The findings indicate, for example, that many manufacturers are succeeding with their service strategies, that they are attracted to these as a source of customer focus and revenue growth, and that such strategies require less organizational change than might be expected.

The paper by Tim Baines and others (*Baines et al. 2009*) illustrates the form of a real-life servitization process model and summarizes the key challenges that a typical manufacturer experiences in supporting such a model. The paper is based on an in-depth case study with a leading provider of industrial products and related services, presenting an illustration of its servitized process model and the implications that supporting this model has on the wider manufacturing enterprise.

The paper by Tim Baines and others (*Baines et al. 2011*) identifies the key facilities and practices that successful servitizing manufacturers appear to be deploying and the underlying rationale behind their configuration.

The paper by Tim Baines and others (*Baines et al. 2012*) goes further, presenting in-depth case studies of a range of organizations including Caterpillar, Xerox, MAN, and Alstom and exploring a variety of topics including facilities, ICT, vertical integration and supplier relationships, performance measurement systems, and business processes.

The paper by Howard Lightfoot and others (*Lightfoot et al. 2011*) examines the enabling ICT that successful servitized manufacturers appear to be adopting, concluding that ICT capabilities adopted by successfully servitizing manufacturers can differ significantly from those in conventional production operations because of the differences in business pressures and subsequent performance measures. Production tends to focus on cost, quality, and delivery, whereas advanced services contracts center on performance, availability, reliability, and cost.

The presentation by Andreas Schroeder and Julia Kotlarsky (*Schroeder and Kotlarsky 2014*) explores how the digital economy affects servitization, defining "digital servitization" as both the provision of IT-enabled services by relying on digital components embedded in physical products and the creation of digital ecosystems that help build relationships.

The paper by Tim Baines and others (*Baines et al. 2013*) questions how manufacturers configure their operations to deliver advanced services, focusing on the practices and technologies that underpin the capability to successfully deliver such advanced services.

Jim Reed's presentation (*Reed 2013*) looks at servitization from the customer's point of view, emphasizing the need to be an intelligent customer and understand the true drivers for outsourcing, the intent and details of contracts, and the supplier's aims.

The paper by Alex Bill (*Bill 2014*) describes the approach Alstom has taken to servitize its large gas turbine/combined cycle power plant business.

The paper by Jamie Burton and others (*Burton et al. 2015*) investigates the impact of territorial tensions on manufacturer, intermediary, and customer perspectives during servitization efforts, concluding that business leaders need to both understand how servitization might be perceived as territorial aggression and provide strong leadership in order to position servitization activity with sufficient transparency to make the process appear less threatening.

The paper by Ian Machan (*Machan 2015*) expands the traditional lean manufacturing value stream mapping into a tool that can be applied to the design and management of servitized commercial relationships.

The paper by Victoria Uren and Panagiotis Petridis (*Uren and Petridis 2015*) describes a servitization game in which players must make service design decisions based on information provided by non-players.

The paper by Shaun West and Adriano Pascual (*West and Pascual 2015*) reviews the use of equipment life cycle analysis to identify new service opportunities.

Other UK Universities

Other UK academic groups with strong servitization research capabilities include the universities of Cranfield (including the commercially sponsored Integrated Vehicle Health Monitoring (IVHM) Centre: www.cranfield.ac.uk/ivhm), Exeter, and Warwick (http://warwick.ac.uk/fac/sci/wmg/).

The presentations by Jim Angus (*Angus 2015*) and Ip-Shing Fan (*Fan 2015*) discuss the development, operation, and outputs from the Cranfield IVHM Centre in detail.

Howard Lightfoot's presentation (*Lightfoot 2011*) gives an excellent overview of PSSs, looking at how a manufacturer can compete by producing the best total solution for the customer rather than the traditional battlegrounds of operational excellence or product leadership. The presentation discusses business conditions, service types, and risks and includes several supporting case studies.

This point is reinforced by Irene Ng in her paper (*Ng 2010*) which introduces the concept of value co-creation by manufacturers and their customers which can unlock and enhance more value than either party can generate on its own.

The discussion paper by Laura Smith and others (*Smith et al. 2010*) investigates value delivery in equipment-based service, addressing the questions of where and how value is delivered, either via recovery of equipment function, guaranteeing equipment availability, or delivering the outcome the equipment produces.

The research paper by Irene Ng and others (*Ng et al. 2010*) introduces the concept of outcome-based contracting (OBC) as the mechanism for firms to focus on delivering value-in-use as the driver for value co-creation as firms jointly deliver outcomes with the customer. The paper analyzes two OBC-type contracts between the UK Ministry of Defence and two of its industrial partners, finding that in delivering outcomes and achieving value-in-use firms need to reevaluate the way they are structured to deliver a better service.

The discussion paper by Irene Ng and others (*Ng et al. 2011*) argues that the use of technology by value co-creators is determined by the degree of variability of the contexts that they face and could be means-driven rather than goals-driven where goals are unknown.

The presentation by Martin Spring and Katy Mason (*Spring and Mason 2010*) provides a high-level overview of the need to redesign business models to combine product and service elements and align with customers' businesses.

The PhD thesis by Richard Clayton (*Clayton 2011*) investigates whether the approaches to PSS development reported within the PSS literature reflect the actual PSS development practice of servitized manufacturers. It concludes that there are a number of significant differences between the practice of servitized manufacturers and the literature and proposes a new PSS development model that reflects this.

The presentation by Matthew Cook (*Cook 2015*) looks at whether PSS can make consumer markets more sustainable, concluding that sustainability is driven principally by individual consumer attitudes to ownership and behavior which will swamp any PSS benefits.

The presentation by John Erkoyuncu (*Erkoyuncu 2015*) looks at the issues surrounding the modelling of uncertainty and its impact on the evaluation of PSSs.

The presentations by Glenn Parry (*Parry 2014*) and Irene Ng (*Ng 2014b*) discuss the broad principles behind servitization and value co-creation, looking at the delivery of customer experiences rather than products or services.

The PhD thesis by Miguel Puche Alonso (*Puche Alonso 2007*) surveys a number of companies to gather their views on the benefits and barriers to introducing PSS to their businesses. The companies surveyed saw clear benefits from increases in differentiation, customer intimacy, market dominance, and revenues but were seeing problems with lack of experience in services design and cultures focused on the product.

The article by Amir Toossi and others (*Toossi et al. 2010*) considers PSS as an emerging approach to creating a win-win situation for OEMs and their customers and discusses maintenance outsourcing as a step toward applying this new concept. Two important elements in successful maintenance outsourcing are understanding the risks to the service providers in taking responsibility for their customers' maintenance activities and understanding the value that is delivered to the customers by the maintenance service provider. The concept of "value-in-use" is introduced as an improved decision criterion for maintenance outsourcing, and the need for a tool to assess value-in-use is explained.

The presentation by Nabil Sultan (*Sultan 2014*) looks at the disruptive effects of servitization on IT and education via cloud services, including massive open online courses (MOOCs) which could be considered as cloud services for education.

The Engineering and Physical Sciences Research Council (EPSRC) in the UK has recognized the need to support technology developments in this field and has established the EPSRC Centre for Innovative Manufacturing in Through-life Engineering Services, hosted by Cranfield and Durham Universities (www.through-life-engineering-services.org, www.cranfield.ac.uk/sas/tesi). The proceedings of its international conferences (*Roy et al. 2012, 2014*) present a number of papers addressing several aspects of the technological and operational challenges involved in service provision.

The book edited by Louis Redding and Rajkumar Roy (*Redding and Roy 2015*) takes a wide-ranging view of through-life engineering services (TES), their supporting functions, servitization, and service delivery, with contributions from many experts in this field. Louis Redding's PhD thesis (*Redding 2012*) describes strategies and methods for manufacturing companies seeking to embrace servitization, identifying IVHM as a key enabler.

The paper by Raj Metha (*Metha 2015*) shows the importance of the support and service industry to the UK economy, which generates £11bn in UK revenue and £23bn globally.

Universities Outside the United Kingdom

Rogelio Oliva's presentation (*Oliva 2009*) looks at evidence of companies making the transition from products to services, the difficulties they may face and the steps they need to go through.

Charlotta Windahl's PhD thesis (*Windahl 2007*) discusses and analyzes the challenges of developing and commercializing integrated solutions in the capital goods sector, building on case studies of firms experimenting with integrated-solution offerings. Her thesis shows that the development and commercialization of integrated solutions represents a multifaceted, iterative, and complex process for the firms under study, who need to combine product, service, and business innovations and create new business structures and new relationships with customers and possible partners. The three activities of innovating, organizing, and building relationships are dependent on changing market structures, customer demands, and business cycles.

The paper by Heiko Gebauer and Thomas Friedli (*Gebauer and Friedli 2005*) attempts to provide an understanding of behavioral processes and their impact on the transition from products to services, focusing on German and Swiss machinery and medical equipment manufacturing industries. The important managerial implications and recommendations are to establish "value-added" managerial service awareness, change managerial role understanding from traditional customer support to business manager, establish a "value-added" employee service awareness, and change employees' understanding of their roles, from selling products to providing services.

The report by Ulf Karlsson (*Karlsson 2007*) describes three industries in Sweden (trucks, fast trains, and medical devices) that have implemented aftermarket-based manufacturing strategies successfully.

The article by Saara Brax and Katrin Jonsson (*Brax and Jonsson 2009*) describes two manufacturing firms entering the condition-based maintenance business and reveals the complex nature of establishing integrated solutions where value is created incrementally through the customer-provider coproduction process. Building an integrated solutions business requires managing the interdependence of the offerings both within the provider company and between the provider and the client to enable this collaborative process.

The presentation by Arnold Tukker (*Tukker 2014*) discusses PSSs and the literature describing them, concluding that results-oriented offerings are better than use-oriented offerings, that PSS need to be evaluated like any other business models, and that all PSS designers should ask, "Why has a smart entrepreneur not yet put my idea on the market?"

Long-term, service-driven engineering projects demand that maintenance and maintainability issues need to be considered at a much earlier stage by the manufacturer, who was never impacted directly by such issues before. The paper by Chris Ivory and others (*Ivory et al. 2001*) concludes that project managers are finding that it is not enough simply to tie together the various packages of goods and services to form a physical product, but that is necessary to fully integrate service and physical elements into a unified service-led product offering. Project managers are under pressure to find ways to ensure that the design of equipment and the organizational structures left behind after the delivery of the assets reflect the need to maintain the equipment for maximum availability over the long term.

The thesis by Alessandro Bertoni (*Bertoni 2012*) investigates the early stages of aerospace product development, proposing methods and tools that help engineers and designers understand the value impact that different design alternatives make to PSSs. He also shows the need to complement requirement information with value assessments and proposes color-coded computer-aided design (CAD) models to help communication of value outcomes.

The PhD thesis by Saara Brax (*Brax 2013*) reviews existing service definitions, showing that eight different definitional approaches are recognized (which indicates a lack of any common definition of services) and that authors in operations management in many journals usually do not analyze or develop a robust definition of services.

Jorge Eduardo Parada Puig's PhD thesis (*Parada Puig 2015*) studies the role of serviceability of capital assets in the practices of large acquisition projects, using trains as an example.

References

* denotes sources that the author found particularly helpful and/or inspiring.

Aberdeen Group, *Best Practices in Strategic Services Management: Integrating the Service Chain to Drive Profits and Competitive Advantage* (Boston, MA: Aberdeen Group, 2005). Pg. 27

Angehrn, P., Siepen, S., Lässig, R., and Herweg, O., *Evolution of Service* (Hamburg, Germany: Roland Berger Strategy Consultants GmbH, 2013). Pg. 27

Angus, J., "Integrated Vehicle Health Management Centre," *Integrated Vehicle Health Management Event*, CleanTech Business Network, Geldards, Pride Park, Derby, UK, April 30, 2015 (Cranfield, UK: Cranfield IVHM Centre, Cranfield University). Pg. 35

Arthur D Little, *Profitable Growth through Service* (London, UK: Arthur D Little, 2005). Pg. 27

Aston Business School, *Proceedings of the 2013 Spring Servitization Conference*, Aston Business School, Birmingham, UK, May 20-22, 2013a (Birmingham, UK: Aston Business School, Aston University). Pg. 33

Aston Business School, "Servitization Impact Study: How UK Based Manufacturing Organisations Are Transforming Themselves to Compete through Advanced Services," Birmingham, UK: Aston Business School, Aston University, 2013b. Pg. 33

Aston Business School, *Proceedings of the 2014 Spring Servitization Conference*, Aston Business School, Birmingham, UK, May 12-14, 2014 (Birmingham, UK: Aston Business School, Aston University). Pg. 33

Aston Business School, *Delivering Growth: Exploring the Potential of Advanced Services within the Road Transport Industry* (Birmingham, UK: Aston Business School, Aston University, 2015a). Pg. 15

Aston Business School, "Made to Serve: An Executive Education Programme in Servitization and Advanced Services," Birmingham, UK: Aston Business School, Aston University, 2015b. Pg. 33

Aston Business School, *Proceedings of the 2015 Spring Servitization Conference*, Aston Business School, Birmingham, UK, May 18-19, 2015c (Birmingham, UK: Aston Business School, Aston University). Pg. 33

Aston Business School, "Servitization Explained," Birmingham, UK: Aston Business School, Aston University, 2015d. Pg. 33

Baines, T., "Operations Strategy: Systems Engineering in the Business Process for Technology-Based Companies," Royal Academy of Engineering, London, UK, July 19, 2012 (Birmingham, UK: Aston Business School, Aston University). Pg. 33

Baines, T., "Made to Serve - What It Takes to Compete through Servitization and Product-Service Systems," *Spring Servitization Conference 2013*, Aston Business School, Birmingham, UK, May 20-22, 2013a (Birmingham, UK: Aston Business School, Aston University). Pg. 33

Baines, T., "Servitization Explained," *The Manufacturer* 16, no. 9 (November 2013b): 54-55. Pg. 33

Baines, T., "Servitization Impact Study - Lessons from Industrial Leaders in the UK," *Spring Servitization Conference 2013*, Aston Business School, Birmingham, UK, May 20-22, 2013c (Birmingham, UK: Aston Business School, Aston University). Pg. 33

Baines, T., "Sustainability and Growth - The Impact of Servitization," *The Manufacturer* 16, no. 10 (December/January 2014a): 44-45 (London, UK: The Manufacturer). Pg. 33

Baines, T., *Competing through Servitization* (Birmingham, UK: Aston Business School, Aston University, 2014b). Pgs. 10, 11, 12, 33

Baines, T., "Made to Serve: What It Takes to Compete through Servitization and Product-Service Systems," *Aftermarket Europe Conference*, Grand Hotel Huis ter Duin, Noordwijk, the Netherlands, October 22-24, 2014c (Birmingham, UK: Aston Business School, Aston University). Pg. 33

Baines, T., "More Companies Are Now Made to Serve," UK Manufacturing, March 3, 2014d (London, UK: Raconteur/CIMdata/Times Newspapers Ltd). Pg. 33

Baines, T., "Serve Yourself," *The Manufacturer* 14, no. 4 (May 2014e): 54-55. Pg. 33

Baines, T., "The Small Serving Big," *The Manufacturer* 14, no. 3 (April 2014f): 44-45. Pg. 33

Baines, T., "Servitization: Organisational Structure and Culture," *ESRC/BAE Systems Servitization Seminar*, Aston Business School, Birmingham, UK, February 23, 2015a, Unpublished presentation (Birmingham, UK: Aston Business School, Aston University). Pg. 33

Baines, T., "Welcome to the New Age…," *Field Service News*, no. 5 (March/April 2015b): 36. Pg. 33

* Baines, T. and Lightfoot, H., *Made to Serve* (Chichester, UK: John Wiley & Sons Ltd, 2013). Pg. 25

Baines, T., Clark, L., King, L., Lamb, J. et al., *Product Lifecycle Management* (London, UK: Raconteur/CIMdata/Times Newspapers Ltd., January 27, 2014). Pg. 26

Baines, T., Lightfoot H., Benedettini, O., Whitney, D. et al., "The Adoption of Servitization Strategies by UK Manufacturers," *Proceedings of the IMechE, Part B: Journal of Engineering Manufacture* 224 (May 1, 2010): 815-829. Pg. 34

Baines, T., Lightfoot, H., and Kay, J., "Servitized Manufacture - Practical Challenges of Delivering Integrated Products and Services," *Proceedings of the IMechE, Part B: Journal of Engineering Manufacture* 223 (September 1, 2009): 1207-1215. Pg. 34

Baines, T., Lightfoot, H., and Smart, P., "Servitization within Manufacturing Operations," *Proceedings of the IMechE, Part B: Journal of Engineering Manufacture* 226 (December 13, 2011): 377-380. Pg. 34

Baines, T., Lightfoot, H., and Smart, P., "Servitization of Manufacture: Exploring the Deployment and Skill of People Critical to the Delivery of Advanced Services," Journal of Manufacturing Technology Management 24, no. 4 (April 26, 2013a): 10. Pg. 33

Baines, T., Lightfoot, H., Benedettini, O., and Kay, J., *The Servitization of Manufacturing: A Review of Literature* (Birmingham, UK: Aston Business School, Aston University, 2008a). Pg. 34

Baines, T., Lightfoot, H., Evans, S., Neely, A. et al., "State of the Art in Product-Service Systems," *Proceedings of the IMechE, Part B: Journal of Engineering Manufacture* 221 (October 1, 2007): 1543-1552. Pg. 34

Baines, T., Lightfoot, H., Peppard, J., Johnson, M. et al., *Towards an Operations Strategy for Product Centric Servitization* (Cranfield, UK: Cranfield Innovative Manufacturing Research Centre, Cranfield University, 2008b). Pg. 34

Baines, T., Lightfoot, H., Smart, P., and Fletcher, S., *Servitization of the Manufacturing Firm: Exploring the Operations Practices and Technologies That Deliver Advanced Services* (Birmingham, UK: Aston Business School, Aston University, 2013). Pg. 34

Barclays, *Servitization and the Future of Manufacturing* (London, UK: Barclays Bank plc, 2011). Pg. 26

* Baréz-Brown, C., *Free! Love Your Work, Love Your Life* (London, UK: Penguin Group, 2014). Pg. 17

Barrett, M., Velu, C., Kohli, R., Salge, T. et al., "Making the Transition to Collaborative Innovation," Cambridge, UK: Cambridge Service Alliance, Cambridge University/NESTA, 2011. Pg. 30

Barringer, H., "Life Cycle Cost and Good Practices," *NPRA Maintenance Conference*, San Antonio Convention Center, San Antonio, TX, May 19-22, 1998 (Humble, TX: Barringer & Associates, Inc.). Pg. 23

Barringer, H., "A Life Cycle Cost Summary," *International Conference of Maintenance Societies (ICOMS®-2003)*, Sheraton Hotel, Perth, Western Australia, May 20-23, 2003 (Humble, TX: Barringer & Associates, Inc.). Pg. 23

Barringer, H. and Weber, P., "Life Cycle Cost Tutorial," *Fifth International Conference on Process Plant Reliability*, Marriott Houston Westside, Houston, TX, October 2-4, 1996 (revised December 2, 1996) (Humble, TX: Barringer & Associates, Inc.). Pg. 23

Bayley, S. and Mavity, R., *Life's a Pitch* (London, UK: Bantam Press, 2007). Pg. 17

Belanger, D., "A Vision of Enterprise Reliability: Looking into the Future of the Industry," *Uptime* 5, no. 40 (April/May 2010): 18-21. Pg. 19

Benedettini, O. and Neely, A., "Complexity in Services: An Interpretive Framework," *23rd Annual POMS Conference*, Chicago, IL, April 20-23, 2012a (Cambridge, UK: Cambridge Service Alliance, Cambridge University). Pg. 29

Benedettini, O. and Neely, A., "Factors Influencing Service Complexity: The Perspective of Servitized Manufacturers," *19th International European Operations Management Association Conference*, Amsterdam, the Netherlands, July 1-5, 2012b (Cambridge, UK: Cambridge Service Alliance, Cambridge University). Pg. 29

Benedettini, O., Davies, J., and Neely, A., "A Capability-Based View of Service Transitions," Cambridge, UK: Cambridge Service Alliance, Cambridge University, 2015. Pg. 33

Benedettini, O., Swink, M., and Neely, A., "Firm's Characteristics and Servitization Performance: A Bankruptcy Perspective," Cambridge, UK: Cambridge Service Alliance, Cambridge University, 2013. Pg. 30

Benedettini, O., Swink, M., and Neely, A., "Impact of Firm Characteristics on Survival: An Empirical Analysis in the Context of Service Strategies," Cambridge, UK: Cambridge Service Alliance, Cambridge University, 2014. Pg. 31

Bertoni, A., "Value Assessment Capabilities in Early PSS Development: A Study in the Aerospace Industry," Licentiate thesis, Department of Business Administration, Technology and Social Sciences, Luleå University of Technology Luleå, Sweden, 2012. Pg. 38

Besanko, D., Dranove, D., Shanley, M., and Schaefer, S., *Economics of Strategy*, 5th ed. (Hoboken, NJ: John Wiley & Sons Inc., 2010). Pg. 18

Bigdeli, A. and Musson, E., "Advanced Services in the Road Transport Industry...," *Field Service News*, no. 7 (July/August 2015): 20 (London, UK: 1927 Media Ltd. Pg. 15

Bill, A., "Competing through Advanced Services: Gas Service Solutions," *Spring Servitization Conference 2014*, Aston Business School, Birmingham, UK, May 12-14, 2014 (Birmingham, UK: Aston Business School, Aston University). Pg. 35

Block, P., *Flawless Consulting: A Guide to Getting Your Expertise Used*, 3rd ed. (San Francisco, CA: Pfeiffer/Wiley, 2011). Pg. 18

Bolton, R., Gustaffsson, A., McColl-Kennedy, J., Sirianni, N. et al., "Small Details That Make Big Differences: A Radical Approach to Consumption Experience as a Firms Differentiating Strategy," Cambridge, UK: Cambridge Service Alliance, Cambridge University, 2014. Pg. 32

Bombardier, "Bombardier Launches Ground-Breaking Innovation in Predictive Maintenance," Press release, September 20, 2006 (Montréal, Québec: Bombardier Inc.). Pg. 14

Bombardier, "Bombardier Inc. Annual Report Fiscal Year Ended, December 31, 2012" (Montréal, Québec: Bombardier Inc., 2013). Pg. 14

Brax, S., "The Process Based Nature of Services," Ph.D. thesis, Department of Industrial Engineering and Management, Aalto University, Helsinki, Finland, 2013. Pg. 38

Brax, S. and Jonsson, K., "Developing Integrated Solution Offerings for Remote Diagnostics," *International Journal of Operations and Production Management* 29, no. 5 (2009): 539-560. Pg. 37

* Brealey, R. and Myers, S., *Principles of Corporate Finance*, 3rd ed. (New York: McGraw-Hill, 1988). Pg. 19

Brintrup, A., McFarlane, D., Ranasinghe, D., Sánchez López, T. et al., "Towards Self-Serving Aircraft - Revolutionising the Service Supply Chain," Cambridge, UK: Cambridge Service Alliance, Cambridge University/Boeing Co., 2012. Pg. 30

Brownlow, J., Zaki, M., Neely, A., and Urmetzer, F., "Data and Analytics - Data-Driven Business Models: A Blueprint for Innovation," Cambridge, UK: Cambridge Service Alliance, Cambridge University, 2015. Pg. 32

Burton, J., Story, V., Zolkiewski, J., Raddats, C. et al., "Identifying (Territorial) Tensions in the Servitized Value Shain," *Spring Servitization Conference 2015*, Aston Business School, Birmingham, UK, May 18-19, 2015 (Birmingham, UK: Aston Business School, Aston University). Pg. 35

Buschak, D., "Benefits and Sacrifices of After-Sales Services in the German Machine Building Industry," *Spring Servitization Conference 2014*, Aston Business School, Birmingham, UK, May 12-14, 2014 (Birmingham, UK: Aston Business School, Aston University). Pg. 28

Buzan, T., *The Ultimate Book of Mind Maps®* (London, UK: Thorsons/HarperCollins, 2005). Pg. 18

Cambridge Service Alliance, *Engineering Asset Management: Issues and Challenges - Delivering Business Objectives by Extracting Value from Assets* (Cambridge, UK: Cambridge Service Alliance, Cambridge University, 2013a). Pg. 29

Cambridge Service Alliance, *KT-Box Handbook: Diagnostic and Management Tools to Support Engineering Service Operations* (Cambridge, UK: Cambridge Service Alliance, Cambridge University, 2013b). Pg. 33

Chan Kim, W. and Mauborgne, R., "Blue Ocean Strategy," *Harvard Business Review* 82, no. 10 (2004): 76-84. Pg. 18

Chan Kim, W. and Mauborgne, R., "Blue Ocean Strategy: From Theory to Practice," *California Management Review* 47, no. 3 (Spring 2005a): 105-121. Pg. 18

Chan Kim, W. and Mauborgne, R., *Blue Ocean Strategy: How to Create Uncontested Market Space and Make the Competition Irrelevant* (Boston, MA: Harvard Business School Press, 2005b). Pg. 18

Cheng, S., Azarian, M., and Pecht, M., "Sensor Systems for Prognostics and Health Management," *Sensors* 10, no. 6 (2010): 5774-5797. Pg. 24

Cinquini, L., Di Minin, A., and Varaldo, R. eds., *New Business Models and Value Creation: A Service Science Perspective* (Milan, Italy: Springer-Verlag, 2013). Pg. 26

Clark, T., Osterwalder, A., and Pigneur, Y., *Business Model You®* (Hoboken, NJ: John Wiley & Sons Inc., 2012). Pg. 18

Clayton, R., "Investigating the Development and Delivery of Integrated Product-Service Systems," Ph.D. thesis, Systems Engineering Innovation Centre, Loughborough University, Loughborough, UK, 2011. Pg. 36

Clayton, R., Backhouse, C., Provost, M., Dani, S. et al., "Applying Systems Engineering to Optimise the Operation and Maintenance of Railway Vehicles throughout the Value Chain," *7th Annual Conference on Systems Engineering Research 2009 (CSER 2009)*, Loughborough, UK, April 20-23, 2009 (Loughborough, UK: Research School of Systems Engineering, Loughborough University). Pg. 14

* Clemen, R. and Reilly, T., *Making Hard Decisions with DecisionTools*, 3rd ed. (Mason, OH: South-Western/Cengage Learning, 2014). Pg. 20

Cohen, M., Agrawal, N., and Agrawal, V., "Winning in the Aftermarket," *Harvard Business Review* 11, no. 5 (2006): 129-138. Pg. 27

Conklin, J., *Dialogue Mapping: Building Shared Understanding of Wicked Problems* (Chichester, UK: John Wiley & Sons Ltd, 2006). Pg. 18

Cook, M., "Implementing More Sustainable Product Service Systems in Consumer Markets: Exploring the Challenges and Opportunities," *ESRC/BAE Systems Servitization Seminar*, Aston Business School, Birmingham, UK, February 23, 2015, Unpublished presentation Milton Keynes, UK: Open University). Pg. 36

Cook, M. and Muiter, M., "Estimating Failure Avoidance Costs," *Uptime* 7, no. 49 (October/November 2011): 44-49. Pg. 22

Cook, S., *Delivering Service through Engineering and Product Expertise* (Derby, UK: Rolls-Royce plc, 2004). Pg. 14

Copeland, T., Koller, T., and Murrin, J., *Valuation: Measuring and Managing the Value of Companies*, 2nd ed. (New York: John Wiley & Sons Inc., 1996). Pg. 19

Cordner, K. ed., *The Modern Railway: The Definitive Guide to the UK's Railway Industry in 2014* (Stamford, UK: Key Publishing Ltd, 2013). [Pg. 15]

Cordner, K. ed., *The Modern Railway: The Definitive Guide to the UK's Railway Industry in 2015* (Stamford, UK: Key Publishing Ltd, 2014). [Pg. 15]

Côrtes, J., "Maintenance Engineering and Engineering Economics," *Asset Management & Maintenance Journal* 24, no. 2 (April 2011): 6-8. [Pg. 22]

Côrtes, J., "Maintenance Engineering and Engineering Economics," *Maintenance & Engineering* 14, no. 4 (July/August 2014): 49-51. [Pg. 22]

* Crundwell, F., *Finance for Engineers: Evaluation and Funding of Capital Projects* (London, UK: Springer-Verlag London Ltd, 2008). [Pg. 19]

Cullen, J., "Servitization: A Journey to Better Sell Service Value," *Spring Servitization Conference 2015*, Aston Business School, Birmingham, UK, May 18-19, 2015 (Birmingham, UK: Aston Business School, Aston University). [Pg. 16]

Cuthbert, R., McFarlane, D., and Neely, A., "The Impact of Contract Type on Service Provider Information Requirements," Cambridge, UK: Cambridge Service Alliance, Cambridge University, 2012. [Pg. 30]

Data Systems & Solutions, "Are Your Engines Really as Healthy as They Seem?," in: Copping, P. ed., *The Engine Yearbook 2005* (London, UK: Aviation Industry Press Ltd, 2005), 54-59. [Pg. 12]

Davenport, T. and Harris, J., *Competing on Analytics: The New Science of Winning* (Boston, MA: Harvard Business School Press, 2007). [Pg. 19]

Davenport, T., Harris, J., and Morison, R., *Analytics at Work: Smarter Decisions, Better Results* (Boston, MA: Harvard Business School Press, 2010). [Pg. 19]

de Neufville, R. and Scholtes, S., *Flexibility in Engineering Design* (Cambridge, MA: MIT Press, 2011). [Pg. 23]

Dinges, V., Urmetzer, F., Martinez, V., Zaki, M. et al., *The Future of Servitization: Technologies That Will Make a Difference* (Cambridge, UK: Cambridge Service Alliance, Cambridge University, 2015). [Pg. 33]

* Dixit, A. and Nalebuff, B., *Thinking Strategically: The Competitive Edge in Business, Politics and Everyday Life* (New York: Norton, 1991). [Pg. 18]

Dixit, A. and Pindyck, R., *Investment under Uncertainty* (Princeton, NJ: Princeton University Press, 1994). [Pg. 19]

Easwapillai, P., "ICT: A Key Enabler for Servitization," *Spring Servitization Conference 2013*, Aston Business School, Birmingham, UK, May 20-22, 2013 (Noida, India: HCL Technologies). [Pg. 19]

Economist, "Britain's Lonely High-Flier," *The Economist*, January 8, 2009a. [Pg. 14]

Economist, "Coming in from the Cold," *The Economist*, January 8, 2009b. [Pg. 14]

Economist, "A Tale of Two Industries," *The Economist*, July 30, 2011. [Pg. 14]

Ellis, S., "Integrated Vehicle Health Management: Rail Industry," *Integrated Vehicle Health Management Event*, CleanTech Business Network, Derby, UK, April 30, 2015 (Derby, UK: Bombardier Transportation). [Pg. 14]

Erkoyuncu, J., "Uncertainty in Industrial Product-Service Systems," *ESRC/BAE Systems Servitization Seminar*, Aston Business School, Birmingham, UK, February 23, 2015, Unpublished presentation (Cranfield, UK: Cranfield University). [Pg. 36]

Erskine, A., "Whole Life Costing," *2nd Annual Predictive Asset Management Seminar*, Royal Garden Hotel, London, UK, September 11-13, 2012 (London, UK: IBC Energy/Informa UK Ltd). Pg. 23

Essential Energy, "Value Selling Handbook - Telecom," Internal presentation, Bangalore, India: Essential Energy India Pvt. Ltd, 2014. Pg. 16

Eurich, M. and Burtscher, M., "The Business-to-Consumer Lock-In Effect," Cambridge, UK: Cambridge Service Alliance, Cambridge University, 2014. Pg. 32

European Rail Outlook, "Predictive Maintenance Delivers Great Results," *European Rail Outlook*, Spring 2007. Pg. 14

Europump/Hydraulic Institute, "Pump Life Cycle Costs: A Guide for LCC Analysis for Pumping Systems," Brussels, Belgium/Parsippany, NJ: Europump/Hydraulic Institute, 2001. Pg. 23

Europump/Hydraulic Institute/US DoE, "Pump Life Cycle Costs: A Guide to LCC Analysis for Pumping Systems - Executive Summary," Brussels, Belgium/Parsippany, NJ/Washington DC: Europump/Hydraulic Institute/US Department of Energy, 2000. Pg. 23

Evans, D., *Switching the Lights On*. Swindon, UK: MAN Truck & Bus Ltd (2008). Pg. 15

Evans, D., "MAN Enough for Servitization?," *Field Service News*, no. 6 (May/June 2015a): 20. Pg. 15

Evans, D., "Servitisation for Trucks!," *Service Community Event*, Fujitsu, Stevenage, UK, April 16, 2015b (Swindon, UK: MAN Truck & Bus Ltd). Pg. 15

Fan, I.-S., "IVHM Business Transformation Opportunities," *Integrated Vehicle Health Management Event*, CleanTech Business Network, Geldards, Pride Park, Derby, UK, April 30, 2015 (Cranfield, UK: Cranfield IVHM Centre, Cranfield University). Pg. 35

Farr, J., *System Life Cycle Costing: Economic Analysis, Estimation and Management* (Boca Raton, FL: CRC Press/Taylor & Francis, 2011). Pg. 23

Feldman, K., Jazouli, T., and Sandborn, P., "A Methodology for Determining the Return on Investment Associated with Prognostics and Health Management," *IEEE Transactions on Reliability* 58, no. 2 (June 2009): 305-316. Pg. 24

Feldman, K., Sandborn, P., and Jazouli, T., "The Analysis of Return on Investment for PHM Applied to Electronic Systems," *Prognostics and Health Management (PHM), 2008 IEEE Conference*, Marriott Tech Center, Denver, CO, October 6-9, 2008 (Red Hook, NY: Curran Associates Inc). Pg. 24

Fielder, P., Roper, A., Walby, B., Fuse, J. et al., "Product Safety in a World of Services - Through-Life Accountability," Cambridge, UK: Cambridge Service Alliance, Cambridge University, 2014. Pg. 30

Foresight, "The Future of Manufacturing: A New Era of Opportunity and Challenge for the UK - Project Report," London, UK: Government Office for Science, 2013a. Pg. 27

Foresight, "The Future of Manufacturing: A New Era of Opportunity and Challenge for the UK - Summary Report," London, UK: Government Office for Science, 2013b. Pg. 27

Forrest, P., Provost, M., and Lovell, J., "Diagnostic system and method for monitoring a rail system," World Intellectual Property Organisation International Publication Number WO 2008/034583 A1, Published March 27, 2008. Pg. 14

Forrest, P., Provost, M., and Lovell, J., "Diagnostic system and method for monitoring a rail system," US Patent Application Publication 2010/0204857 A1, Published August 12, 2010. Pg. 14

Frank, N., "Research Findings: Growth through Services," *Spring Servitization Conference 2014*, Aston Business School, Birmingham, UK, May 12-14, 2014 (Birmingham, UK: Aston Business School, Aston University). Pg. 27

Frank, N., "Good-Bye Products, Hello Services? The Road to Monetising the iotS...," *Field Service News*, no. 6 (May/June 2015a): 54. Pg. 27

Frank, N., "Service Thinking," *The Manufacturer* 18, no. 4 (May 2015b): 61. Pg. 27

* Fraser, D., *Relationships Made Easy* (Evesham, UK: HotHive Books, 2010). Pg. 17

Friend, G. and Zehle, S., *Guide to Business Planning* (London, UK: The Economist/Profile Books, 2004). Pg. 20

Gaiardelli, P., Resta, B., Martinez, V., Pinto, O. et al., "A Classification Model for Product-Service Offerings," Cambridge, UK: Cambridge Service Alliance, Cambridge University, 2014. Pg. 32

Galinsky, A. and Schweitzer, M., *Friend & Foe: When to Cooperate, When to Compete and How to Succeed at Both* (London, UK: Random House Business Books, 2015). Pg. 17

Gebauer, H. and Friedli, T., *Behavioural Implications of the Transition Process from Products to Services* (St Gallen, Switzerland: Institute of Technology Management, St Gallen University, 2005). Pg. 37

Geropp, B., "You Can Make It Happen! Clear Communication as Key for a Successful Condition Based Maintenance Program," *Condition Monitoring and Analysis Seminar*, Dexter House, London, UK, February 5-6, 2014 (London, UK: IBC Energy/Informa UK Ltd). Pg. 17

Glueck, J., Koudal, P., and Vaessen, W., "The Service Revolution," *Deloitte Review*, August 2007, 23-33. Pg. 27

* Godin, S., *Linchpin: Are You Indispensable?* (London, UK: Piatkus, 2010). Pg. 17

Gopalan, R. and Kumar, S., "Asset Management & Servitization," *Uptime* 10, no. 69 (February/March 2015): 54-56. Pg. 19

Grantham, A., "ORBITA™ Shows the Way to Optimised Maintenance," *Railway Gazette International*, July 2007. Pg. 14

Gray, D., Brown, S., and Macanufo, J., *Gamestorming: A Playbook for Innovators, Rulebreakers and Changemakers* (Sebastopol, CA: O'Reilly Media, 2010). Pg. 18

Gulati, R., "Does It Pay to Design for Maintainability and Reliability?," *Asset Management & Maintenance Journal* 26, no. 3 (May 26, 2013): 39-40. Pg. 23

Haddad, G., Sandborn, P., and Pecht, M., "A Real Options Optimization Model to Meet Availability Requirements for Offshore Wind Turbines," *MFPT: The Applied Systems Health Management Conference 2011*, Sheraton Oceanfront Hotel, Virginia Beach, VA, May 10-12, 2011a (Dayton, OH: Society for Machinery Failure Prevention Technology). Pg. 25

Haddad, G., Sandborn, P., and Pecht, M., "Using Real Options to Manage Condition-Based Maintenance Enabled by PHM," *Prognostics and Health Management (PHM), 2011 IEEE Conference*, Hyatt Regency Denver Tech Center, Denver, CO, June 20-23, 2011b (Red Hook, NY: Curran Associates Inc). Pg. 25

* Hadfield, C., *An Astronaut's Guide to Life on Earth* (London, UK: Pan Macmillan, 2013). Pg. 17

Hamel, G. and Prahalad, C., *Competing for the Future: Breakthrough Strategies for Seizing Control of Your Industry and Creating the Markets of Tomorrow* (Boston, MA: Harvard Business School Press, 1994). Pg. 18

Harrison, A., *Designing for Service - A Brief Introduction* (Derby, UK: Rolls-Royce plc, 2012). Pg. 14

Harrison, A., "Service Data and Knowledge," *Service Community Birmingham Special Event*, Aston Business School, Birmingham, UK, September 30, 2015 (Derby, UK: Rolls-Royce plc). Pg. 14

Hart, D., Lynch, C., Zeilinski, R., Lightfoot, H. et al., "2013-14 European Services and Trends Report," London, UK: Worldwide Business Research, 2013. Pg. 28

Hegarty, T, "Driving a Services Culture," *Spring Servitization Conference 2014*, Aston Business School, Birmingham, UK, May 12-14, 2014 (Birmingham, UK: Aston Business School, Aston University). Pg. 14

Heller, R. and Hindle, T., *Essential Manager's Manual* (London, UK: Dorling Kindersley Ltd, 1998). Pg. 18

Higgins, R., *Analysis for Financial Management* (New York: McGraw-Hill/Irwin, 2009). Pg. 19

HM Treasury, "Public Competition and Purchasing Unit Guidance No. 35: Life Cycle Costing," London, UK: HM Treasury, 1992. Pg. 23

Hodges, N., *The Economic Management of Physical Assets* (London, UK: Mechanical Engineering Publications Ltd, 1996). Pg. 20

Holland, S., "Integrated Vehicle Health Management in the Automotive Industry," *Electronics System-Integration Technology Conference 2008*, Greenwich, UK, September 1-4, 2008 (New York: Institute of Electrical and Electronics Engineers). Pg. 15

Hou, J. and Neely, A., "Barriers of Servitization: Results of a Systematic Literature Review," Cambridge, UK: Cambridge Service Alliance, Cambridge University, 2013. Pg. 30

Hou, J. and Neely, A., "Analysing the Effects of Social Capital on Risks Taken by Suppliers in Outcome-Based Contracts," Cambridge, UK: Cambridge Service Alliance, Cambridge University, 2014. Pg. 30

Houghton, D. and Lea, G., "Managing and Supporting Ship Availability," *Maintenance & Engineering* 9 no. 2 (March/April 2009): 36-43. Pg. 28

* Hubbard, D., *How to Measure Anything: Finding the Value of Intangibles in Business – Conference Presentation* (Glen Ellyn, IL: Hubbard Decision Research, 2007). Pg. 26

Hubbard, D., *How to Measure Anything: Finding the Value of Intangibles in Business*, 3rd ed. (Hoboken, NJ: John Wiley & Sons Inc, 2014). Pg. 26

Hunt, A., "Reducing the Risk of Asset Failure through Cost Risk Optimisation," *1st Annual Predictive Asset Management Seminar*, Mayfair Conference Centre, London, UK, October 19-20, 2011 (London, UK: IBC Energy/Informa UK Ltd). Pg. 24

Ivory, C., Thwaites, A., and Vaughan, R., "Design for Maintainability: The Innovation Process in Long Term Engineering Projects," *The Future of Innovation Studies Conference*, Eindhoven University of Technology, Eindhoven, the Netherlands, September 20-23, 2001 (Eindhoven, the Netherlands: Eindhoven Centre for Innovation Studies). Pg. 37

Jazouli, T. and Sandborn, P., "A Design for Availability Approach for Use with PHM," *Annual Conference of the Prognostics and Health Management Society 2010*, Hilton Portland and Executive Tower, Denver, CO, October 10-16, 2010 (New York: Prognostics and Health Management Society). Pg. 25

Jazouli, T. and Sandborn, P., "Using PHM to Meet Availability-Based Contracting Requirements," *Prognostics and Health Management (PHM), 2011 IEEE Conference*, Hyatt Regency Denver Tech Center, Denver, CO, June 20-23, 2011 (Red Hook, NY: Curran Associates Inc). Pg. 25

Jennions, I., "5 Steps to Heaven – A Rolls-Royce Perspective," *Innovating with Information*, IBM Amalden Institute, May 7, 2008 (San Jose, CA: IBM Corp). Pg. 14

* Jennions, I. ed., *Integrated Vehicle Health Management: Perspectives on an Emerging Field* (Warrendale, PA: SAE International, 2011). Pg. 18

* Jennions, I. ed., *Integrated Vehicle Health Management: Business Case Theory and Practice* (Warrendale, PA: SAE International, 2012). Pg. 18

* Jennions, I. ed., *Integrated Vehicle Health Management: The Technology* (Warrendale, PA: SAE International, 2013). Pg. 18

* Jennions, I. ed., *Integrated Vehicle Health Management: Implementation and Lessons Learned* (Warrendale, PA: SAE International, 2014). Pg. 18

Jennions, I., Walthall, R., and O'Dell, R., *Taking Data to New Heights: How Airlines, Plane Manufacturers and Suppliers Are Shaping the Future of Integrated Vehicle Health Management* (Warrendale, PA: SAE International, 2013). Pg. 18

Kabukin, D., "Reviewing the Blue Ocean Strategy: Is the Blue Ocean Strategy Valid and Reliable?," MBA thesis, University of Twente, Enschede, the Netherlands, 2014. Pg. 18

Karlsson, U., *Service Based Manufacturing Strategies* (Göteborg, Sweden: Göteborg University, 2007). Pg. 37

Keller, K., *Health Management Technology Integration* (St Louis, MO: Boeing Co, 2006). Pg. 15

King, S., "IVHM in the Aerospace Sector: Benefits of Health Monitoring," *Integrated Vehicle Health Management Event*, CleanTech Business Network, Derby, UK, April 30, 2015 (Derby, UK: Rolls-Royce Controls and Data Services Ltd). Pg. 14

Kirkwood, C., *System Dynamics Methods: A Quick Introduction* (Tempe, AZ: College of Business, Arizona State University, 1998). Pg. 19

Kirkwood, C., "Decision Tree Primer," Tempe, AZ: Department of Supply Chain Management, Arizona State University, 2002. Pg. 21

Kotter, J., "Leading Change: Why Transformation Efforts Fail," *Harvard Business Review* 85, no. 1 (2007): 96-103. Pg. 17

Kotter, J. and Schlesinger, L., "Choosing Strategies for Change," *Harvard Business Review* 86, no. 7 (2008): 130-139. Pg. 17

Lay, G. ed., *Servitization in Industry* (Cham, Switzerland: Springer International Publishing, 2014). Pg. 26

Lee, R., "Rolls Keeps Engine Ticking over to Ride Out the Storm," *The Times*, February 10, 2012. Pg. 14

Lerch, C., Gotch, M., Weidner, N., and Jäger, A., "Service Offers as Competitive Strategy in Industrial Firms," *Spring Servitization Conference 2014*, Aston Business School, Birmingham, UK, May 12-14, 2014 (Birmingham, UK: Aston Business School, Aston University). Pg. 28

Lightfoot, H., *Partnership in Practice* (Cranfield, UK: Cranfield University, 2011). Pg. 35

Lightfoot, H., Baines, T., and Smart, P., "Examining the Information and Communication Technologies Enabling Servitized Manufacture," *Proceedings of the IMechE, Part B: Journal of Engineering Manufacture* 225 (November 9, 2011): 1964-1968. Pg. 34

Lucas, R., "Data Connections," *Professional Engineering*, December 2013, 47-48. Pg. 28

Machan, I., "Stepping from Lean to Servitization," *Spring Servitization Conference 2015*, Aston Business School, Birmingham, UK, May 18-19, 2015 (Birmingham, UK: Aston Business School, Aston University). Pg. 35

MacKenzie, G., *Orbiting the Giant Hairball: A Corporate Fool's Guide to Surviving with Grace* (New York: Viking, 1996). Pg. 18

Makri, C. and Neely, A., *Through-Life Accountability: Managing Complex Services* (Cambridge, UK: Cambridge Service Alliance, Cambridge University, 2015). Pg. 33

Martinez, V. and Turner, T., "Designing Competitive Service Models," Cambridge, UK: Cambridge Service Alliance, Cambridge University, 2014. Pg. 30

Martinez, V., Pouthas, V. and the 'Shift to Service' Team, *Making and Sustaining the Shift to Services in the Animal Health Industry* (Cambridge, UK: Cambridge Service Alliance, Cambridge University, 2015). Pg. 33

McCormick, S., "Advancing the Connected Vehicle," *Automotive Megatrends USA*, The Henry, Fairlane Plaza, Dearborn, MI, March 17, 2015 (Plymouth, MI: Connected Vehicle Trade Association). Pg. 15

Meeker, M., "Internet Trends Report 2014 - Code Conference," San Francisco, CA: Kleiner Perkins Caufield & Byers, 2014. Pg. 28

Meeker, M., "Internet Trends Report 2015 - Code Conference," San Francisco, CA: Kleiner Perkins Caufield & Byers, 2015. Pg. 28

Metha, R., *UK Support and Service Industry a High Value Employer and a Net Exporter* (Cranfield, UK: EPSRC Centre for Innovative Manufacturing in Through-life Engineering Services, Cranfield University, 2015). Pg. 37

Miller, I., "Life Cycle Costing: Art or Science?," *Maintenance & Engineering* 14, no. 2 (March/April 2014): 42-44. Pg. 24

Mills, J., Parry, G., and Purchase, V., "Towards Understanding the Value of the Client's Aspirations and Fears," Cambridge, UK: Cambridge Service Alliance, Cambridge University, 2012a. Pg. 31

Mills, J., Purchase, V., and Parry, G., "Enterprise Imaging - Representing Complex Multi-Organizational Service Enterprises," Cambridge, UK: Cambridge Service Alliance, Cambridge University, 2012b. Pg. 31

* Moran, K., *Investment Appraisal for Non-Financial Managers* (London, UK: Pitman/Institute of Management, 1995). Pg. 19

Morecroft, J., *Strategic Modelling and Business Dynamics: A Feedback Systems Approach* (Chichester, UK: John Wiley & Sons Ltd, 2007). Pg. 19

Morrison, M., *Rolling on Innovation* (London, UK: Flight International, 2007). Pg. 14

NASA, "NASA Cost Estimating Handbook: Version 4.0," Washington DC: National Aeronautics and Space Administration, 2015. Pg. 23

Neely, A., *The Servitization of Manufacturing: A Longitudinal Study of Global Trends* (Cambridge, UK: Cambridge Service Alliance, Cambridge University, 2009). Pg. 28

Neely, A., *Society's Grand Challenges: What Role for Services?* (Cambridge, UK: Cambridge Service Alliance, Cambridge University, 2012). Pg. 29

Neely, A., "Servitization in Germany - An International Comparison," Cambridge, UK: Cambridge Service Alliance, Cambridge University, 2013. Pg. 31

Neely, A., "Innovate Your Business Model by Making the Shift to Services: New Roles for Big Data and Analytics," *2nd AIMP Congress on Marketing in a Disruptive World*, Eindhoven, the Netherlands, April 2014a (Cambridge, UK: Cambridge Service Alliance, Cambridge University). Pg. 32

Neely, A., *The Servitization of Manufacturing* (Cambridge, UK: Cambridge Service Alliance, Cambridge University, 2014b). Pg. 32

Neely, A., "The Shift to Services: Insights from the Cambridge Service Alliance," *ESRC/BAE Systems Servitization Seminar*, Royal Society, London, UK, October 22, 2014c, Unpublished presentation (Cambridge, UK: Cambridge Service Alliance, Cambridge University). Pg. 32

Neely, A., "Servitization and Service Integration in China: Reflections from Shanghai," *Field Service News*, no. 6 (May/June 2015): 26. Pg. 32

Neely, A. and Benedettini, O., "The Challenges of Successful Servitization," *AIM Social Science Week Event: The Servitization of Manufacturing*, Woburn House Conference Centre, London, UK, 2010 (Cambridge Service Alliance, Cambridge University). Pg. 29

Neely, A., Benedettini, O., and Visnjic, I., "The Servitization of Manufacturing: Further Evidence," *18th International European Operations Management Association Conference*, Cambridge, UK, July 3-6, 2011 (Cambridge, UK: Cambridge Service Alliance, Cambridge University). Pg. 28

Ng, I., *The Pricing and Revenue Management of Services: A Strategic Approach* (Abingdon, UK: Routledge/Taylor & Francis Group, 2009). Pg. 26

Ng, I., "The Move to Service: Value Co-creation and Delivering on Outcomes," *AIM Social Science Week Event: The Servitization of Manufacturing*, Woburn House Conference Centre, London, UK, March 17, 2010 (Exeter, UK: University of Exeter Business School/Cambridge Service Alliance, Cambridge University). Pg. 35

Ng, I., *Value and Worth: Creating New Markets in the Digital Economy* (Cambridge, UK: Innovorsa Press, 2013). Pg. 26

Ng, I., *Creating New Markets in the Digital Economy: Value and Worth* (Cambridge, UK: Cambridge University Press, 2014a). Pg. 26

Ng, I., "Servitisation: A Customer-Based Approach," *ESRC/BAE Systems Servitization Seminar*, Royal Society, London, UK, October 22, 2014b, Unpublished presentation (Coventry, UK: Warwick Manufacturing Group, Warwick University). Pg. 36

Ng, I., Guo, L., and Ding, Y., *The Use of Information Technology as Value Co-creation* (Exeter, UK: Exeter University Business School, 2011). Pg. 36

Ng, I., Nudurupati, S., and Tasker, P., *Value Co-creation in the Delivery of Outcome-Based Contracts for B2B Service* (Cranfield, UK: Advanced Institute of Management Research, Cranfield University, 2010). Pg. 35

Ng, I., Parry, G., Wild, P., McFarlane, D. et al. eds., *Complex Engineering Service Systems: Concepts and Research* (London, UK: Springer-Verlag London Ltd, 2011). Pg. 26

NVable, *NVable Case Study: Bombardier ORBITA™: The Future of Maintenance* (Glasgow, UK: NVable Ltd, 2009). Pg. 14

Oldland, K., "Servitization: The Next Industrial Revolution?," *Field Service News*, no. 7 (July/August 2015): 52-54. Pg. 33

Oldland, K. and Gordon, D., "Viva la Revolution," *Field Service News*, no. 3 (October 2014): 44-45. Pg. 14

Oldland, K. and Bill, A., "The Power of Services," *Field Service News*, no. 6 (May/June 2015): 28. Pg. 16

Oldland, K., Baines, T., Viggers, B. and Dyaeyer, K., "We're Talking Servitization," *Field Service News*, no. 4 (January/February 2015): 18-20. Pg. 16

Oliva, R., "Managing the Transition from Products to Services," *Transformations to Servitized Organisational Forms Industry Day*, Cranfield School of Management, June 17, 2009 (College Station, TX: Mays Business School, Texas A&M University). Pg. 37

Osterwalder, A. and Pigneur, Y., *Business Model Generation* (Hoboken, NJ: John Wiley & Sons Inc, 2010). Pg. 18

Owen, J., *How to Influence* (Harlow, UK: Prentice Hall Business, 2010). Pg. 17

Oxford Economics, "Manufacturing Transformation: Achieving Competitive Advantage in a Changing Global Marketplace," Oxford, UK: Oxford Economics, 2013. Pg. 27

Papazian, A., *Space Exploration and Money Mechanics: An Evolutionary Challenge* (Exeter, UK: Keipr Ltd, 2012). Pg. 20

Papazian, A., *Space Value of Money: The Spatial Dynamics of Monetary Value* (Exeter, UK: Keipr Ltd, 2014). Pg. 20

Parada Puig, J., "Serviceability of Passenger Trains during Acquisition Projects," Ph.D. thesis, University of Twente, Enschede, the Netherlands, 2015. Pg. 38

Parmar, R., Mackenzie, I., Cohn, D., and Gann, D., "The New Patterns of Innovation: How to Use Data to Drive Growth," *Harvard Business Review* 92, no. 1 (2014): 86–95. Pg. 28

Parry, G., "Servitization," *ESRC/BAE Systems Servitization Seminar*, Royal Society, London, UK, October 22, 2014, Unpublished presentation (Bristol, UK: University of the West of England). Pg. 36

* Payne, M., *How to Kill a Unicorn... and Build Bold Ideas That Make It to Market, Transform Industries and Deliver Growth* (London, UK: Nicholas Brealey Publishing, 2014). Pg. 18

Porter, M., *The Competitive Advantage of Nations* (Basingstoke, UK: Macmillan Press, 1990). Pg. 18

Powley, T., *UK Manufacturers Find Value in Services* (London, UK: Financial Times Ltd, 2015). Pg. 12

Professional Engineering, "Data Monitoring Cuts Rail Running Costs," November 8, 2006, 50, London, UK: Professional Engineering Publishing Ltd. Pg. 14

Provost, M., "COMPASS™: A Generalized Ground-Based Monitoring System," in: Rao, B. and Hope, A. eds., *COMADEM 89 International: Proceedings of the First International Congress on Condition Monitoring and Diagnostic Engineering Management* (London, UK: Kogan Page, 1989), 74-87. Pg. 12

Provost, M., "Effective Monitoring of High-Value Assets," *Railway Condition Monitoring - Why? What? How?*, Pride Park Stadium, Derby, UK, February 25, 2004 (Stevenage, UK: Institution of Electrical Engineers Railway Professional Network). Pg. 19

Provost, M., "Beyond Condition Monitoring: From Data to Business Value," *Intelligent Trains – Use of Rail Vehicle Condition Monitoring*, Institution of Mechanical Engineers, London, UK, March 20, 2008 (London, UK: Institution of Mechanical Engineers). Pg. 14

* Provost, M., "Bombardier ORBITA™: Railway Asset Management for the 21st Century," *Open Conference on Availability and Resource Utilization*, Swedish Institute of Computer Science, Kista, Sweden, December 2, 2009 (Derby, UK: Bombardier Transportation). Pg. 14

* Provost, M., "Bombardier ORBITA™: Railway Asset Management for the 21st Century," *Journal of the Safety and Reliability Society* 30, no. 1 (2010): 46-56. Pg. 14

Provost, M, "Calculating the Value of Monitoring," Internal presentation, Loughborough, UK: Intelligent Energy Ltd., 2013a. Pg. 21

Provost, M., "Markov Chains," Unpublished presentation, Nottingham, UK, 2013b. Pg. 22

Provost, M., "Servitization? Asset Management? Why Should I Bother?," *Spring Servitization Conference 2015*, Aston Business School, Birmingham, UK, May 18-19, 2015 (Birmingham, UK: Aston Business School, Aston University). Pg. 12

PTC, "Manufacturing Transformation: Achieving Competitive Advantage in a Changing Global Marketplace," Needham, PA: PTC Inc., 2013a. Pg. 27

PTC, "The Service Imperative: Key Takeaways for Service Executives from the Oxford Economics Report on Manufacturing Transformation," Needham, PA: PTC Inc., 2013b. Pg. 27

PTC, "Proving the Service Continuum," Needham, PA: PTC Inc., 2015. Pg. 27

Puche Alonso, M., "Product Service System Benefits and Barriers," M.Sc. thesis, Cranfield University, Cranfield, UK, 2007. Pg. 36

Rackley, S., "21st Century Maintenance for a 21st Century Railway," *The Rail Engineer*, December 2006. Pg. 14

Rail Engineer, "Reliability Goes into ORBITA™," *The Rail Engineer*, June 2009. Pg. 14

Read, B., "Flying Doctor," *Aerospace International*, August 2006: 16-17. Pg. 15

Redding, L., "A Strategy Formulation Methodology for Companies Seeking to Compete through IVHM Enabled Service Delivery Systems," Ph.D. thesis, School of Applied Sciences, Cranfield University, Cranfield, UK, 2012. Pg. 36

Redding, L. and Roy, R. eds., *Through-Life Engineering Services: Motivation, Theory and Practice* (Cham, Switzerland: Springer International Publishing, 2015). Pg. 36

Reed, J., "How to Gain and Keep Control of Your Service," *Spring Servitization Conference 2013*, Aston Business School, Birmingham, UK, May 20-22, 2013 (Nottingham, UK: Nottingham University). Pg. 35

Reekie, B., Whitehall, K., and Hunt, A., "SALVO Project: Scottish Water," *Uptime* 10, no. 68 (December/January 2015): 12-16. Pg. 24

Ren, G. and Gregory, M., "Growing Services in Manufacturing: Lessons from Leading Industrial Companies," Cambridge, UK: Institute for Manufacturing, Cambridge University, 2009. Pg. 30

Riis, T., "Quantifying the Value of Information," *Petroleum Engineer International* (Houston, TX: Hart Energy, June 1999). Pg. 21

Robinson, T., "Derby - We Have a Problem," *Aerospace International*, April 2006: 34-35. Pg. 12

Roland Berger, "Evolution of Service," *Asset Management & Maintenance Journal* 27, no. 2 (March 2014): 48-52. Pg. 27

Rolls-Royce, "Rolls-Royce Celebrates 50th Anniversary of Power-by-the-Hour™," Press release, October 30, 2012. Pg. 12

Roy, R., Shehab, E., Hockley, C., and Khan, S. eds., *Enduring and Cost-Effective Engineering Support Solutions: Proceedings of the 1st International Conference on Through-Life Engineering Services* (Cranfield, UK: Cranfield University Press, 2012). Pg. 35

Roy, R., Tomiyama, T., Tiwari, A., Tracht, K. et al. eds., *Proceedings of the 3rd International Conference in Through-Life Engineering Services: Procedia CIRP*, Vol. 22 (Amsterdam, the Netherlands: ScienceDirect/Elsevier BV, 2014), 1-298. Pg. 36

RSSB, "Research Programme Strategy Support: Mapping Current Remote Condition Monitoring Activities to the System Reliability Framework," London, UK: Rail Safety and Standards Board Ltd, 2009. Pg. 14

Sampson, S., *Introduction to PCN Analysis* (Provo, UT: Marriott School of Management, Brigham Young University, 2011a). Pg. 25

Sampson, S., *PCN Diagrams: A New Tool for Service Design and Innovation* (Provo, UT: Marriott School of Management, Brigham Young University, 2011b). Pg. 25

Sampson, S., "Service Performance Measurement," *Spring Servitization Conference 2014*, Aston Business School, Birmingham, UK, May 12-14, 2014. Birmingham, UK: Aston Business School, Aston University. Pg. 25

Sandborn, P., "A Decision Support Model for Determining the Applicability of Prognostic Health Management (PHM) Approaches to Electronic Systems," *Reliability and Maintainability Symposium (RAMS)*, Arlington, VA, January 24-27, 2005 (New York: Institute of Electrical and Electronics Engineers). Pg. 24

Sandborn, P., "PHM Return on Investment (ROI) Analysis and the Use of PHM in Maintenance Planning," Unpublished presentation, College Park, MD: Maryland University, 2010. Pg. 24

Sandborn, P., "PHM Enabled Logistics (Return on Investment and Availability Management)," *2011 Prognostics and System Health Management Conference (PHM-2011 Shenzhen)*, Intercontinental Shenzhen Hotel, Shenzhen, China, May 24-25, 2011 (Red Hook, NY: Institute of Electrical and Electronics Engineers). Pg. 24

Sandborn, P. and Feldman, K., "The Economics of Prognostics and Health Management," in: *Electronic System Prognostics and Health Management* (Hoboken, NJ: John Wiley & Sons Inc., 2008). Pg. 25

Sandborn, P. and Wilkinson, C., "A Maintenance Planning and Business Case Development Model for the Application of Prognostics and Health Management (PHM) to Electronic Systems," *Microelectronics Reliability* 47, no. 12 (December 2007): 1889-1901. Pg. 25

Sankey, P. and Gaughan, C., "Securing Investment for Assets in the Current Economic Climate," *1st Annual Predictive Asset Management Seminar*, Mayfair Conference Centre, London, UK, October 19-20,2011 (London, UK: IBC Energy/Informa UK Ltd). Pg. 23

* Savage, S., *Decision Making with Insight*, 2nd ed. (Belmont, CA: Thomson Brooks/Cole, 2003). Pg. 21

Sawyer, D., "Another Year of Progress," *Modern Railways*, January 2009. Pg. 14

Scanff, E., Feldman, K., Ghelam, S., Sandborn, P. et al., "Life Cycle Cost Impact of Using Prognostic Health Management (PHM) for Helicopter Avionics," *Microelectronics Reliability* 47, no. 12 (December 2007): 1857-1864. Pg. 25

Schlosser, M., *Corporate Finance*, 2nd ed. (Hemel Hempstead, UK: Prentice Hall, 1992). Pg. 20

Schroeder, A. and Kotlarsky, J., "Servitization and IT-Driven Innovation," *Spring Servitization Conference 2014*, Aston Business School, Birmingham, UK, May 12-14, 2014 (Birmingham, UK: Aston Business School, Aston University). Pg. 34

Schuh, L. and Schiefer, M., "Developing Value Added Services," *Asset Management & Maintenance Journal* 25, no. 2 (March 2012): 20-24. Pg. 26

Secrett, M., *Mastering Spreadsheet Budgets and Forecasts* (London, UK: Pitman/Institute of Management, 1993). Pg. 20

Secured by Design, *An Overview and Opportunities for EV Telematics* (Milton Keynes, UK: Secured by Design Ltd, 2010). Pg. 15

Senior, C., *Challenging Diagnostics Approaches in Automotive* (Reading, UK: Snap-on Business Solutions Ltd, 2014). Pg. 16

Simon, P., *How to Get over Your Inaction on Big Data* (Watertown, MA: Harvard Business Review, 2014). Pg. 19

Slack, N., *Patterns of Servitization* (Cambridge, UK: Institute for Manufacturing, Cambridge University, 2005). Pg. 28

Slater, P., "Don't Get Even, Get M.A.D.: Making the Decision to Make a Difference," *Uptime* 5, no. 37 (October/November 2009): 38-41. Pg. 17

Smith, L., Ng, I., and Maull, R., *The Three Value Cycles of Equipment Based Service* (Exeter, UK: Exeter University Business School, 2010). Pg. 35

Smith, R., "Looking Up: What to Do after an Economic Crisis," *Uptime* 5, no. 37 (October/November 2009): 18-20. Pg. 19

Spring, M. and Mason, K., "Business Model Innovation," *AIM Social Science Week Event: The Servitization of Manufacturing*, Woburn House Conference Centre, London, UK, 2010 (Lancaster, UK: Lancaster University Management School). Pg. 36

Stickdorn, M. and Schneider, J., *This Is Service Design Thinking: Basics - Tools - Cases* (Amsterdam, the Netherlands: BIS Publishers, 2014). Pg. 26

Sultan, N., "The Servitization of IT and Education through Clouds and MOOCS," *Spring Servitization Conference 2014*, Aston Business School, Birmingham, UK, May 12-14, 2014 (Birmingham, UK: Aston Business School). Pg. 36

Tennent, J. and Friend, G., *Guide to Business Modelling*, 2nd ed. (London, UK: The Economist/Profile Books, 2005). Pg. 20

Tersago, S. and Visnjic, I., "Business Model Innovations in Healthcare," Cambridge, UK: Cambridge Service Alliance, Cambridge University, 2011. Pg. 31

The Week, "The Sharing Economy," Digital edition, June 14, 2014 (London, UK: Dennis Publishing Ltd). Pgs. 15, 16

Thomas, S., "Uncover the Hidden Value: Internal Consulting in Your Organization," *Uptime* 4, no. 34 (April/May 2009): 46-49. Pg. 18

Thompson, A., *The SALVO Project: Innovative Approaches to Decision-Making for the Management of Ageing Physical Assets* (Kingsclere, UK: The Woodhouse Partnership Ltd, 2011). Pg. 24

Thompson, A., "The SALVO Project: Innovative Approaches to Decision-Making in the Management of Ageing Assets," *Asset Management & Maintenance Journal* 25, no. 5 (October 2012a): 36-43. Pgs. 24

Thompson, R. *10 Steps to a Successful Smart Services Evaluation* (Foxboro, MA: Axeda Corp, 2012b). Pg. 28

Thompson, A., "Managing Ageing Assets and Optimal Replacement Timing," *3rd Annual Predictive Asset Management Seminar*, Dexter House, London, UK, December 4-5, 2013 (London, UK: IBC Energy/Informa UK Ltd). Pg. 24

Tomczykowski, W., *Physics Based Modelling* (Beltsville, MD: DfR Solutions LLC, 2011). Pg. 23

Toossi, A., Lockett, H., Macdonald, E., Greenough, R. et al., "Maintenance Outsourcing – A Step towards Product Service Systems," *Maintenance & Engineering* 10, no. 3 (May/June 2010): 38-41. Pg. 36

Trevor, J. and Williamson, P., "A Blueprint for the Next Generation Organisation: Reconciling Agility, Efficiency and Purpose," Cambridge, UK: Cambridge Service Alliance, Cambridge University, 2014. Pg. 32

Tukker, A., "Product Services for a Resource Efficient and Circular Economy," *Spring Servitization Conference 2014*, Aston Business School, Birmingham, UK, May 12-14, 2014 (Birmingham, UK: Aston Business School, Aston University). Pg. 37

Turunen, T. and Neely, A., *Organising Servitization: An In-Depth Case Study* (Cambridge, UK: Cambridge Service Alliance, Cambridge University, 2012). Pg. 29

Uren, V. and Petridis, P., "This Beer Is Off: Building a Dialogue Game for Servitization," *Spring Servitization Conference 2015*, Aston Business School, Birmingham, UK, May 18-19, 2015 (Birmingham, UK: Aston Business School, Aston University). Pg. 35

Urmetzer, F., Parlikad, A., Pearson, C., and Neely, A., "Key Considerations in Asset Management System Design," Cambridge, UK: Cambridge Service Alliance, Cambridge University, 2014. Pg. 32

Valerdi, R., *The Constructive Systems Engineering Cost Model (COSYSMO) – Quantifying the Costs of Systems Engineering Effort in Complex Systems* (Saarbrücken, Germany: VDM Verlag Dr Müller Aktiengesellschaft and Co. KG, 2008). Pg. 24

Velu, C. and Stiles, P., "Managing Decision Making and Cannibalization for Parallel Business Models," Cambridge, UK: Cambridge Service Alliance, Cambridge University, 2013. Pg. 31

Velu, C., Barrett, M., Kohli, R., and Salge, T., "Thriving in Open Innovation Ecosystems - Towards a Collaborative Market Orientation," Cambridge, UK: Cambridge Service Alliance, Cambridge University, 2013. Pg. 31

Viljakainen, A., Toivonen, M., and Aikala, M., "Industry Transformation towards Service Logic - A Business Model Approach," Cambridge, UK: Cambridge Service Alliance, Cambridge University, 2013. Pg. 31

Visnjic Kastalli, I. and Neely, A., "Collaborate to Innovate - How Business Ecosystems Unleash Business Value," Cambridge, UK: Cambridge Service Alliance, Cambridge University, 2013. Pg. 31

Visnjic Kastalli, I. and Van Looy, B., "Servitization: Disentangling the Impact of Service Business Model Innovation on Manufacturing Firm Performance," *Journal of Operations Management* 31, no. 4 (May 2013): 169-180. Pg. 29

Visnjic, I. and Neely, A., "From Processes to Promise: How Complex Service Providers Use Business Model Innovation to Deliver Sustainable Growth (presentation)," Cambridge, UK: Cambridge Service Alliance, Cambridge University, 2012a. Pg. 29

Visnjic, I. and Neely, A., "From Processes to Promise: How Complex Service Providers Use Business Model Innovation to Deliver Sustainable Growth (white paper)," Cambridge, UK: Cambridge Service Alliance, Cambridge University, 2012b. Pg. 29

Visnjic, I. and Van Looy, B., "Can a Product Manufacturer Be a Successful Service Provider? In Pursuit of a Business Model That Fosters Complementarity between Product and Service Activities Perspectives," *Academy of Management Conference*, San Antonio Hotel, San Antonio, TX, August 1-16, 2011 (Cambridge, UK: Cambridge Service Alliance, Cambridge University). Pg. 29

Visnjic, I. and Van Looy, B., "Successfully Implementing a Service Business Model in a Manufacturing Firm," Cambridge, UK: Cambridge Service Alliance, Cambridge University, 2013. Pg. 29

Visnjic, I., Neely, A., and Weingarten, F., "Another Performance Paradox? A Refined View on the Performance Impact of Servitization," Cambridge, UK: Cambridge Service Alliance, Cambridge University, 2012. Pg. 30

Visnjic, I., Turunen, T., and Neely, A., "When Innovation Follows Promise: Why Service Innovation Is Different and Why That Matters," Cambridge, UK: Cambridge Service Alliance, Cambridge University, 2013. Pg. 29

Warren, K., *Strategic Management Dynamics* (Chichester, UK: John Wiley & Sons Ltd, 2008). Pg. 19

Waters, N., "Engine Health Management," *Ingenia*, no. 39 (June 2009): 37-42. Pg. 14

Weiller, C. and Neely, A., "Business Model Design in an Ecosystem Context," Cambridge, UK: Cambridge Service Alliance, Cambridge University, 2013. Pg. 31

West, S. and Pascual, A., "The Use of Equipment Life-Cycle Analysis to Identify New Service Opportunities," *Spring Servitization Conference 2015*, Aston Business School, Birmingham, UK, May 18-19, 2015 (Birmingham, UK: Aston Business School, Aston University). Pg. 35

Westerman, G., Calméjane, C., Bonnet, D., Ferraris, P. et al., "Digital Transformation - A Roadmap for Billion-Dollar Organizations," Paris, France: Capgemini Consulting/MIT Center for Digital Business, 2011. Pg. 28

Windahl, C., "Integrated Solutions in the Capital Goods Sector: Exploring Innovation, Service and Network Perspectives," Ph.D. thesis, Linköping University, Linköping, Sweden, 2007. Pg. 37

Winkler, A., *Overbeck Analitica PAM Executive Summary* (Rome, Italy: Overbeck Analitica, 2015a). Pg. 24

Winkler, A., *Using Survival Analysis and Simulation for Predictive Asset Management: Overbeck Analitica's PAM System* (Rome, Italy: Overbeck Analitica, 2015b). Pg. 24

Woodall, P., Borek, A., and Parlikad, A., "Data Quality Assessment - The Hybrid Approach," Cambridge, UK: Cambridge Service Alliance, Cambridge University, 2013. Pg. 31

* Woodhouse, J., *Asset Management Decision Making: The SALVO Process* (Kingsclere, UK: The Woodhouse Partnership Ltd., 2014). Pg. 24

Woodhouse, J., "Decisions in Asset Management," *Asset Management & Maintenance Journal* 28, no. 1 (January 2015): 4-6. Pg. 24

Wynn, G., "How to Get People to Do What You Want Them to Do," *Uptime* 9, no. 63 (February/March 2014): 34-35. Pg. 17

Zaki, M. and Neely, A., "Optimising Asset Management within Complex Service Networks: The Role of Data," Cambridge, UK: Cambridge Service Alliance, Cambridge University, 2014. Pg. 32

* Zeithaml, V., Brown, S., Bitner, M., and Salas, J., *Profiting from Services and Solutions: What Product-Centric Firms Need to Know* (New York: Business Expert Press, 2014). Pg. 26

Zenger, J., Folkman, J., and Edinger, S., "Making Yourself Indispensable," *Harvard Business Review* 89, no. 10 (2011): 84-92. Pg. 17

Websites

Aberdeen Group Pg. 27	www.aberdeen.com
Airbnb Pg. 16	www.airbnb.com
Arthur D Little Pg. 27	www.arthurdlittle.com
Aston Business School Pg. 33	www.aston-servitization.com
Bernd Geropp Pg. 17	www.more-leadership.com
Blue Ocean Strategy Pg. 18	www.blueoceanstrategy.com
BMW Pg. 15	www.bmw.com
Boeing Co. Pg. 15	www.boeing.com
Bombardier Inc. Pg. 14	www.bombardier.com
Business Model Generation Pg. 18	www.businessmodelgeneration.com
Business Model You® Pg. 18	http://businessmodelyou.com
Business Smart International Pg. 19	www.business-smart.com
Cambridge Service Alliance Pg. 28	www.cambridgeservicealliance.org
Capgemini Consulting Pg. 28	www.capgemini.com
Car2Go Pg. 15	www.car2go.com
Centre for Advanced Life Cycle Engineering, Maryland University Pg. 24	www.calce.umd.edu
CogNexus Institute Pg. 18	http://cognexus.org/cognexus_institute
Connected Vehicle Trade Association Pg. 15	www.connectedvehicle.org
Controls and Data Services Pg. 14	www.controlsdata.com
David Fraser Pg. 17	www.drdavidfraser.com
Decision Support Tools Ltd. Pg. 24	www.decisionsupporttools.com
EPSRC Centre for Innovative Manufacturing: Through-life Engineering Services Pg. 36	www.through-life-engineering-services.org
Fahrenheit 212 Pg. 18	www.fahrenheit-212.com
Finoptek Pg. 20	www.finoptek.com
Fleetmatics Ltd. Pg. 15	www.fleetmatics.co.uk
FleetStar Pg. 15	www.fleetstar-online.com
Fraunhofer Institute Pg. 28	www.fraunhofer.de
General Motors Pg. 15	www.gm.com

Harvard Business Review Pg. 27	http://hbr.org
Home Depot Pg. 16	www.homedepot.com
Hubbard Decision Research Pg. 26	www.hubbardresearch.com
Integrated Vehicle Health Monitoring Centre, Cranfield University Pg. 35	www.cranfield.ac.uk/ivhm
Intelligent Energy Ltd. Pg. 16	www.intelligent-energy.com
Kleiner Perkins Caufield & Byers Pg. 28	www.kpcb.com
KT-Box Pg. 33	www.cambridgeservicealliance.org/kt-box-learn.html
Lyft Pg. 16	www.lyft.com
MACRO Project Pg. 24	www.macroproject.org
MAN Trucks Pg. 15	www.man.eu
Microlise Ltd. Pg. 15	www.microlise.com
NavStar Pg. 15	www.navstar.co.uk
Noventum IT Management Consulting Pg. 27	www.noventum.com
Oracle Crystal Ball Pg. 20	www.oracle.com/us/products/applications/crystalball
Oxford Economics Pg. 27	www.oxfordeconomics.com
Palisade Corp. (@Risk®, Decision Tools suite) Pg. 20	www.palisade.com
PAM Analytics Pg. 24	www.pamanalytics.com
Paul Barringer & Associates Inc. Pg. 23	www.barringer1.com
PTC Inc. Pg. 27	www.ptc.com
Quartix Ltd. Pg. 15	www.quartix.net
Rolls-Royce plc. Pg. 13	www.rolls-royce.com
SALVO Project Pg. 24	www.salvoproject.org
Sampson on Services Pg. 25	http://services.byu.edu/wp
Service in Industry Pg. 27	www.serviceinindustry.com
Simplytrak Ltd. Pg. 15	www.simplytrak.co.uk
Snap-on Inc. Pg. 16	www.snapon.com
Strategy Dynamics Pg. 19	http://strategydynamics.com
Sysdea Pg. 19	http://sysdea.com
Through-life Engineering Services Institute, Cranfield University Pg. 36	www.cranfield.ac.uk/sas/tesi
Uber Pg. 16	www.uber.com
Warwick Manufacturing Group, Warwick University Pg. 35	http://warwick.ac.uk/fac/sci/wmg/

Additional Material

References

* denotes sources that the author found particularly helpful and/or inspiring.

Aston Business School, *Servitization Case Study: Alstom Transport* (Birmingham, UK, Aston Business School, Aston University, 2018).

Bal, A., *Introduction to SmartFleet* (Utrecht, the Netherlands: Ricardo Rail Nederland, 2016).

Beckford, J., *Information for Change* (Newbury, UK: Beckford Consulting, 2010).

Beckford, J., *The Intelligent Organisation: Realising the Value of Information* (Abingdon, UK: Routledge/Taylor & Francis Group, 2015).

Bennett Stewart, G., *The Quest for Value: A Guide for Senior Managers* (New York, NY: HarperCollins Inc., 1991).

Cordner, K. ed., *The Modern Railway: The Definitive Guide to the UK's Railway Industry in 2016* (Stamford, UK: Key Publishing Ltd., 2015).

Cordner, K. ed., *The Modern Railway: The Definitive Guide to the UK's Railway Industry in 2017* (Stamford, UK: Key Publishing Ltd., 2016).

Cordner, K. ed., *The Modern Railway: The Definitive Guide to the UK's Railway Industry in 2018* (Stamford, UK: Key Publishing Ltd., 2017).

Damodaran, A., *Damodaran on Valuation: Security Analysis for Investment and Corporate Finance* (New York: John Wiley & Sons Inc., 1964).

Dixit, A. and Nalebuff, B., *The Art of Strategy: A Game Theorist's Guide to Success in Business and Life* (New York: Norton, 2008).

Dohrmann, K., Hölter, K., Bigdeli, A., and Musson, E., *Servitization and Supply Chains: A DHL Perspective on Future Engineering & Manufacturing Supply Chains* (Bonn, Germany: DHL Customer Solutions & Innovation, 2018).

Dubé, E., *Bombardier ORBITA™ Predictive Asset Management* (Victoria, Australia: Bombardier Transportation Australia, 2007).

Dutta, P., *Strategies and Games: Theory and Practice* (Cambridge, MA: MIT Press, 2001).

Eisenführ, F., Weber, M., and Langer, T., *Rational Decision Making* (Berlin, Germany: Springer-Verlag, 2010).

Engelbrecht, C., *Using Asset Condition Monitoring Involving into Predictive Maintenance* (Kelvin, South Africa: Bombela Concession Company (Pty) Ltd., 2016).

* Fischer, T., Gebauer, H., and Fleisch, E., *Service Business Development: Strategies for Value Creation in Manufacturing Firms* (Cambridge, UK: Cambridge University Press, 2012).

* Fisher, N., *Analytics for Leaders: A Performance Measurement System for Business Success* (Cambridge, UK: Cambridge University Press, 2013).

* Francis, J., *Investments: Analysis and Management – International Edition*, 4th ed. (Singapore: McGraw-Hill Book Co, 1987).

Frank, N., Service Transformation. Introduction to Service Innovation and Technologies Free Online Course: Module 3, Cranfield, UK: Operations Excellence Institute, Cranfield University, 2016, http://opex-institute.co.uk/index.php/service-innovation-and-operations.

* Galar, D., Sandborn, P., and Kumar, U., *Maintenance Costs and Life Cycle Cost Analysis* (Boca Raton, FL: CRC Press/Taylor & Francis, 2017).

Ganz, C., "Enabler for Advanced Services - IoT and Analytics for New Service Offerings," *Creating Value through Customer Services*, Møller Centre, Churchill College, Cambridge, UK, October 6, 2015 (Zürich, Switzerland: ABB Group).

Gill, A., Hogan, D., Haden, C., and Urmetzer, F., *Service Transformation and Ecosystem Mapping* (Cambridge, UK: Cambridge University, 2018).

Goodson, S., "Second Tier Operators – First Class Service," *Rolls-Royce Magazine*, no. 151 (December 2016): 27-29.

Harrison, A., Designing for Service Lifecycle Cost. Introduction to Service Innovation and Technologies Free Online Course: Module 2, Cranfield, UK: Operations Excellence Institute, Cranfield University, 2016, http://opex-institute.co.uk/index.php/ service-innovation-and-operations.

Hou, J. and Neely, A., *Investigating Risks of Outcome-Based Service Contracts from a Provider's Perspective* (Cambridge, UK: Cambridge Service Alliance, Cambridge University, 2017).

Howie, D., "Data Driven Aviation: How Rolls-Royce Is Digitally Enabling Airlines to Improve Efficiency," *Rolls-Royce Magazine*, no. 150 (September 2016): 58-61.

* Hubbard, D., *The Failure of Risk Management: Why It's Broken and How to Fix It* (Hoboken, NJ: John Wiley & Sons Inc, 2009).

* Hubbard, D., *How to Measure Anything Workbook: Finding the Value of Intangibles in Business* (Hoboken, NJ: John Wiley & Sons Inc., 2014b).

Hughes, A. and Hughes, J., *Through-Life Engineering Services (TES): Market and Data Review* (Cranfield, UK: ESPRC Centre for Innovative Manufacturing - Through-life Engineering Services, 2016).

Kohtamäki, M., Baines, T., Rabetino, R., and Bigdeli, A., *Practices and Tools for Servitization: Managing Service Transition* (Cham, Switzerland: Palgrave Macmillan/Springer International Publishing, 2018).

* Kowalkowski, C. and Ulaga, W., *Service Strategy in Action: A Practical Guide for Growing Your B2B Service and Solutions Business* (Linköping, Sweden/Tempe, AZ: Service Strategy in Action LLC, 2017).

Kunneke, K. and Engelbrecht, C. *Advance Asset Condition Assessment* (Kelvin, South Africa: Bombela Concession Company (Pty) Ltd., 2012).

Lightfoot, H., Competing through Services. Introduction to Service Innovation and Technologies Free Online Course: Module 1, Cranfield, UK: Operations Excellence Institute, Cranfield University, 2016a, http://opex-institute.co.uk/index.php/service-innovation-and-operations.

* Lightfoot, H., "Why Engineers Must Protect and Serve," in: Stirling, W. ed., *UK Manufacturing Review 2015-2016* (Teddington, UK: Stirling Media Ltd, 2016b), 148-150.

Martinez, V., Bastl, M., Kingston, J. and Evans, S., "Challenges in Transforming Manufacturing Organisations into Product-Service Providers," *Journal of Manufacturing Technology Management* 21, no. 4 (2010): 2.

Martinez, V., Neely, A., Urmetzer, F., Alison, N. et al. *Seven Critical Success Factors in the Shift to Services* (Cambridge, UK: Cambridge University, 2016).

Mehnen, J., Service Enabling Technologies. Introduction to Service Innovation and Technologies Free Online Course: Module 4, Cranfield, UK: Operations Excellence Institute, Cranfield University, 2016, http://opex-institute.co.uk/index.php/service-innovation-and-operations.

Morris, M., *TotalCare® Accounting* (London, UK: Rolls-Royce plc, 2014).

* Neely, A., "Understanding the Customer: New-Service-Based Business Models," in: Stirling, W. ed., *UK Manufacturing Review 2015-2016* (Teddington, UK: Stirling Media Ltd., 2016), 151-152.

O'Sullivan, B., "A Service Mentality," *Rolls-Royce Magazine*, no. 148 (March 2016): 25-29.

* PA Consulting, *From Products to Services: Creating Sustainable Growth in Industrial Manufacturing through Servitization* (London, UK: PA Knowledge Ltd., 2017).

Palmer, T., "Rolls-Royce Civil Aerospace - Better Services for a Changing World," *Creating Value through Customer Services*, Møller Centre, Churchill College, Cambridge, UK, October 6, 2015 (Derby, UK: Rolls-Royce plc).

Palmer, T., "The Power of Digital," *Farnborough International Airshow Official Show Catalogue 2016* (Farnborough, UK: Farnborough International Ltd., 2016a), 38-39.

Palmer, T., "The Shape of Services to Come," *Rolls-Royce Magazine*, no. 151 (December 2016b): 24-26.

PAM Analytics, *PAM Analytics Predictive Asset Management System* (Radlett, UK: PAM Analytics Ltd., 2016).

Pistoni, A. and Songini, L., *Servitization Strategy and Managerial Control* (Bingley, UK: Emerald Publishing Ltd., 2017).

Provost, M., "Introduction to Servitization," Unpublished presentation, Nottingham, UK, 2016a.

Provost, M., "The Value of Servitization," Unpublished presentation, Nottingham, UK, 2016b.

Roy, R., "Engineering for Life," in: Stirling, W. ed., *UK Manufacturing Review 2015-2016* (Teddington, UK: Stirling Media Ltd., 2016), 216-219.

Ryals, L., *Rolls-Royce TotalCare®: Meeting the Needs of Key Customers* (Cranfield, UK: Cranfield School of Management, Cranfield University, 2010).

* Savage, S., *The Flaw of Averages: Why We Underestimate Risk in the Face of Uncertainty* (Hoboken, NJ: John Wiley & Sons Inc., 2012).

Tasker, P., Shaw, A., Hughes, A., Gill, A. et al., *A National Strategy for Engineering Services: Delivering UK Economic Growth by Making Things Work Better for Longer* (Cranfield, UK: ESPRC Centre for Innovative Manufacturing - Through-life Engineering Services, 2016a).

Tasker, P., Shaw, A., Hughes, A., Gill, A. et al., *A National Strategy for Engineering Services: Delivering UK Economic Growth by Making Things Work Better for Longer*. (Cranfield, UK: ESPRC Centre for Innovative Manufacturing - Through-life Engineering Services, 2016b).

Tasker, P., Shaw, A., Hughes, A., Gill, A. et al., *A National Strategy for Engineering Services: Delivering UK Economic Growth by Making Things Work Better for Longer – Sector Report* (Cranfield, UK: ESPRC Centre for Innovative Manufacturing - Through-life Engineering Services, 2016c).

TES, *Making Things Work: Engineering for Life – Developing a Strategic Vision* (Cranfield, UK: ESPRC Centre for Innovative Manufacturing - Through-life Engineering Services, 2015).

TES, "*A National Strategy for Engineering Services: Delivering UK Economic Growth by Making Things Work Better for Longer*," Cranfield, UK: ESPRC Centre for Innovative Manufacturing - Through-life Engineering Services, 2016.

The Manufacturer, *Moving Up the Value Chain: UK Manufacturing's Servitization Journey* (London, UK: The Manufacturer, 2017).

Vanson Bourne and ServiceMax, *Asset and Service Data Gravity* (Newbury, UK and Pleasanton, CA: Vanson Bourne Ltd. and ServiceMax Inc., 2018).

* Vose, D., *Risk Analysis: A Quantitative Guide*, 3rd ed. (Chichester, UK: John Wiley & Sons Ltd., 2008).

Vose, D., *Ten Steps to Incorporating Risk Analysis into Your Business* (Sint-Amandsberg, Belgium: Vose Software, 2018).

Walsh, C., *Key Management Ratios*, 4th ed. (Harlow, UK: Financial Times/Prentice Hall, 2006).

* Warren, K., *The Trouble with Strategy: The Brutal Reality of Why Business Strategy Doesn't Work and What to Do about It* (Seattle WA: CreateSpace/Amazon.com Inc., 2012).

* Warren, K. *Strategy Dynamics Essentials* (Seattle WA: CreateSpace/Amazon.com Inc., 2015).

Warren, K., *System Dynamics: A Core Systems Engineering Capability* (Princes Risborough, UK: Strategy Dynamics Ltd., 2018).

West, S., "Advanced Services - Contracts and Modelling," *Aston MBA Programme*, Aston Business School, Birmingham, UK, March 1, 2016a, Birmingham, UK: Aston Business School, Aston University.

West, S., "Creating and Measuring Value: Alternative Operations and Maintenance Business Models," *The Future of Gas Turbine Technology, European Turbine Network 8th International Gas Turbine Conference*, Hotel Le Plaza, Brussels, Belgium, October 12-13, 2016c (Brussels, Belgium: European Turbine Network).

West, S., "Service Pricing Strategies in Maintenance Services," *23rd EurOMA Conference*, Norwegian University of Science and Technology, Trondheim, Norway, June 17-22, 2016d (Brussels, Belgium: European Operations Management Association).

West, S., "Value Based Pricing in the Capital Equipment Aftermarket," *10th Annual Aftermarket Business Platform*, Hotel Dorint Pallas, Wiesbaden, Germany, October 19-21, 2016e (Stockholm, Sweden: Copperberg AB).

West, S., Gaiardelli, P., Bigdeli, A., and Baines, T., "Exploring Operational Challenges for Servitization - A European Survey," *Spring Servitization Conference 2018*, Copenhagen Business School, Frederiksberg, Denmark, May 14-16 2018 (Birmingham, UK: Aston Business School, Aston University).

Wood, T., *TotalCare®* (London, UK: Rolls-Royce plc, 2014).

Xerox, *Servitization: The Shift to Technology Enabled Services* (Norwalk, CT: Xerox Corp, 2013).

Websites

ABB Group	www.abb.com
Beckford Consulting	www.beckfordconsulting.com
Copperberg A.B.	www.copperberg.com
European Operations Management Association (EurOMA)	www.euroma-online.org
Flaw of Averages	www.flawofaverages.com
PA Consulting	www.paconsulting.com
Riversimple Movement Ltd.	www.riversimple.com
Service Strategy in Action	www.servicestrategyinaction.com
Si2 Partners	www.si2partners.com
The Manufacturer	www.themanufacturer.com
Vose Software	www.vosesoftware.com
Xerox Corp.	www.xerox.com, www.xerox.co.uk

CHAPTER 3

Modelling and Simulation

Introduction

This chapter discusses the advantages of mathematical modelling of assets and its extension to the simulation of the maintenance and operation of assets to create predictions of whole-life technical and financial performance of both single assets and whole fleets. Some useful sources of information about general mathematical modelling and simulation tools, as well as specific equipment types, are also discussed. The modelling of a fuel cell is described in Appendix B.

Why Is Modelling and Simulation Useful?

Modelling and simulation is an extremely powerful capability for any asset producer or user to have, for a number of reasons. Clearly, trying things out (and making all the mistakes in), a computer instead of hardware provides significant opportunities to increase quality and reduce costs, timescales, and risk. A simulation and modelling capability can also:

- Improve communication within a project, across both time and space.
- Provide baselines to compare real hardware performance with design intent.

FIGURE 3.1 Where simulation and modelling help.

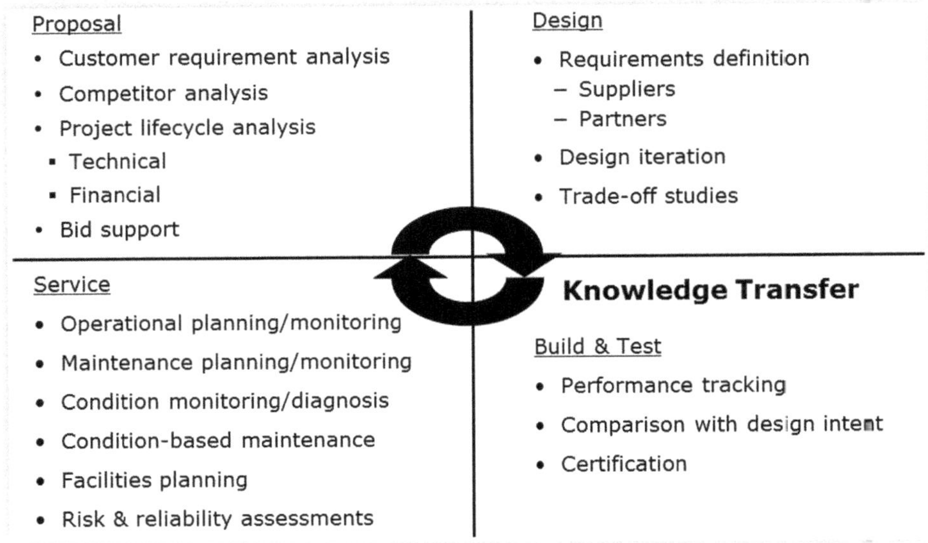

- Enable the rapid assessment of hardware performance in different operational conditions.
- Enable "what-if" studies on customer requirements and understanding of competitors' market offerings.
- Enable optimization of business performance.
- Vastly improve condition monitoring and asset management data processing efficiency and knowledge creation.

Simulation and modelling are keys to rational predictions and the move to an "information culture." Figure 3.1 summarizes how a simulation and modelling capability can help a project.

Modelling supports further developments discussed in Chapters 4, 5, and 6 of this book, such as observability analysis to optimize sensor sets, Kalman Filtering for sensor bias detection, and time series analysis of differences between actual and expected performance. A simulation and modelling capability enables an organization to move from a culture of "build, test, analyze, then understand" to "use test results to refine previous predictions and confirm understanding," which has profound positive effects on cost, speed, and efficiency of hardware development and in-service support.

Levels of Modelling

It is useful to be able to model an asset at a number of levels of detail, in order to make the process more efficient. The results of detailed modelling of individual components can be described in one or more component characteristics, in which high-level relationships between component performance parameters are summarized. Expressing parameters as appropriate nondimensional groups by using Buckingham's rule, an approach widely used in aerospace (see *Whittle 1981*; *Walsh and Fletcher 1998*; *Saravanamuttoo et al. 2001*; *Jackson et al. 1992*) and wind turbines (see *Burton et al. 2001*; *Hau 2006*), can

FIGURE 3.2 Physical asset modelling pyramid (Intelligent Energy Ltd.: reproduced with permission).

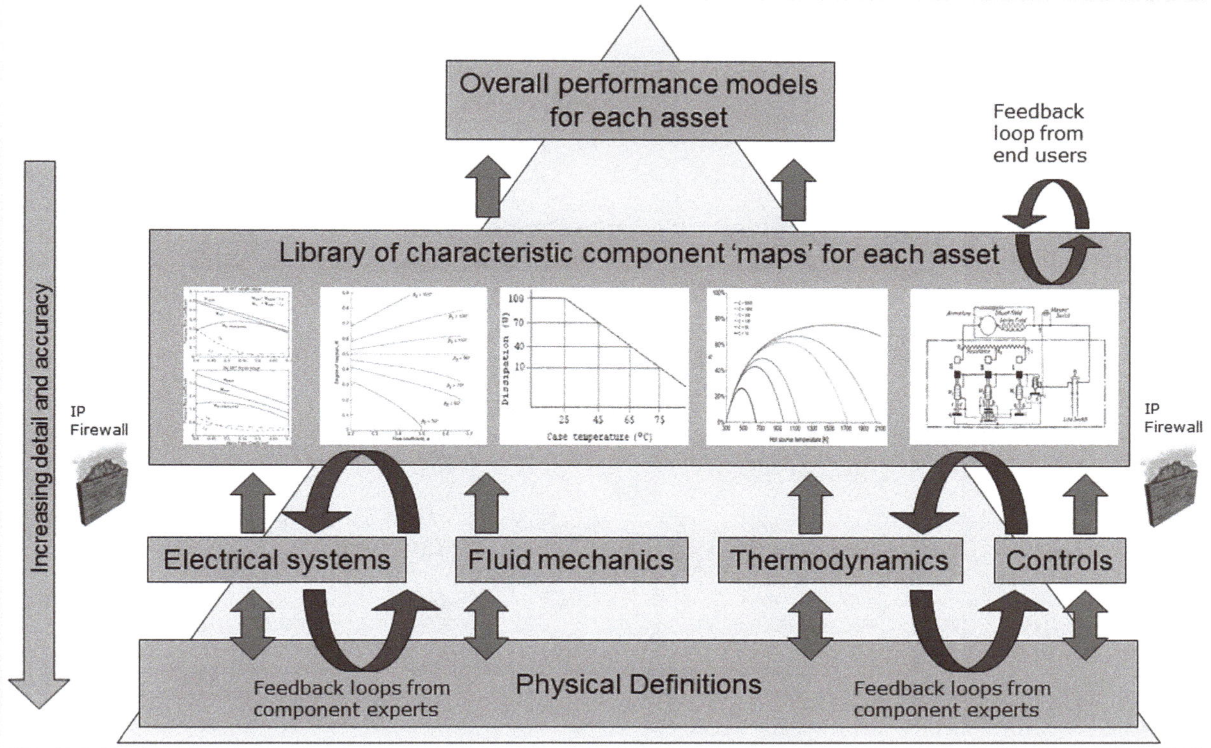

be very powerful. These component performance characteristics can then be used, using appropriate physical laws (such as conservation of energy, conservation of mass, conservation of current, etc.) to combine the components into a model of the complete asset; see Figure 3.2.

Note how the generation of component performance characteristics also provides a "firewall" that protects the organization's intellectual property by concealing the high-fidelity details of component configuration and how component performance is physically realized.

Simulation

Since the performance of an asset is defined by the performance of each of its components, overall asset performance can be determined at any given time by feeding the state of those components (expressed as levels of component performance parameters) into the model described earlier. The operation of that asset can then be simulated, including environmental and operation variation, consumable usage, and any degradation of component performance due to wear and tear resulting from the operation, to determine the new (possibly degraded) asset performance after that operation. Since the purpose of maintenance, in its broadest sense, is to replenish consumables and ensure the asset and its components are in a useable state for the next operation, operations and maintenance continually change the states of an asset and its components throughout their lives, as shown diagrammatically in Figure 3.3.

FIGURE 3.3 Asset operation and maintenance simulation (*Provost 2012*).

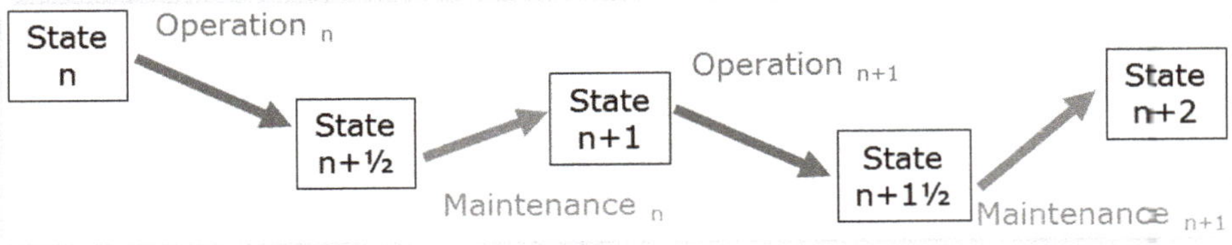

FIGURE 3.4 Project simulation (*Provost 2012*).

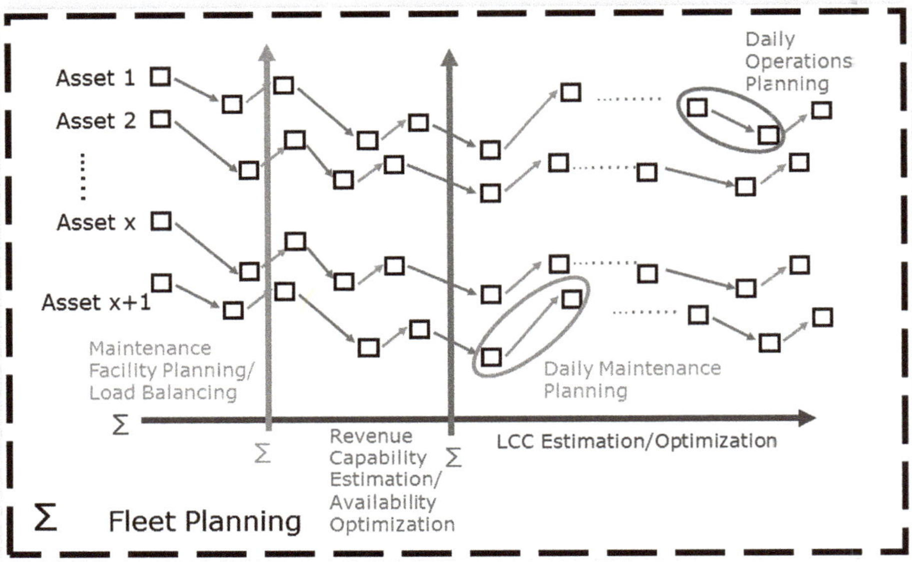

By adding the financial effects of operation and maintenance to the asset performance model, a complete project planning simulation can be created, as shown in Figure 3.4 by:

- Adding up the results from the operation and maintenance simulations for each individual asset over its life, to provide a life-cycle cost for that asset.
- Adding up the operation simulations for every asset in a fleet at any point in time, to provide an assessment of fleet operational performance and availability at that time.
- Adding up the maintenance simulations for every asset in a fleet at any point in time, to provide an assessment of fleet maintenance performance and availability at that time.
- Adding up the results from the operation and maintenance simulations for all assets in a fleet (or fleets) over their lives, to provide a project life-cycle cost.

Figure 3.4 shows how operation and maintenance modelling tools can also be used for daily operation and maintenance planning for each asset. Optimization of fleet technical and financial performance over the whole life of the fleet is also possible.

The reader should now be able to appreciate the advantages of possessing a modelling and simulation capability:

- Optimization of asset technical and economic performance is much easier.
- The capability provides a solid base for dialogue with customers and understanding of competitors.

- The capability provides consistent and traceable prediction of asset performance, from individual units to whole fleets.
- Fast, consistent, and accurate analysis and monitoring of asset performance is made much easier, both during development and in service.
- Knowledge transfer within and between projects is facilitated.

Books and Papers

Jason Merrick's presentation (*Merrick 2013*) gives a high-level overview of simulation, while Mark Matzopoulos' article (*Matzopoulos 2013*) discusses how modelling can accelerate development, eliminating the need for prototypes in particularly complex cases like petrochemical plants. The article by Martin Courtney (*Courtney 2014*) shows how the latest high-performance modelling and simulation tools use powerful math and physics computation to engineer a wide range of products faster and more accurately, while the paper by Carley Jurishica (*Jurishica 2010*) offers some practical advice for organizations new to simulation. The article by Michael Carone (*Carone 2014*) shows how iterating between modelling and simulation can improve the quality of system designs early, reducing the number of errors found later on in the design process. A presentation from the MathWorks (*MathWorks 2007*) shows how modelling and experimental design can refine the simulation of a complex system, while the article by Ben Hargreaves (*Hargreaves 2014*) describes the software and benefits of multi-physics simulation, which brings together the modelling of disparate physical phenomena (such as aerodynamics, heat transfer, structures, and electrics) into an integrated whole.

Finally, Horace Judson's wide-ranging, high-level overview of science and technology aimed at the lay reader (*Judson 1980*) includes a chapter on modelling and simulation, describing it as a rehearsal for reality and a way of trialing ideas that minimizes the penalties for error: good models include only the parts of a system that are relevant to the problem in hand. Readers should also bear in mind the saying attributed to the eminent statistician G.E.P. Box: "All models are wrong, but some are useful."

Some Useful Sources of Information for Specific Disciplines and Equipment Types

Below are some of the books and papers that the author has found helpful over the years. These are obviously not the only sources of information: some of them are now quite old, going back to the author's university days, so they may be superseded by more modern works. Books inevitably are a personal choice: the reader should use his or her own experience, as well as reviews posted on the Internet, to guide any purchases or loans. Remember that even if a book has only one or two good ideas, it will have paid for itself, while a book that is liked and used often is a real treasure. The Internet, particularly Wikipedia (www.wikipedia.com), can be very useful, although much of the material that search engines retrieve may not be peer reviewed and should therefore be checked against other sources before use.

General Mathematical Modelling

Those who find mathematics intimidating may find David MacKay's excellent book about sustainable energy (*MacKay 2009*) helpful. As well as being a very readable book on general energy matters, it shows mathematics being used in a very pragmatic and useful way. Jordan Ellenberg's book (*Ellenberg 2014*) is a good introduction to applying simple mathematical techniques and thought processes to everyday issues, showing how a little mathematical knowledge can reveal hidden structures that lie beneath many physical and social phenomena. Sanjoy Mahajan has produced a book to help those who are less confident about their mathematical thought processes get into the right frame of mind (*Mahajan 2010*). The books by Frank Giordano and others (*Giordano et al. 2003*) and Neil Gershenfeld (*Gershenfeld 2002*) explain a useful range of mathematical techniques. Ka-Kit Tung's book (*Tung 2007*) contains some new ideas that readers may find helpful and interesting. For those who are more academically inclined, the books by Rutherford Aris (*Aris 1994*) and Edward Bender (*Bender 1978*) are worth studying. Readers who need a refresher in matrix algebra are referred to *Matrices: Their Meaning and Manipulation* (*Bickley and Thompson 1964*). The book by Charles Close and others (*Close et al. 2002*) explains the modelling of dynamic systems in some detail, while the book by Jerry Banks and others (*Banks et al. 2010*) looks at the problem of simulating discrete events.

The subjects of accuracy, repeatability, measurement error, and how to deal with them are covered in *Hayward* (*1977*) and *Topping* (*1972*). Finally, the book by Phillip Slater (*Slater 2012*) shows that optimization needs to be conducted in context and with an appreciation for the true constraints and culture that affect the problem in order to be successful. While the example used relates to inventory management, the lessons can be applied to any situation.

Modelling Tools

William Press and others have produced an excellent series of books explaining and detailing a number of useful mathematical algorithms in various computer languages (see *Press et al. 1986* for the FORTRAN version, as well as the Numerical Recipes website (www.nr.com)).

The widely regarded "gold standards" for mathematical modelling tools are MatLab® and Simulink®, produced by the MathWorks Inc. (www.mathworks.com). Training packages are available from the MathWorks (*MathWorks 2004*), and there are a number of excellent introductory and advanced books available (*Pratap 2006*; *Hunt et al. 2001*; *Chaturvedi 2010*; *Gilat and Subramaniam 2008*; and others). There is also a great deal of guidance and information on the Internet, such as Allen Downey's book (*Downey 2011*), as well as course notes and/or presentations available online such as those by Robert Marino and Andrew Pruszynski (*Marino and Pruszynski 2009*). MatLab® Central is a thriving online community offering advice and solutions to common (and not-so-common) issues.

Two good free competitors to MatLab® and Simulink® are Scilab™ and Scicos™, respectively (www.scilab.org) and (www.scicos.org). Stephen Campbell and others have produced a book on Scilab™ and Scicos™ (*Campbell et al. 2006*), and there are online guides available (e.g., *Baudin 2010*, *Scilab Group 2010*, and *Nikoukhah and Steer 2010*). The general approach and syntax is similar, though not identical, to MatLab®. For modelling micro-power systems, the HOMER package (www.homerenergy.com) is comprehensive and easy to use; there are many online guides and examples of its use. Tom Lambert and others have provided an excellent introduction to HOMER (*Lambert et al. 2006*).

The Microsoft Excel® spreadsheet program is also widely used for a host of business and technical modelling tasks. DoneEx (www.doneex.com) produce software that will compile Microsoft Excel® spreadsheets into secured, executable programs which look like spreadsheets: however, all the formulae are hidden, and only cells containing data can be changed by the user.

The books by Sam Savage (*Savage 2003*) and Robert Clemen and Terence Reilly (*Clemen and Reilly, 2014*) provide excellent introductions to a range of useful decision-making tools such as forecasting, decision trees, optimization, and Monte Carlo analysis. Finally, the article by Ben Sampson (*Sampson 2009*) describes several companies that offer modelling and other IT process as a service via the Internet (Software as a Service (SaaS)), while the book by Philip Armour (*Armour 2004*) makes a powerful case for the use of agile software methods for the production and management of software, since software is now a repository of knowledge that must keep pace with user understanding.

Reference Material

Theodore Baumeister and others have edited a standard mechanical engineering handbook, which provides a general overview of almost everything to do with engineering (*Baumeister et al. 1978*). Dan Marghitu has edited a less comprehensive but more detailed coverage of general mechanical engineering knowledge (*Marghitu 2001*). Clifford Matthews has produced two useful data books for aeronautical (*Matthews 2002*) and mechanical (*Matthews 2012*) engineers, while the Gieck brothers have written a concise yet comprehensive book of technical formulae (*Gieck and Gieck 1996*). An interesting and attractive loose-leaf book by Alex Moulton and others (*Moulton et al. 2005*) presents many topics in statics, material properties, dynamics, and vibrations in the form of single sheets for each topic.

Aerodynamics, Fluid Mechanics, and Turbomachinery

John Allen's book (*Allen 1986*) is a comprehensive yet accessible discourse on aerodynamics, covering a range from insects to weather systems, as well as cars, aircraft, and spacecraft. *Kay and Nedderman 1974* provides an introduction to fluid mechanics and heat transfer, while Frank White's book (*White 1999*) goes into more detail on compressible and incompressible flow, while still being very accessible. Sir John Horlock's books (*Horlock 1973a*, *1973b*) are turbomachinery classics, while the books by Archie McKenzie (*McKenzie 2001*) and Geoff Wilde (*Wilde 1999*) discuss the design, development, performance, and performance calculation methods of Rolls-Royce aircraft engine axial compressors in some detail. Nick Cumpsty's book (*Cumpsty 1989*) presents a more-up-to-date view of compressor aerodynamics. A simpler introduction to turbomachinery is provided by Dixon (*Dixon 1975*), while John Wendt has edited a very good introduction to computational fluid dynamics (*Wendt 1995*).

Thermodynamics

The classic thermodynamics textbook is *Rogers and Mayhew 1967*; however, *Çengel and Boles* (*1998*) is more modern and accessible. Richard Haywood's book (*Haywood 1975*) goes into more detail about the modelling of simple and advanced gas turbine, steam turbine, internal combustion engine, and refrigeration cycles. Iain Staffell has provided a useful data sheet of properties of some common fuels (*Staffell 2011*).

Control Systems

Richard Dorf's book (*Dorf 1989*) gives a comprehensive introduction to control theory. The book by Richard Jagacinski and John Flach (*Jagacinski and Flach 2003*) is also useful. John Prentis' book (*Prentis 1970*) provides a fine overview of control systems and their relationship to vibration theory and system dynamics.

Electrical Systems and Batteries

The classic electrical systems textbook is *Hughes Electrical Technology* (*McKenzie-Smith and Hughes 1995*). Arthur Seidman and others have assembled together an excellent selection of electrical system formulae, including some relevant financial algorithms (*Seidman et al. 1996*).

Thomas Reddy and David Linden have produced a very comprehensive book that describes all types of batteries (*Reddy and Linden 2011*). Modelling of lead-acid batteries is discussed in the papers by Henrik Bindner and others (*Binder et al. 2005*), Ahmad Darabi and others (*Darabi et al. 2013*), and Robyn Jackey (*Jackey 2007*), while Ander Tenno (*Tenno 2004*) describes the modelling of a variant of lead-acid batteries known as VRLA (valve-regulated lead-acid) which is commonly used for backup power. The paper by Sukhvinder Badwal and others (*Badwal et al. 2014*) presents an overview of several emerging electrochemical energy conversion and storage technologies, together with a discussion of some of their key technical challenges.

Two papers by Duke Energy and David Lawrence and others (*Duke Energy 2014*; *Lawrence et al. 2014*: www.duke-energy.com, www.elp.com) describe modelling of an electricity grid in some detail, using big data techniques to bring together many disparate datasets.

Heating, Ventilation, and Air Conditioning (HVAC)

The course materials by Matthew Cloutier (*Cloutier 2001*) and PDH Online (*PDH Online 2011*: www.pdhonline.com) are comprehensive introductions to refrigeration cycles, fans, and blowers.

Aeroengines and Gas Turbines

The modelling of gas turbines has a long history, dating back to the days of Frank Whittle, who produced accurate models of his jet engines using his slide rule. The Rolls-Royce plc book *The Jet Engine* (*Rolls-Royce 2005*) is a well-written and easy-to-understand introduction to the world of aeroengines, with a minimum of formulae. Nick Cumpsty has provided a more detailed book, which focuses on the main drivers of aeroengine performance and design (*Cumpsty 2003*), while Klaus Hünecke discusses mechanical design issues as well as performance (*Hünecke 1997*). Claire Soares' comprehensive book (*Soares 2008*) goes into more detail across a wide range of gas turbine topics covering all applications, including history, thermodynamics, mechanical design, operations, and business issues, drawing upon much material published elsewhere. The classic UK gas turbine textbook, which has been available for over sixty years, is by Saravanamuttoo and others (*Saravanamuttoo et al. 2001*). The U.S. classics, with more focus on aeroengines, are by Jack Mattingly and others (*Mattingly 1996; Mattingly et al. 2002*). David Wilson and Theodosios Korakianitis (*Wilson and Korakianitis 1998*) provide another comprehensive description of the thermodynamics and aerodynamics of these fascinating machines. Zak Razak has produced both course notes (*Razak 2007a*) and a book (*Razak 2007b*) focusing on the performance of industrial gas turbines. Mention should also be made of Sir Frank Whittle's book (*Whittle 1981*), given his monumental contributions to the industry.

The GasTurb program produced by Joachim Kurzke, and now marketed by GasTurb GmbH (www.gasturb.de), is an extremely comprehensive software package for modelling all types of gas turbines. Philip Walsh and Paul Fletcher's book goes into considerable detail on gas turbine performance calculations, giving the reader enough material to produce his/her own simulations (*Walsh and Fletcher 1998*). The Rolls-Royce plc *Jackson Course* ran for many years and was given by its authors both internally and at several airframers and academic institutions (*Jackson et al. 1992*).

Jane's Aero Engines, edited by Mark Daly (*Daly 2015*), describes every aeroengine currently in service and includes introductory chapters on engine technology. A very detailed and fascinating history of the design, development, and entry into service of the Rolls-Royce RB211-22B turbofan has been written by Phil Ruffles (*Ruffles 2014*); technical, financial, and commercial aspects are covered. The Rolls-Royce Heritage Trust (www.rolls-royce.com/about/heritage-trust.aspx) have also published histories of other Rolls-Royce engines, such as the Eagle, Merlin, Dart, Tyne, Spey, and Olympus; details of these can be found on the website. The report by Stanley Hooker and others (*Hooker et al. 1997*) is a classic description of how to calculate the performance of supercharged aircraft piston engines.

Aircraft

Mark Davies has edited a comprehensive handbook (*Davies 2003*) which provides a brief introduction to all of the important aeronautical subjects. *Kermode* (*1972*) provides a very simple introduction to aircraft flight. Darrol Stinton's book (*Stinton 1998*) goes into more detail, showing why aircraft configurations vary so much, while the book by Laurence Loftin (*Loftin 2014*) discusses the evolution of modern aircraft, showing by way of numerical comparisons of key airframe configuration, engine power, and overall performance and aerodynamic metrics how their performance has evolved.

John Fielding's book (*Fielding 1999*) also looks at the fundamentals of civil and military aircraft design, while Ray Whitford's books (*Whitford 2000, 2007*) are very comprehensive yet accessible accounts of the development of fighters and civil aircraft.

Shannon Ackert has written a simple introduction to civil aircraft payload-range analysis (*Ackert 2013*), while Lloyd Jenkinson and others cover civil aircraft performance in more detail (*Jenkinson et al. 1999*), and John Anderson's book covers both civil and military aircraft performance and design (*Anderson 1999*). The handbook by Airbus Customer Services (*Airbus Customer Services 2002*) goes into the fine detail of defining aircraft performance over all phases of flight, including emergency situations. Egbert Torenbeek's book (*Torenbeek 1982*) is a very detailed discussion of subsonic aircraft design. Reg Austin's book (*Austin 2010*) covers many of the technical and nontechnical aspects of unmanned aircraft ("drones"), while the book by Ian Moir and Allan Seabridge (*Moir and Seabridge 2008*) describes aircraft systems (electrics, hydraulics, pneumatics, etc.) in detail.

Automotive

Hans-Hermann Braess and Ulrich Sieffert have edited a comprehensive handbook (*Braess and Sieffert 2005*) which provides a brief introduction to all of the important automotive subjects. Allen Fuhs' book (*Fuhs 2009*) presents many of the calculations and design choices made in the design of a passenger vehicle, while road vehicle powertrain and overall performance is covered in some detail in *Lucas* (*1986*). The PhD theses by Martin Passmore (*Passmore 1990*) and Andrew Simpson (*Simpson 2005*) discuss automotive vehicle drag and energy consumption in detail, while the paper by Tony Markel and others (*Markel et al. 2002*) describes a commercial systems analysis tool (ADVISOR®) for advanced vehicle modelling. A compilation of papers discussing the performance of Rolls-Royce pre-WWII motor cars, including many graphs and charts and information on performance methods, has been published by the Rolls-Royce Heritage Trust (*Hives et al. 2005*).

Diesels and Turbocharging

Richard Stone's book (*Stone 1999*) is an excellent introduction to the design and performance of both spark ignition and compression ignition engines, as well as fuels, turbocharging, engine modelling, and mechanical design. John Lumley's book (*Lumley 1999*) goes into more detail and complements the Stanford engine simulation program (ESP), while the report by Dan Cordon and others (*Cordon et al. 2007*) describes a commercial internal combustion engine simulation package called WAVE, available from Ricardo plc. Bernard Challen's and Rodica Baranescu's book (*Challen and Baranescu 1999*) is a comprehensive introduction to all aspects of diesel engine design and operation. The book by Robert Bosch GmbH (*Bosch 2005*) discusses diesel engine management in more detail, while Jesper Ritzén's thesis (*Ritzén 2003*) covers the modelling of a turbocharged diesel engine. Those who need details of diesel engine transient performance need look no further than *Rakopoulos and Giakoumis* (*2009*). Nick Baines has written a comprehensive account of turbocharging (*Baines 2005*), while the paper by Paul Moraal and Ilya Kolmanovsky (*Moraal and Kolmanovsky 1999*) and thesis by Oskar Leufvén (*Leufvén 2010*) present turbocharger modelling in detail.

Railways

Considering the importance of the rail industry to the smooth functioning of our society, there are remarkably few technical books on the subject. Clifford Bonnett has produced a general overview of railway engineering (*Bonnett 1996*), while John Glover's book (*Glover 2013*) looks at the general principles governing railway operation. David Clark's course (*Clark 2009*) gives an overview of the UK railway business and the technical aspects of rail vehicles. An excellent general overview of the rail business, including management, operation, infrastructure, and vehicles, is provided by Profillidis (*2006*). David Clough's and Martin Beckett's book (*Clough and Beckett 1988*), which was originally written for enthusiasts, contains several equations and performance curves for UK diesel traction. A spreadsheet is available from an enthusiast group, the Railway Performance Society (www.railperf.org.uk), that calculates track gradients and models traction horsepower. Alan Wickens has produced the definitive guide on rail vehicle dynamics (*Wickens 2003*), while the book by Ingo Hansen and Jörn Pachl (*Hansen and Pachl 2008*) looks at analysis, modelling, and simulation of railway timetables and traffic.

Ships

The excellent book by Eric Tupper (*Tupper 2013*) is an accessible introduction to the fundamentals of all aspects of naval architecture, while the book by Bryan Barrass (*Barrass 2004*) goes into more detail on ship design and performance. The paper by MAN Diesel & Turbo (*MAN Diesel & Turbo 2011*: http://dieselturbo.man.eu) discusses the basic principles of ship propulsion, while the course notes by Rod Sampson (*Sampson 2008*), the book by Anthony Molland and others (*Molland et al. 2011*), and the report by Hans Otto Kristensen and Marie Lützen (*Kristensen and Lützen 2012*) cover the subjects of ship propulsion and resistance in considerable detail. Finally, the book by Bryan Barrass and Captain D. Derrett (*Barras and Derrett 2012*) provides an exhaustive coverage of ship stability.

Combined Heat and Power (CHP)

Milton Meckler and Lucas Hyman have edited a book (*Meckler and Hyman 2010*) that covers the basics, feasibility studies, design, construction, and operation of CHP plants, including some case studies.

Fuel Cells

Sharon Thomas and Maria Zalbowitz have produced an excellent pamphlet that concisely introduces fuel cells (*Thomas and Zalbowitz 1999*). A good overview of fuel cell characteristics and technology types is provided in a datasheet produced by Fuel Cell Today (*Fuel Cell Today 2012*: www.fuelcelltoday.com); they have also produced factsheets on fuel cell applications which can be downloaded from the website. The basics of fuel cells are also discussed in a paper by Ram Krishna and others (*Krishna et al. 2012*), while the article by Chris Jackson (*Jackson 2008*) gives a good overview of fuel cell technology and its applications.

James Larminie's and Andrew Dicks' book (*Larminie and Dicks 2003*) is an excellent introduction to the theory, performance, and design of the various types of fuel cells. Frano Barbir has written an introductory chapter on fuel cells (*Barbir 2006*) and a book that goes into more detail (*Barbir 2005*), as does the book by O'Hayre (*O'Hayre et al. 2009*). The report by the U.S. Department of Energy (*US DOE 2004*) is widely regarded as a definitive reference. Mark Matzopoulos' article (*Matzopoulos 2007*) discusses how fuel cell modelling can accelerate development, while Colleen Spiegel's book (*Spiegel 2008*) enables the reader to produce a detailed MatLab® model of a proton exchange membrane (PEM) fuel cell. *Gou et al.* (*2010*) and *Pukrushpan et al.* (*2010*) discuss the modelling of fuel cells and their control systems in considerable detail.

Pumps

Good introductions to pumps and pumping systems can be found in the presentation by Mark Hemeyer and Adam Mudge (*Hemeyer and Mudge 2012*), the series of articles by Hans Vogelesang (*Vogelesang 2008a, 2008b, 2008c, 2008d, 2009*), and a presentation by the author (*Provost 2015*). The article by Roland McKinney (*McKinney 2010*) illustrates the benefits that can be gained from a pump system optimization program aimed at reducing energy use and operating cost; since these can extend well beyond energy use, savings from optimization programs tend to be underestimated. The article by James Boyle (*Boyle 2003*) summarizes the issues that govern the successful selection of centrifugal pumps.

The detailed handbook by Europump (*Europump 2006*) looks at selecting pumps for maximizing energy efficiency, while the executive summary by Europump, the Hydraulic Institute, and the U.S. Department of Energy (*Europump/Hydraulic Institute/US DoE 2004*) and the corresponding handbook (*Europump/Hydraulic Institute 2004*) look at successful applications of variable speed pumping. More detailed help and guidance on pump selection, best practices, expected pump efficiency levels, and other technical issues is given in *GAMBICA/BPMA* (*2003*), *Europump* (*2008*), *GAMBICA/Europump* (*2008*), *Europump* (*2003*), *ETSU/AEAT* (*2001*), *Europump* (*1999a*), *Europump* (*1999b*), and *Europump* (*2000*). Crane Ltd. have produced a detailed, regularly updated guide on fluid flow through pipes and fittings (*Crane 1986, 1988*).

Finally, the guides produced by Sustainability Victoria in Australia (*Sustainability Victoria 2009*) and the U.S. Environmental Protection Agency (*US EPA 2013*) look at some of the practical issues and best practices around pumping energy efficiency in water and wastewater facilities, supported by various case studies.

Wind Turbines, Solar, and Renewable Energy

The excellent book by Paul Lynn on marine energy systems (*Lynn 2014*) discusses wave and tidal power (including a useful section on the basics of AC power), while his other books (*Lynn 2010, 2012*) are very accessible introductions to photovoltaics and wind energy, respectively.

Volker Quaschning's book (*Quaschning 2005*) offers a good general introduction to renewable energy systems. Christopher Martin's and Yogi Goswami's pocketbook (*Martin and Goswami 2005*) contains formulae, tables, and other reference material for solar power systems, while the book written by Peter Jensen and others (*Jensen et al. 2007*), which is part of the same series, covers wind energy in the same way.

Donald Wroblewski has written a very simple guide to wind turbines (*Wroblewski 2013*), while more detailed discussions about the technical and nontechnical aspects of wind turbines can be found in *Burton et al. (2001)* and *Hau (2006)*. The report by David Corbus and Mark Meadors (*Corbus and Meadors 2005*) describes the results of testing a small (10 kW) research wind turbine, while a factsheet from the Minnesota Municipal Power Agency (*MMPA 2014*) discusses the importance of wind turbine tip speed ratio.

References

* denotes sources that the author found particularly helpful and/or inspiring.

Ackert, S., *Aircraft Payload-Range Analysis for Financiers* (San Francisco, CA: Jackson Square Aviation, 2013). Pg. 72

Airbus Customer Services, *Getting to Grips with Aircraft Performance* (Blagnac, France: Airbus SAS, 2002). Pg. 72

Allen, J., *Aerodynamics: The Science of Air in Motion*, 5th ed. (Blythburgh, UK: Allen Brothers & Father, 1986). Pg. 69

Anderson, J., *Aircraft Performance and Design* (Singapore: McGraw-Hill, 1999). Pg. 72

Aris, R., *Mathematical Modelling Techniques* (Mineola, NY: Dover Publications, 1994). Pg. 68

Armour, P., *The Laws of Software Process: A New Model for the Production and Management of Software* (Boca Raton, FL: Auerbach Publications/CRC Press, 2004). Pg. 69

Austin, R., *Unmanned Aircraft Systems: UAVS Design Development and Deployment* (Chichester, UK: John Wiley & Sons Ltd, 2010). Pg. 72

Badwal, S., Giddey, S., Munnings, C., Bhatt, A. et al., "Emerging Electrochemical Energy Conversion and Storage Technologies," *Frontiers in Chemistry* 2 (September 2014). Pg. 70

Baines, N., *Fundamentals of Turbocharging* (White River Junction, VT: Concepts NREC, 2005). Pg. 72

Banks, J., Carson, J., Nelson, B., and Nicol, D., *Discrete-Event System Simulation*, 5th ed. (Upper Saddle River, NJ: Pearson Education, 2010). Pg. 68

Barbir, F., *PEM Fuel Cells: Theory and Practice* (Burlington, MA: Elsevier Academic, 2005). Pg. 74

Barbir, F., "PEM Fuel Cells," in: Sammes, N. ed., *Fuel Cell Technology: Reaching towards Commercialization* (London, UK: Springer-Verlag London Ltd, 2006). Pg. 74

Barras, C., *Ship Design and Performance for Masters and Mates* (Oxford, UK: Elsevier Butterworth-Heinemann, 2004). Pg. 73

Barras, C. and Derrett, D., *Ship Stability for Masters and Mates*, 7th ed. (Oxford, UK: Elsevier Butterworth-Heinemann, 2012). Pg. 73

Baudin, M., *Introduction to Scilab™* (Le Chesnay, France: Scilab™ Consortium, 2010). Pg. 68

Baumeister, T., Avallone, E., and Baumeister, T. eds., *Mark's Standard Handbook for Mechanical Engineering*, 8th ed. (New York: McGraw-Hill, 1978). Pg. 69

Bender, E., *An Introduction to Mathematical Modelling* (Mineola, NY: Dover Publications, 1978). Pg. 68

Bickley, W. and Thompson, R., *Matrices: Their Meaning and Manipulation* (London, UK: Unibooks/English Universities Press Ltd, 1964). Pg. 68

Binder, H., Cronin, T., Lundsager, P., Manwell, J. et al., *Lifetime Modelling of Lead Acid Batteries* (Roskilde, Denmark: Risø National Laboratory, 2005). Pg. 70

Bonnett, C., *Practical Railway Engineering* (London, UK: Imperial College Press, 1996). Pg. 73

Bosch, *Diesel-Engine Management: Systems and Components* (Plochingen, Germany: Robert Bosch GmbH, 2005). Pg. 72

Boyle, J., "Specifying Centrifugal Pumps – For Energy, Efficiency and Effective Performance," *Maintenance & Engineering* 15, no. 4 (July/August 2015): 19-20. Pg. 74

Braess, H. and Sieffert, U. eds., *Handbook of Automotive Engineering* (Warrendale, PA: SAE International, 2005). Pg. 72

Burton, T., Sharpe, D., Jenkins, N., and Bossanyi, E., *Wind Energy Handbook* (Chichester, UK: John Wiley & Sons Ltd, 2001). Pgs. 64, 75

Campbell, S., Chancelier, J., and Nikoukhah, R., *Modeling and Simulation in Scilab™/Scicos™* (New York: Springer-Verlag, 2006). Pg. 68

Carone, M., *Using Modeling and Simulation to Test Designs and Requirements* (Natick, MA: MathWorks Inc., 2014). Pg. 67

Çengel, Y. and Boles, M., *Thermodynamics: An Engineering Approach*, 3rd ed. (Boston, MA: WCB/McGraw-Hill, 1998). Pg. 70

Challen, B. and Baranescu, R. eds., *Diesel Engine Reference Book*, 2nd ed. (Oxford, UK: Butterworth-Heinemann, 1999). Pg. 72

Chaturvedi, D., *Modeling and Simulation of Systems Using MatLab® and Simulink®* (Boca Raton, FL: CRC Press/Taylor & Francis, 2010). Pg. 68

Clark, D., *Get on the Right Track: Issue B* (Derby, UK: Bombardier Transportation/Catalis Ltd, 2009). Pg. 73

* Clemen, R. and Reilly, T., *Making Hard Decisions with DecisionTools*, 3rd ed. (Mason, OH: South-Western/Cengage Learning, 2014). Pg. 69

Close, C., Frederick, D., and Newell, J., *Modeling and Analysis of Dynamic Systems*, 3rd ed. (New York: John Wiley & Sons Inc., 2002). Pg. 68

Clough, D. and Beckett, M., *BR Motive Power Performance* (Shepperton, UK: Ian Allan, 1988). Pg. 73

Cloutier, M., "Refrigeration Cycles – Course Notes," Tuscaloosa, AL: Alabama University, 2001. Pg. 70

Corbus, D. and Meadors, M., "Small Wind Research Turbine," Technical Report NREL/TP-500-38550, Golden, CO: National Renewable Energy Laboratory, 2005 Pg. 75

Cordon, D., Dean, C., Steciak, J., and Beyerlein, S., "One-Dimensional Engine Modeling and Validation Using Ricardo WAVE," Moscow, ID: National Institute for Advanced Transportation Technology, Idaho University, 2007. Pg. 72

Courtney, M., "Better by Design," *Engineering and Technology* 9, no. 3 (April 2014): 71-76. Pg. 67

Crane, "Technical Paper 410: Flow of Fluids through Valves, Fittings and Pipe (Metric Edition - SI Units)," London, UK: Crane Ltd., 1986. Pg. 74

Crane, "Technical Paper 410: Flow of Fluids through Valves, Fittings and Pipe (US Units)," Chicago, IL: Crane Co., 1988. Pg. 74

Cumpsty, N., *Compressor Aerodynamics* (Harlow, UK: Longman Scientific & Technical, 1989). Pg. 69

Cumpsty, N., *Jet Propulsion: A Simple Guide to the Aerodynamic Design and Performance of Jet Engines*, 2nd ed. (Cambridge, UK: Cambridge University Press, 2003). Pg. 71

Daly, M. ed., *IHS Jane's Aero Engines 2015-2016* (Coulsdon, UK: IHS Global Ltd, 2015). Pg. 71

Darabi, A., Hosseina, M., Gholami, H., and Khakzad, M., "Modeling of Lead-Acid Battery Bank in the Energy Storage Systems," *International Journal of Emerging Technology and Advanced Engineering* 3, no. 3 (March 2013): 932-937. Pg. 70

Davies, M. ed., *The Standard Handbook for Aeronautical and Astronautical Engineers* (New York: McGraw-Hill, 2003). Pg. 71

Dixon, S., *Fluid Mechanics, Thermodynamics of Turbomachinery*, 2nd ed. (Oxford, UK: Pergamon Press, 1975). Pg. 69

Dorf, R., *Modern Control Systems*, 5th ed. (Menlo Park, CA: Addison-Wesley, 1989). Pg. 70

Downey, A., *Physical Modeling in MatLab®* (Needham, MA: Green Tea Press, 2011). Pg. 68

Duke Energy, *Duke Energy Emerging Technologies Organization: Data Modeling and Analytics Initiative* (Charlotte, NC: Duke Energy Corp, 2014). Pg. 70

* Ellenberg, J., *How Not to Be Wrong: The Hidden Maths of Everyday Life* (London, UK: Penguin Books/Allen Lane, 2014). Pg. 68

ETSU/AEAT, "Study on Improving the Energy Efficiency of Pumps," Brussels, Belgium: European Commission, 2001. Pg. 74

Europump, *Attainable Efficiencies of Volute Casing Pumps: Europump Guide to Advanced Pumping Technology No. 2* (Kidlington, UK: Elsevier Advanced Technology, 1999a). Pg. 74

Europump, *NPSH for Rotodynamic Pumps: Europump Guide to Advanced Pumping Technology No. 1* (Kidlington, UK: Elsevier Advanced Technology, 1999b). Pg. 74

Europump, *Operating Rotodynamic Pumps Away from Design Conditions: Europump Guide to Advanced Pumping Technology No. 3* (Kidlington, UK: Elsevier Advanced Technology, 2000). Pg. 74

Europump, *European Guide to Pump Efficiency for Single Stage Centrifugal Pumps* (Brussels, Belgium: Europump, 2003). Pg. 74

Europump, *System Efficiency: A Guide for Energy Efficient Rotodynamic Pumping Systems* (Brussels, Belgium: Europump, 2006). Pg. 74

Europump, *Guide to the Selection of Rotodynamic Pumps* (Brussels, Belgium: Europump, 2008). Pgs. 74, 74

Europump/Hydraulic Institute, *Variable Speed Pumping: A Guide to Successful Applications* (Kidlington, UK: Elsevier Advanced Technology, 2004). Pg. 74

Europump/Hydraulic Institute/US DoE, "Variable Speed Pumping: A Guide to Successful Applications - Executive Summary," Brussels, Belgium/Parsippany, NJ/Washington DC: Europump/Hydraulic Institute/US Department of Energy 2004. Pg. 74

Fielding, J., *Introduction to Aircraft Design* (Cambridge, UK: Cambridge University Press, 1999). Pg. 72

Fuel Cell Today, "Fuel Cell Basics: Technology Types," London, UK: Fuel Cell Today Ltd/Johnson Matthey Ltd, 2012. Pg. 73

Fuhs, A., *Hybrid Vehicles and the Future of Personal Transportation* (Boca Raton, FL: CRC Press, 2009). Pg. 72

GAMBICA/BPMA, "Variable Speed Driven Pumps: Best Practice Guide," London, UK/Birmingham, UK: GAMBICA Association Ltd/British Pump Manufacturers Association, 2003. Pg. 74

GAMBICA/Europump, "Variable Speed Electro Submersible Pumps: A Guide to the Specific Application of VSDs for Borehole and Wet-Well Pumps," London, UK/Brussels, Belgium: GAMBICA Association Ltd/Europump, 2008. Pg. 74

Gershenfeld, N., *The Nature of Mathematical Modelling* (Cambridge, UK: Cambridge University Press, 2002). Pg. 68

Gieck, K. and Gieck, R., *Technical Formulae*, 8th ed. (Germering, Germany: Gieck Verlag, 1996). Pg. 69

Gilat, A. and Subramaniam, V., *Numerical Methods for Engineers and Scientists: An Introduction with Applications Using MatLab®* (Hoboken, NJ: John Wiley & Sons Inc., 2008). Pg. 68

Giordano, F., Weir, M., and Fox, W., *A First Course in Mathematical Modelling*, 3rd ed. (Pacific Grove, CA: Brooks/Cole – Thompson Learning, 2003). Pg. 68

Glover, J., *Principles of Railway Operation* (Hersham, UK: Ian Allan Publishing Ltd, 2013). Pg. 73

Gou, B., Na, W., and Diong, B., *Fuel Cells: Modeling, Control and Applications* (Boca Raton, FL: CRC Press/Taylor & Francis, 2010). Pg. 74

Hansen, I. and Pachl, J. eds., *Railway Timetable and Traffic: Analysis, Modelling, Simulation* (Hamburg, Germany: Eurailpress/DVV Media Group GmbH, 2008). Pg. 73

Hargreaves, B., "Better Together," *Professional Engineering*, October 2014, 41-43. Pg. 67

Hau, E., *Wind Turbines: Fundamentals, Technologies, Application, Economics*, 2nd ed. (Berlin, Germany: Springer-Verlag, 2006). Pgs. 64, 75

Hayward, A., *Repeatability and Accuracy* (London, UK: Mechanical Engineering Publications Ltd, 1977). Pg. 68

Haywood, R., *Analysis of Engineering Cycles*, 2nd ed. (Oxford, UK: Pergamon Press, 1975). Pg. 70

Hemeyer, M. and Mudge, A., *Efficient Pumping* (Wixom, MI: Kennedy Industries Inc., 2012). Pg. 74

Hives, E., Lovesey, A., and Robotham, A., *Fundamentals of Car Performance* (Derby, UK: Rolls-Royce Heritage Trust, 2005). Pg. 72

Hooker, S., Reed, H., and Yarker, A., *The Performance of a Supercharged Aero Engine* (Derby, UK: Rolls-Royce Heritage Trust, 1997). Pg. 71

Horlock, Sir J., *Axial Flow Compressors: Fluid Mechanics and Thermodynamics* (Huntington, NY: Krieger Publishing Co, 1973a). Pg. 69

Horlock, Sir J., *Axial Flow Turbines: Fluid Mechanics and Thermodynamics* (Huntington, NY: Krieger Publishing Co, 1973b). Pg. 69

Hünecke, K., *Jet Engines: Fundamentals of Theory, Design and Operation* (Shrewsbury, UK: Airlife Publishing, 1997). Pg. 71

Hunt, B., Lipsman, R., Rosenberg, J., Coombes, K. et al., *A Guide to MatLab® for Beginners and Experienced Users* (Cambridge, UK: Cambridge University Press, 2001). Pg. 68

Jackey, R., *A Simple Effective Lead-Acid Battery Modeling Process for Electrical System Component Selection* (Natick, MA: MathWorks Inc., 2007). Pg. 70

Jackson, A., Bashforth, S., and Provost, M. *A Short Course on Gas Turbine Performance: Course Notes*, 6th ed. (Derby, UK: Rolls-Royce plc, 1992). Pgs. 64, 71

Jackson, C., "Current Thinking on Fuel Cells," *Energy in Buildings and Industry*, March 2008, 29-32. Pg. 74

Jagacinski, R. and Flach, J., *Control Theory for Humans: Quantitative Approaches to Modeling Performance* (Mahwah, NJ: Lawrence Erlbaum Associates Inc., 2003). Pg. 70

Jenkinson, L., Simpkin, P., and Rhodes, D., *Civil Jet Aircraft Design* (London, UK: Arnold, 1999). Pg. 72

Jensen, P., Meyer, N., Mortensen, N., and Oster, F., *Wind Energy Pocket Reference* (Tampa, FL: Earthscan, 2007). Pg. 75

Judson, H., *The Search for Solutions* (London, UK: Hutchinson Publishing Group, 1980). Pg. 67

Jurishica, C., "Practical Advice for Organizations New to Simulation," *Proceedings of the 2010 Winter Simulation Conference*, Marriott Waterfront Hotel, Baltimore, MD, December 5-8, 2010 (New York: Institute of Electrical and Electronics Engineers). Pg. 67

Kay, J. and Nedderman, R., *An Introduction to Fluid Mechanics and Heat Transfer*, 3rd ed. (Cambridge, UK: Cambridge University Press, 1974). Pg. 69

Kermode, A., *Mechanics of Flight* (London, UK: Pitman Publishing, 1972). Pg. 71

Krishna, R., Laxmi, M., and Kumar, N., "Fuel Cell for Telecom Applications: A Low Cost Green Back-Up Power System Alternative to Diesel Generators," New Delhi, India: Department of Telecommunications, Government of India, 2012. Pg. 73

Kristensen, H. and Lützen, M., "Prediction of Resistance and Propulsion Power of Ships," Technical Report, Technical University of Denmark/University of Southern Denmark, Copenhagen, Denmark/Odense, Denmark, 2012. Pg. 73

Lambert, T., Gilman, P., and Lilienthal, P., "Micropower System Modeling with HOMER," in: Farret, F. and Simões, M. eds., *Integration of Alternative Sources of Energy* (Hoboken, NJ: Wiley-IEEE Press, 2006), 379-418. Pg. 69

Larminie, J. and Dicks, A., *Fuel Cell Systems Explained*, 2nd ed. (Chichester, UK: John Wiley & Sons Ltd, 2003). Pg. 74

Lawrence, D., Mulder, D., and Canders, Z., "Duke Energy's Data Modeling and Analytics Initiative," *Powergrid International*, August 13, 2014. Pg. 70

Leufvén, O., "Compressor Modeling for Control of Automotive Two Stage Turbochargers," Ph.D. thesis, Linköping University, Linköping, Sweden, 2010. Pg. 72

Loftin, L., *Quest for Performance: The Evolution of Modern Aircraft* (Washington, DC: National Aeronautics & Space Administration, 2014). Pg. 71

Lucas, G., *Road Vehicle Performance* (New York: Gordon and Breach, 1986). Pg. 72

Lumley, J., *Engines: An Introduction* (Cambridge, UK: Cambridge University Press, 1999). Pg. 72

* Lynn, P., *Electricity from Sunlight: An Introduction to Photovoltaics* (Chichester, UK: John Wiley & Sons Ltd, 2010). Pg. 74

* Lynn, P., *Onshore and Offshore Wind Energy: An Introduction* (Chichester, UK: John Wiley & Sons Ltd, 2012). Pg. 74

* Lynn, P., *Electricity from Wave and Tide: An Introduction to Marine Energy* (Chichester, UK: John Wiley & Sons Ltd, 2014). Pg. 74

* MacKay, D., *Sustainable Energy – Without the Hot Air* (Cambridge, UK: UIT Cambridge Ltd, 2009). Pg. 68

Mahajan, S., *Street-Fighting Mathematics: The Art of Educated Guessing and Opportunistic Problem Solving* (Cambridge, MA: MIT Press, 2010). Pg. 68

MAN Diesel & Turbo, *Basic Principles of Ship Propulsion* (Copenhagen, Denmark: MAN Diesel & Turbo, 2011). Pg. 73

Marghitu, D. ed., *Mechanical Engineer's Handbook* (San Diego, CA: Academic Press, 2001). Pg. 69

Marino, R. and Pruszynski, A., *Non-Accredited MatLab® Tutorial Sessions for Beginner to Intermediate Level Users* (Kingston, Ontario: Queen's University, 2009). Pg. 68

Markel, T., Brooker, A., Hendricks, T., Johnson, V. et al., "ADVISOR: A Systems Analysis Tool for Advanced Vehicle Modelling," *Journal of Power Sources* 110 (2002): 255–266. Pg. 72

Martin, C. and Goswami, D., *Solar Energy Pocket Reference* (London, UK: Earthscan, 2005). Pg. 75

MathWorks, *MatLab® Fundamentals and Programming Techniques* (Natick, MA: MathWorks Training Services, 2004). Pg. 68

MathWorks, *Model-Based Calibration Master Class* (Natick, MA: MathWorks Inc., 2007). Pg. 67

Matthews, C., *Aeronautical Engineer's Data Book* (Oxford, UK: Butterworth- Heinemann, 2002). Pg. 69

Matthews, C., *IMechE Engineers' Databook*, 4th ed. (Chichester, UK: John Wiley & Sons Ltd, 2012). Pg. 69

Mattingly, J., *Elements of Gas Turbine Propulsion* (New York: McGraw-Hill, 1996) Pg. 71

Mattingly, J., Heiser, W., and Pratt, D., *Aircraft Engine Design*, 2nd ed. (Reston, VA: American Institute of Aeronautics and Astronautics, 2002). Pg. 71

Matzopoulos, M., "Advanced Modelling Accelerates Fuel Cell Development," *Fuel Cell Focus*, September 2007, 44-47. Pg. 74

Matzopoulos, M., "Model-to-Market," *The Chemical Engineer*, October 2013, 42-44. Pg. 67

McKenzie, A., *Axial Compressor Development at Rolls-Royce Derby, 1946-1962* (Derby, UK: Rolls-Royce Heritage Trust, 2001). Pg. 69

McKenzie-Smith, I. and Hughes, E., *Hughes Electrical Technology*, 7th ed. (Upper Saddle River, NJ: Prentice Hall, 1995). Pg. 70

McKinney, R., "Conserve Your Energy: The Benefits of Pump System Optimization," *Uptime* 6, no. 43 (October/November 2010): 48-55. Pg. 74

Meckler, M. and Hyman, L. eds., *Sustainable On-Site CHP Systems* (New York: McGraw-Hill, 2010). Pg. 73

Merrick, J., *What Is Simulation?* (Morgantown, WV: West Virginia University, 2013). Pg. 67

MMPA, "Wind Energy Math Calculations: Calculating the Tip Speed Ratio of Your Wind Turbine," Arlington, MN: Minnesota Municipal Power Agency, 2014. Pg. 75

Moir, I. and Seabridge, A., *Aircraft Systems: Mechanical, Electrical and Avionics Subsystem Integration*, 3rd ed. (Chichester, UK: John Wiley & Sons Ltd, 2008). Pg. 72

Molland, A., Turnock, S., and Hudson, D., *Ship Resistance and Propulsion: Practical Estimation of Ship Propulsive Power* (Cambridge, UK: Cambridge University Press, 2011). Pg. 73

Moraal, P. and Kolmanovsky, I., "Turbocharger Modeling for Automotive Control Applications," *SAE International Congress and Exposition*, Cobo Center, Detroit, MI, March 1-4, 1999 (Warrendale, PA: Society of Automotive Engineers). Pg. 72

Moulton, A., Grosjean, J., and Owen, G., *The Moulton Formulae and Methods* (London, UK: Professional Engineering Publishing Ltd, 2005). Pg. 69

Nikoukhah, R. and Steer, S., *Scicos™ - A Dynamic System Builder and Simulator* (Paris, France: Institut National de Recherche en Informatique et en Automatique (INRIA), 2010). Pg. 68

O'Hayre, R., Cha, S., Coletta, W., and Prinz, F., *Fuel Cell Fundamentals*, 2nd ed. (Hoboken, NJ: John Wiley & Sons Inc., 2009). Pg. 74

Passmore, M., "The Measurement and Analysis of Road Vehicle Drag Forces," Ph.D. thesis, Department of Transport Technology, Loughborough University, Loughborough, UK, 1990. Pg. 72

PDH Online, "HVAC - Characteristics and Selection Parameters of Fans and Blower Systems: PDH Course M213," Fairfax, VA: PDH Online, 2011. Pg. 70

Pratap, R., *Getting Started with MatLab® 7* (New York: Oxford University Press, 2006). Pg. 68

Prentis, J., *Dynamics of Mechanical Systems* (London, UK: Longman, 1970). Pg. 70

Press, W., Flannery, B., Teukolsky, S., and Vetterling, W., *Numerical Recipes: The Art of Scientific Computing* (Cambridge, UK: Cambridge University Press, 1986). Pg. 68

Profillidis, V., *Railway Management and Engineering* (Aldershot, UK: Ashgate, 2006). Pg. 73

* Provost, M., "Planes, Trains and Clean Energy: A Life of Simulation, Analysis, Monitoring and Asset Management," Unpublished presentation, Loughborough, UK: Intelligent Energy Ltd, 2012. Pgs. 66, 66

Provost, M., "Reducing Pumping System Costs: An Overview for Enzen Global Ltd.," Unpublished presentation Nottingham, UK, 2015. Pg. 74

Pukrushpan, J., Stefanopoulou, A., and Peng, H., *Control of Fuel Cell Power Systems: Principles, Modeling, Analysis and Feedback Design* (London, UK: Springer-Verlag London Ltd, 2010). Pg. 74

Quaschning, V., *Understanding Renewable Energy Systems* (London, UK: Earthscan, 2005). Pg. 75

Rakopoulos, C. and Giakoumis, E., *Diesel Engine Transient Operation: Principles of Operation and Simulation Analysis* (London, UK: Springer-Verlag London Ltd, 2009). Pg. 72

Razak, A., *Industrial Gas Turbine Performance Engineering: Course Notes* (Houston, TX: Decatur Professional Development, LLC, 2007a). Pg. 71

Razak, A., *Industrial Gas Turbines: Performance and Operability* (Cambridge, UK: Woodhead/CRC Press, 2007b). Pg. 71

Reddy, T. and Linden, D. eds., *Linden's Handbook of Batteries*, 4th ed. (New York: McGraw-Hill, 2011). Pg. 70

Ritzén, J., "Modelling and Fixed Step Simulation of a Turbo Charged Diesel Engine," Master's thesis, Linköping University, Linköping, Sweden, 2003. Pg. 72

Rogers, G. and Mayhew, Y., *Engineering Thermodynamics: Work & Heat Transfer*, 2nd ed. (London, UK: Longman, 1967). Pg. 70

Rolls-Royce, *The Jet Engine* (Stamford, UK: Key Publishing Ltd, 2005). Pg. 71

* Ruffles, P., *The History of the Rolls-Royce RB211 Turbofan Engine* (Derby, UK: Rolls-Royce Heritage Trust, 2014). Pg. 71

Sampson, B., "Heads in the Cloud," *Professional Engineering*, August 19, 2009, 33-34. Pg. 69

Sampson, R., "Resistance & Propulsion (1): Course Summary," Newcastle, UK: Newcastle University, 2008. Pg. 73

Saravanamuttoo, H., Rogers, G., and Cohen, H., *Gas Turbine Theory*, 5th ed. (Harlow, UK: Prentice Hall, 2001). Pgs. 64, 71

* Savage, S., *Decision Making with Insight*, 2nd ed. (Belmont, CA: Thomson Brooks/Cole, 2003). Pg. 69

Scilab Group, *Introduction to Scilab™* (Le Chesnay, France: Scilab™ Consortium, 2010). Pg. 68

Seidman, A., Beaty, H., and Mahrous, H., *Handbook of Electric Power Calculations*, 2nd ed. (New York: McGraw-Hill, 1996). Pg. 70

Simpson, A., "Parametric Modelling of Energy Consumption in Road Vehicles," Ph.D. thesis, University of Queensland, St Lucia, Queensland, Australia, 2005. Pg. 72

Slater, P., *The Optimization Trap: Avoiding the Pitfalls That Limit Operational Performance* (Melbourne, Australia: Initiate Action Pty Ltd, 2012). Pg. 68

Soares, C., *Gas Turbines: A Handbook of Air, Sea and Land Applications* (Amsterdam, The Netherlands: Butterworth-Heinemann, 2008). Pg. 71

Spiegel, C., *PEM Fuel Cell Modeling and Simulation Using MatLab®* (Amsterdam, The Netherlands: Academic Press/Elsevier BV, 2008). Pg. 74

Staffell, I., *The Energy and Fuel Data Sheet – W1P1 Revision 1* (Birmingham, UK: University of Birmingham, 2011). Pg. 70

Stinton, D., *The Anatomy of the Aeroplane*, 2nd ed. (Oxford, UK: Blackwell Science, 1998). Pg. 71

Stone, R., *Introduction to Internal Combustion Engines*, 3rd ed. (Basingstoke, UK: Macmillan Press, 1999). Pg. 72

Sustainability Victoria, "Energy Efficiency Best Practice Guide: Pumping Systems," Melbourne, Australia: Sustainability Victoria, 2009. Pg. 74

Tenno, A., "Modelling and Evaluation of Valve-Regulated Lead-Acid Batteries," Ph.D. thesis, Control Engineering Laboratory, Helsinki University of Technology, Helsinki, Finland, 2004. Pg. 70

Thomas, S. and Zalbowitz, M., *Fuel Cells: Green Power* (Los Alamos, NM: Los Alamos National Laboratory, 1999). Pg. 73

Topping, J., *Errors of Observation and Their Treatment*, 4th ed. (London, UK: Chapman & Hall/Science Paperbacks, 1972). Pg. 68

Torenbeek, E., *Synthesis of Subsonic Airplane Design* (Dordrecht, The Netherlands: Delft University Press/Kluwer Academic Publishers, 1982). Pg. 72

Tung, K., *Topics in Mathematical Modelling* (Princeton, NJ: Princeton University Press, 2007). Pg. 68

* Tupper, E., *Introduction to Naval Architecture*, 5th ed. (Oxford, UK: Elsevier Butterworth-Heinemann, 2013). [Pg. 73]

US DOE, *Fuel Cell Handbook*, 7th ed. (Morgantown, WV: US Department of Energy/ EG&G Technical Services, Inc., 2004). [Pg. 74]

US EPA, "Energy Efficiency in Water and Wastewater Facilities: A Guide to Developing and Implementing Greenhouse Gas Reduction Programs," Washington DC: US Environmental Protection Agency, 2013. [Pg. 74]

Vogelesang, H., "An Introduction to Energy Consumption in Pumps," *World Pumps*, January 2008a, 28-31. [Pg. 74]

Vogelesang, H., "Energy Consumption in Pumps - Friction Losses," *World Pumps*, April 2008b, 20-24. [Pg. 74]

Vogelesang, H., "Energy Savings in Pump Systems," *World Pumps*, August 2008c, 26-30. [Pg. 74]

Vogelesang, H., "Pump Choice to Optimize Energy Consumption," *World Pumps*, December 2008d, 20-24. [Pg. 74]

Vogelesang, H., "Two Approaches to Capacity Control," *World Pumps*, April 2009, 26-29. [Pg. 74]

* Walsh, P. and Fletcher, P., *Gas Turbine Performance* (Malden, MA: Blackwell Science, 1998a). [Pgs. 64, 71]

Wendt, J. ed., *Computational Fluid Dynamics: An Introduction*, 2nd ed. (Berlin, Germany: Springer-Verlag, 1995). [Pg. 69]

White, F., *Fluid Mechanics*, 4th ed. (Boston, MA: WCB/McGraw-Hill, 1999). [Pg. 69]

Whitford, R., *Fundamentals of Fighter Design* (Shrewsbury, UK: Airlife Publishing Ltd, 2000). [Pg. 72]

Whitford, R., *Evolution of the Airliner* (Marlborough, UK: Crowood Press, 2007). [Pg. 72]

Whittle, Sir F., *Gas Turbine Aero-Thermodynamics, with Special Reference to Aircraft Propulsion* (Oxford, UK: Pergamon Press, 1981). [Pgs. 64, 71]

Wickens, A., *Fundamentals of Rail Vehicle Dynamics* (Lisse, The Netherlands: Swets & Zeitlinger BV, 2003). [Pg. 73]

Wilde, G., *Flow Matching of the Stages of Axial Compressors* (Derby, UK: Rolls-Royce Heritage Trust, 1999). [Pg. 69]

Wilson, D. and Korakianitis, T., *The Design of High-Efficiency Turbomachinery and Gas Turbines* (London, UK: Prentice Hall, 1998). [Pg. 71]

Wroblewski, D., "Wind Turbines: Introduction," Boston, MA: University of Boston, Department of Mechanical Engineering, 2013. [Pg. 75]

Websites

DoneEx [Pg. 69]	www.doneex.com
Duke Energy [Pg. 70]	www.duke-energy.com
Electric Light and Power [Pg. 70]	www.elp.com
Fuel Cell Today [Pg. 73]	www.fuelcelltoday.com
GasTurb GmbH [Pg. 71]	www.gasturb.de
HOMER Energy [Pg. 69]	www.homerenergy.com

MAN Diesel & Turbo Pg. 73	http://dieselturbo.man.eu
MathWorks Pg. 68	www.mathworks.com
Numerical Recipes Pg. 68	www.nr.com
PDH Online Pg. 70	www.pdhonline.com
Railway Performance Society Pg. 73	www.railperf.org.uk
Rolls-Royce Heritage Trust Pg. 71	www.rolls-royce.com/about/heritage-trust.aspx
Scicos™ Pg. 68	www.scicos.org
Scilab™ Pg. 68	www.scilab.org
Wikipedia Pg. 67	www.wikipedia.com

Additional Material

References

* denotes sources that the author found particularly helpful and/or inspiring.

Allen, J., *Aerodynamics: A Space-Age Survey* (London, UK: Hutchinson and Co, 1963).

Angel, W., *HVAC Design Sourcebook* (New York: McGraw-Hill, 2012).

Becker, L. ed., *Pushing the Envelope: A NASA Guide to Engines* (Cleveland, OH: Glenn Research Center, National Aeronautics & Space Administration, 2007).

Boyce, M., *Gas Turbine Engineering Handbook*, 2nd ed. (Woburn, MA: Gulf Professional Publishing/Butterworth-Heinemann, 2002).

Dennis, R. ed., *The Gas Turbine Handbook* (Morgantown, WV: National Energy Technology Laboratory, US Department of Energy, 2006).

Giampaolo, T., *Gas Turbine Handbook: Principles and Practice*, 5th ed. (Lilburn, GA: Fairmont Press Inc., 2014).

Harris, K. and Clarke, J. eds., *Jane's World Railways 2008-2009* (Coulsdon, UK: Jane's Information Group Ltd, 2008).

Hughes, A., *Electric Motors and Drives: Fundamentals, Types and Applications*, 3rd ed. (Oxford, UK: Newnes/Elsevier, 2006).

Hull, D., *Fundamentals of Airplane Flight Mechanics* (Berlin, Germany: Springer-Verlag, 2007).

Hunter, J. ed., *IHS Jane's All the World's Aircraft: In Service 2016-2017* (Coulsdon, UK: IHS Global Ltd, 2015).

IET, *IET Factfiles and Influencing Policy Makers* (Stevenage, UK: Institution of Engineering and Technology, 2015).

Jackson, P., Munson, K., and Peacock, L. eds., *IHS Jane's All the World's Aircraft: Development & Production 2015-2016* (Coulsdon, UK: IHS Global Ltd, 2015).

Jennions, I. eds., *Integrated Vehicle Health Management: Essential Reading* (Warrendale, PA: SAE International, 2013).

* Kurzke, J. and Halliwell, I., *Propulsion and Power: An Exploration of Gas Turbine Performance Modelling* (Cham, Switzerland: Springer International Publishing, 2018).

* Macchi, E. and Astolfi, M., *Organic Rankine Cycle (ORC) Power Systems: Technologies and Applications* (Duxford, UK: Woodhead Publishing/Elsevier, 2017).

* MacIsaac, B. and Langton, R., *Gas Turbine Propulsion Systems* (Chichester, UK: John Wiley & Sons Ltd, 2011).

* Martyr, A. and Plint, M., *Engine Testing*, 3rd ed. (Oxford, UK: Elsevier Butterworth-Heinemann, 2007).

Meadows, D., *Thinking in Systems: A Primer* (White River Junction, VT: Chelsea Green Publishing, 2008).

Narkiewicz, M., "Digital Twins," *Our World*, no. 3, (June 2018), 17.

Otis, C., *Jeppesen Aircraft Gas Turbine Powerplants Workbook*, 3rd ed. (Englewood, CO: Jeppesen, 2010).

Otis, C. and Vosbury, P., *Jeppesen Aircraft Gas Turbine Powerplants Textbook*, 3rd ed. (Englewood, CO: Jeppesen, 2010).

Rill, G., "Vehicle Dynamics: Short Course," Regensberg, Germany: Ostbayerische Technische Hochschule (OTH) Regensburg, 2007.

* Rill, G., *Road Vehicle Dynamics: Fundamentals and Modelling* (Boca Raton, FL: CRC Press/Taylor & Francis, 2012).

Seal, D., "The System Engineering "V": Is It Still Relevant in the Digital Age?" In *Model-Based Enterprise Summit 2018*, NIST Campus, Gaithersburg, MD, April 2-5, 2018. Gaithersburg, MD: National Institute of Standards and Technology (NIST), U.S. Department of Commerce.

* Spiryagin, M., Cole, C., Quan Sun, Y., McClanachan, M. et al., *Design and Simulation of Rail Vehicles* (Boca Raton, FL: CRC Press/Taylor & Francis, 2014).

* Spiryagin, M., Wolfs, P., Cole, C., Spiryagin, V. et al., *Design and Simulation of Heavy Haul Locomotives and Trains* (Boca Raton, FL: CRC Press/Taylor & Francis, 2016).

* Young, T., *Performance of the Jet Transport Airplane: Analysis Methods, Flight Operations and Regulations* (Chichester, UK: John Wiley & Sons Ltd, 2018).

Websites

Energy in Buildings and Industry (part of the Energy Institute)	www.eibi.com, www.energyzine.co.uk
Institution of Engineering and Technology (IET)	www.theiet.org
Institution of Mechanical Engineers (IMechE)	www.imeche.org

CHAPTER 4

Analysis

Introduction

This chapter (based on *Provost 1994*) discusses the various methods that can be used to analyze the performance of assets, both during development and in service. Analysis of a fuel cell system using these methods is described in Appendix B.

Facets of Analysis

The objectives of modelling, discussed in Chapter 3, are clear: use component performance parameter characteristics and physical relationships (such as the laws of conservation of mass and energy) to build a mathematical model of an asset from its component parts. Analysis, on the other hand, can have different objectives depending on how the results will be used. If, for instance, the analysis is being done to determine if an asset is acceptable to the customer, then the calculations to determine its acceptability (or otherwise) will almost certainly be laid down in a contract specification. Depending on the skills and sophistication of the parties involved, this analysis may bear little relation to current understanding of the asset in question. Analysis of assets undergoing performance development testing will have as its objective the detailed understanding of the behavior of all the components. This understanding and the methods used to achieve it develop as analysis proceeds and should not be constrained by contract specification niceties. Using one set of methods to achieve another set of objectives is usually a recipe for confusion and misunderstanding. The performance analysis methods discussed in this chapter address the detailed performance understanding requirement but can also be applied to monitoring of in-service assets.

It should be noted that the performance analyst is rarely asked to explain the absolute levels of overall and component performance that are calculated. Usually, performance of assets and their components is expressed relative to some appropriate datum. For example, the statement that the Specific Fuel Consumption (SFC) of a gas turbine is 10 g/kNs is not nearly as useful as the statement that the SFC is 3% worse than expected; a compressor efficiency of 89% sounds quite good, until comparison with the results of a test rig shows it to be 2% lower than design. Ultimately the purpose of analysis of asset measurements is to identify physical (hardware) faults in the asset by looking at deviations in measurable parameters from expectation.

Difficulties with the Analysis Process

Superficially, the analysis process looks easy: calculate appropriate component performance parameter deviations from measurement deviations relative to a datum, and then use the results to guide the search for hardware faults. In practice, the analysis process can be corrupted by errors in the measurements taken to determine how the asset is behaving. Errors can have serious effects on any analysis, because they result in incorrect component performance parameter calculations which can then lead to misleading hardware fault diagnoses. Failure to detect measurement errors can have serious consequences, as time and effort are spent searching for nonexistent faults in one part of the asset while genuine faults elsewhere go undetected. Any analyst who ignores the possibility of corruption of the analysis calculations by erroneous measurements is treading on dangerous ground and is certain to run into problems. Indeed, such individuals soon gain a reputation for recklessness bordering on irresponsibility.

The experienced analyst learns to look for characteristic "signatures" of single measurement errors. These are usually recognized by calculation of better than expected performance of one or more components, accompanied by worse than expected performance on other components (so-called reciprocal changes). However, when more than one error is present, and/or genuine changes in components that make up error "signatures" have happened, the task of finding errors becomes very much more difficult. It is not unusual for even the most experienced analyst to spend days or weeks producing a credible assessment of overall and component behavior when multiple errors are present. Engineering judgement, trial and error calculations, patience, and a certain amount of luck are all required if any sense is to be made of the results.

The "Traditional" Approach to Analysis

The "traditional" analysis approach can be summarized in four steps:

1. Use appropriate measurements to calculate component performance parameters, depending on what measurements are available.
2. Calculate the differences between the component performance parameter values worked out above and the datum values of those same parameters.
3. Calculate the differences between the observed measurement values and the datum values of the measurements.
4. Do a consistency check to see if the differences in component performance parameters account for the differences in overall asset performance.

The above method has the following drawbacks:

- The required calculations may not be possible, due to missing measurements.
- The possibility of measurement errors is not considered, at least initially.
- Any consistency checks done at the end of the analysis process do not directly indicate the presence of errors in the measurements, since the "reciprocal changes" in component performance parameters discussed earlier will conspire to explain all of the measurements.
- The consistency checks also fail if the analyst fails to calculate some of the component performance parameters.
- Rematching effects (in which the change of performance of the set of components causes movements of component working points along their characteristics that alter component performance expectations) are not explicitly accounted for.

The experienced performance analyst is aware of the above pitfalls and makes allowances for them in the calculations. However, as indicated earlier, the process is time-consuming and heavily dependent on skill and judgement.

Iterative Analysis

So far, the discussion of analysis methods has concentrated on the calculation of component performance parameters directly from measurements. While this method generally works well (and, indeed, may form part of a contract between the asset manufacturer and the customer and/or certifying authority that lays down agreed calculations for proving asset acceptability), there are a number of problems that led to the development by the author and others of a technique called Iterative Analysis that uses pretest models of the asset being analyzed as the starting point for the analysis process. These problems include:

- Potential inconsistencies between the analysis and modelling processes as asset modelling becomes more sophisticated.
- The inability of the "traditional" analysis to proceed if significant measurements are missing.
- The reliance on the skill of the performance analyst to detect and remove measurement errors from the analysis. This can be extremely time-consuming and frustrating if several errors are present.

To understand Iterative Analysis, it is instructive to compare analysis and modelling programs:

- An analysis program calculates, from measurements taken at a defined asset operation level and set of environment conditions, individual component performance parameters.
- A modelling program works in the opposite direction: using predictions of component performance parameters, it calculates a series of working points on these characteristics that satisfy the physical laws governing asset operation.

Figure 4.1 summarizes these differences diagrammatically.

It follows that if the component performance parameter assumptions made in a model of the asset are modified so that, when the model is run at the test environmental and asset operation conditions, the measurements taken on that test are reproduced,

FIGURE 4.1 Comparison of modelling and analysis processes.

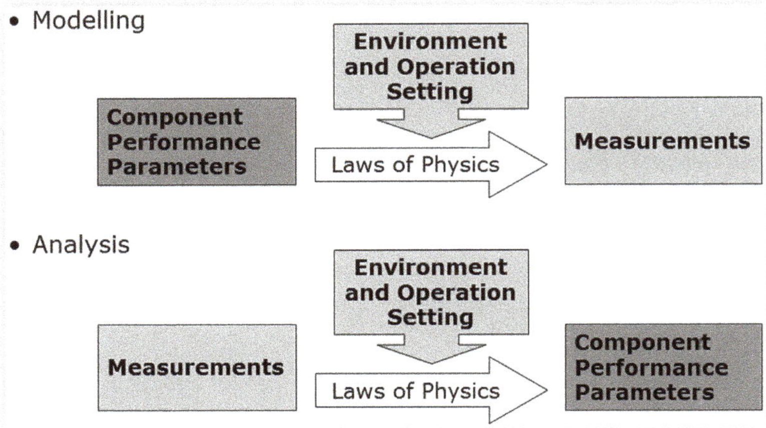

then an analysis will have been done, since at the end of this process all the parameters mentioned above will have been calculated. The process is generally iterative, hence the term Iterative Analysis; see Figure 4.2.

FIGURE 4.2 Iterative Analysis process.

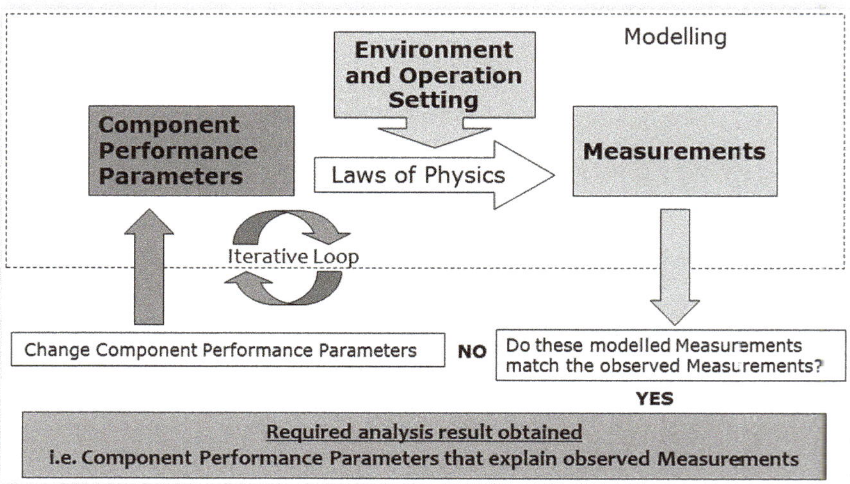

Note that at least one measurement is needed to define the operation level of the asset, while other measurements are needed to define environmental conditions. Any extra measurements that are recorded are reproduced by varying as many component performance parameter assumptions as there are extra measurements.

Note that it is an assumption, based on all the available information, that the components chosen are the ones that have actually changed. In fact, it may have been other components that have changed, but it is impossible to say categorically which ones have changed given the information available. The choice of which component assumptions to vary is a best guess: as more measurements are taken, so more component performance parameter assumptions can be varied, and a more complete picture of the actual asset component performance changes can be built up.

Having converted the analysis process from one of direct calculation to one of varying component performance assumptions within a model, it can be seen that this method of analysis is merely an extension of the calculations done in the modelling process. This brings with it a number of advantages:

- The basic modelling assumptions and data are identical for both modelling and analysis.
- The analysis can proceed if measurements are missing by using component performance parameter assumptions that would be used in a model.
- The changes made to the component performance parameter assumptions in order to reproduce the measurements are known. This means that:
  - The analysis automatically "accounts": the analyzed component parameter performance changes will, by definition, account for the observed measurement values.
  - The changes made to the component performance assumptions can very easily be fed back into a model to produce a model of the asset running at other operational or environment conditions. As an extension of this, the necessity to produce correction factors to correct the measurements to standard environment conditions (a common requirement) is eliminated. The analysis can be done at actual conditions, the resulting component performance parameter changes can be fixed, and then the program can be run again, this time as a model at corrected conditions, to derive what the measurements would have been had the asset been run at standard environment conditions.
- Monitoring of component performance parameter changes, as distinct from absolute levels, eliminates the effects of environmental and operational variability. This enables more subtle changes to be diagnosed.
- Reciprocal changes used to diagnose measurement errors are much easier to spot.
- The effort needed to create a credible analysis is reduced.

As stated previously, each measurement input to an Iterative Analysis program over and above measurement(s) used to define the operating environment of the asset and its operational setting requires a corresponding component performance parameter to be changed in order to reproduce that measurement. Consideration of the functions of the components combined with engineering judgement usually determines which component performance parameter changes are best associated with each measurement. The relationships between measurements and component performance parameters need not be unique but may depend on the presence or absence of other measurements.

An Alternative Approach to Analysis

An alternative approach recognizes that the differences between observed and expected measurements can be caused by up to four distinct causes:

- Genuine component performance changes.
- Biases (shifts of sensor calibration) in the measurements used to define the environment and asset operation conditions. These measurements essentially define the expected levels of the measurements added to the asset for diagnostic purposes; if they are erroneous, then the expectations will also be wrong. Typically, if the asset operation parameter is measured high (say), then all the diagnostic measurements will appear low because the asset is operating at a lower operating point than has been measured. Similarly, if the measurement of ambient pressure is high, then all the diagnostic pressures and other pressure-sensitive measurements

will appear low because the ambient pressure is lower than measured. If the measurement of ambient temperature is high, then all the diagnostic temperatures and other temperature-sensitive measurements will appear low because the ambient temperature is lower than measured.

- Biases (shifts of sensor calibration) in the measurements added to the asset for diagnostic purposes. These affect each observed measurement shift independently.
- Random scatter of all the measurement sensors, both those used to define the environment and asset operation point and those added for diagnostic purposes.

Figure 4.3 shows the above situation in diagrammatic form. Unknown component changes and sensor biases affect the "system" of component performance relationships and measurement dynamics, resulting in a set of measurements which are then corrupted by random noise to produce a set of observed values. The analyst must use these to diagnose component and sensor performance in order to ultimately deduce which hardware problems are affecting both the components making up the asset and the measurement system.

FIGURE 4.3 Diagrammatic representation of the analysis situation.

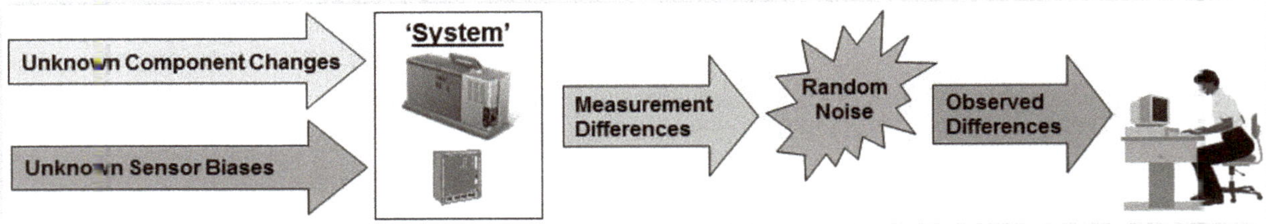

The "system" referred to above can be thought of as a matrix, shown in Figure 4.4. In this matrix, the effects of component changes on the differences between the observed and expected measurements can be represented by a component exchange rate table. Each column of this table represents the effect of a 1% change in each of the component performance parameters in turn. Similarly, the effects of environmental and asset operation measurement sensor biases (environment biases) can be represented by a set of slopes, representing the effects of 1% shifts in the environmental and asset operation measurement calibrations on each of the diagnostic measurements. The final part of the matrix represents the effects of 1% shifts in each of the diagnostic measurement sensor calibrations in turn; since these are independent, they can be represented by a unit matrix of appropriate dimension. It is seen that a 1% shift in the nth diagnostic measurement would alter the nth observed difference by 1%, leaving the others unaltered.

FIGURE 4.4 The system matrix.

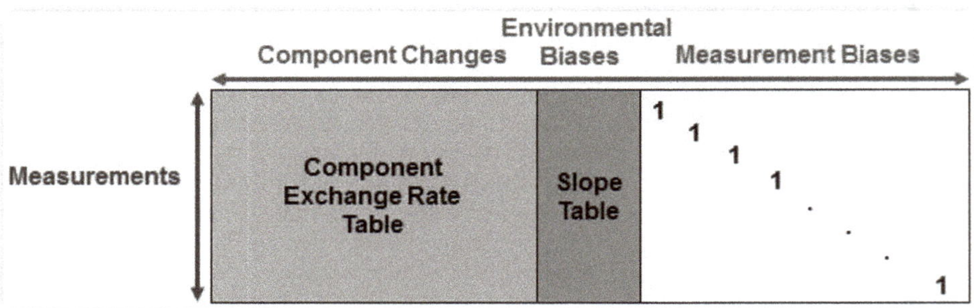

It is immediately obvious from the shape of the matrix represented in Figure 4.4 that there are more unknowns (component performance parameter changes and sensor biases) than knowns (measurement differences from expectation). There are, therefore, an infinite number of possible solutions to the problem of diagnosing the component changes and sensor biases that caused the observed measurement differences. Two of these solutions are apparent by inspection:

- If the sensor bias columns are removed, leaving as many component performance changes to be diagnosed as there are measurement differences (the usual situation in most methods of "traditional" analysis), then a unique solution for the set of component performance changes that produced the set of observed measurement differences can be found.
- If the component change and environmental and asset operation measurement sensor bias columns are removed, leaving only the unit matrix of diagnostic measurement bias effects, then this implies that all the measurement shifts are caused solely by measurement bias and the asset component performance parameters are unchanged.

The true situation is usually somewhere between the two extremes outlined above. Some component performance parameters have changed, while some measurements are biased, and all measurements are scattered. Because there are an infinite number of answers, this looks superficially like an intractable problem.

Optimal Estimation and Kalman Filtering

The history of techniques that can address the above situation begins with the work done by Carl Friedrich Gauss in 1795 using least squares to calculate comet and planetary motion from uncertain astronomical measurements. The maximum likelihood methods of Ronald Fisher in 1912 also provided a stimulus by giving some theoretical and probabilistic backing to the arithmetic of least squares. Andrey Kolmogorov in 1941 and Norbert Weiner in 1942 independently developed a linear least-squares estimation technique which, while somewhat cumbersome, received considerable attention and laid the foundations for the subsequent developments in Kalman Filter theory.

Rudolf Kálmán's first paper on discrete-time, recursive mean-square filtering in 1960 (*Kalman 1960*) was greeted with enthusiasm by engineers and mathematicians, because his general approach to the problem and use of state-space techniques unified known results and the recursive nature of the equations that now bear his name (usually spelt without accents in the literature) are ideally suited to computer implementation. His approach can be related to the work of Bayes in the early eighteenth century, since it (like Bayes' theorem, discussed in Chapter 8) depends very heavily on prior (pre-observation) information which is then combined with observations to produce posterior (post-observation) estimates. Kálmán's ideas produced an explosion of subsequent theoretical work, as well as inspiring engineers in such diverse fields as spacecraft navigation, missile guidance, and control systems analysis to use this new theory to either proceed with problems initially thought intractable or improve the accuracy of existing control and navigation systems by calculation instead of expensive hardware improvements. Louis Urban and Allan Volponi at Hamilton Standard (a division of United Technologies, that also owns Pratt & Whitney aeroengines) began using Kalman Filtering for the analysis of gas turbines in the late 1970s. Their interest was mainly in the analysis of data from civil turbofans gathered in service for maintenance purposes. Their success with the technique resulted in similar initiatives from both General Electric and Rolls-Royce in the field of condition monitoring.

The Kalman Filter is ideally suited to many analysis situations because it treats measurement non-repeatability in a logical fashion (noisy measurements are given less "weight" than more precise ones) and the structure of the matrix equations means that, unlike other analysis methods, it can cope with situations where the number of unknowns in the analysis is different to the number of measurements. However, the output of the Kalman Filter is a best estimate, not a precise answer, because it considers the uncertainties inherent in the analysis situation.

The Kalman Filter Equations

The Kalman Filter, in a simplified form suitable for asset analysis, consists of two equations:

$$\underline{\mathbf{x}}^{\wedge} = \underline{\mathbf{x}}_0^{\wedge} + \underline{\mathbf{K}} \times \left(\underline{\mathbf{y}} - \underline{\mathbf{C}} \times \underline{\mathbf{x}}_0^{\wedge}\right)$$

$$\underline{\mathbf{K}} = \underline{\mathbf{Q}} \times \underline{\mathbf{C}}^{\mathrm{T}} \times \left(\underline{\mathbf{C}} \times \underline{\mathbf{Q}} \times \underline{\mathbf{C}}^{\mathrm{T}} + \underline{\mathbf{R}}\right)^{-1}$$

where

$\underline{\mathbf{x}}^{\wedge}$ = vector of best estimates of component performance parameter changes and sensor biases <u>after</u> measurements taken

$\underline{\mathbf{x}}^{\wedge}_0$ = vector of best estimates of component performance parameter changes and sensor biases <u>before</u> measurements taken (usually zero)

$\underline{\mathbf{y}}$ = vector of measurement differences from expectation

$\underline{\mathbf{K}}$ = Kalman gain matrix

$\underline{\mathbf{C}}$ = system matrix, relating measurement differences to component performance parameter changes and sensor biases

$\underline{\mathbf{Q}}$ = covariance matrix of component performance parameter changes and sensor biases

$\underline{\mathbf{R}}$ = measurement repeatability covariance matrix

Superscript **T** = transpose

Superscript **−1** = inverse

+, −, and × denote matrix addition, subtraction, and multiplication, respectively

If $\underline{\mathbf{x}}^{\wedge}_0 = \underline{\mathbf{0}}$, so all the estimates of the component performance parameter changes and sensor biases before any measurements are taken are zero, the first equation simplifies to $\underline{\mathbf{x}}^{\wedge} = \underline{\mathbf{K}} \times \underline{\mathbf{y}}$.

The Kalman gain matrix $\underline{\mathbf{K}}$ can be considered as a ratio of uncertainties, in the form of an "inverse" of the system matrix $\underline{\mathbf{C}}$. $\underline{\mathbf{K}}$ gives state estimates $\underline{\mathbf{x}}^{\wedge}$ from known observations, while $\underline{\mathbf{C}}$ gives observations from known states $\underline{\mathbf{x}}$. There are two limiting situations (*Provost 2008*):

- As the measurement repeatability covariance matrix $\underline{\mathbf{R}}$ tends to zero, $\underline{\mathbf{K}}$ tends to a true inverse of $\underline{\mathbf{C}}$ if $\underline{\mathbf{C}}$ is square (or a pseudo-inverse of $\underline{\mathbf{C}}$ if $\underline{\mathbf{C}}$ is either over- or underdetermined) and $\underline{\mathbf{x}}^{\wedge}$ is determined entirely by the measurements $\underline{\mathbf{y}}$.

- As the measurement repeatability covariance matrix $\underline{\mathbf{R}}$ becomes large compared to the system covariance matrix $\underline{\mathbf{Q}}$, $\underline{\mathbf{K}}$ tends to zero and the measurements cease to influence the estimate $\underline{\mathbf{x}}^{\wedge}$, which is then set equal to $\underline{\mathbf{x}}^{\wedge}_{0}$.

Various derivations of the Kalman Filter equations are in the references given below. These derivations also produce a third Kalman Filter equation giving the uncertainties in the estimate $\underline{\mathbf{x}}^{\wedge}$, which is of more use in the recursive form of the equations and has therefore been left out of this discussion for simplicity.

Problems with the Kalman Filter

While the Kalman Filter is a very powerful tool that can be (and is) used in a variety of applications, it has several problems which should be considered. One problem of major practical importance when considering applying the Kalman Filter to asset performance analysis is the fact that, given a set of measurement differences from a datum, it calculates changes in all the component performance parameters and sensor biases being considered in the problem, even when it is known that those measurement differences are due to a small subset of the component changes and sensor biases being considered. This "smearing" effect results from the least-squares basis of the Kalman Filter and can produce potentially misleading analyses by indicating that more hardware changes and sensor biases are present in the asset than is actually the case. Underestimation of magnitudes of simulated faults of 50% or more is possible: the analyst, faced with both underestimated genuine faults and spurious indications of hardware changes and sensor biases that are not present, can rapidly lose confidence in the technique. Selecting the single "biggest" change as being the only fault worth investigating is unsatisfactory, since it is perfectly possible for two or more faults to be present simultaneously, all of which require investigation. A method to overcome this is described in the author's PhD thesis (*Provost 1994*), the author's paper on the subject in the book by Mathioudakis and Sieverding (*Mathioudakis and Sieverding 2003*) and the patents by the author and David Nevell (*Provost and Nevell 1992, 1994*). Another problem is ensuring that the inputs to the Kalman Filter are set up and optimized to produce meaningful analyses. The appropriateness of the measurements used in any analysis also needs to be assessed; this is discussed in more detail in Chapter 5.

Books and Papers

The presentations by Jonathan Batson (*Batson 2013*), Tarun Kumar (*Kumar 2012*), and Ajith Parlikad (*Parlikad 2012*) provide high-level overviews of the latest innovations in predictive analytics, while the presentation by Seth Muthuraman (*Muthuraman 2014*) looks at the range of techniques used by a major UK power generation company to look after its assets. The MSc thesis by Seçil Ariduru (*Ariduru 2004*) describes a commonly used technique (the Rainfall Method) for the analysis of fatigue life, while the book by Steven Smith (*Smith 1999*: www.dspguide.com) offers a comprehensive introduction to digital signal processing.

The presentation by Dan Simon (*Simon 2001*) is a good introduction to Kalman Filtering, while the lecture notes by Ramsey Faragher (*Faragher 2012*) provide a simple and intuitive derivation of the Kalman Filter for people without a strong mathematical background. The book edited by Arthur Gelb (*Gelb 1974*) is widely regarded as one of the best books on Kalman Filtering available, while Dan Simon's book (*Simon 2006*) is seen as a worthy successor. Both books discuss the theory in considerable detail, as does the book by Mohinder Grewal and Angus Andrews (*Grewal and Andrews 2008*) which includes a CD containing MatLab® code. Greg Welch and Gary Bishop have also produced some detailed course notes on the subject (*Welch and Bishop 2001a, 2001b, 2006*). A rather simpler book was written by Bozic (*Bozic 1979*). The author has also produced a simple guide to Kalman Filtering (*Provost 2008*) as well as fuller descriptions (*Provost 1994*) and the author's paper on Kalman Filtering in Mathioudakis and Sieverding's work (*Mathioudakis and Sieverding 2003*).

A number of aircraft and aeroengine health monitoring techniques are discussed in the paper by Irem Turner and Anupa Bajwa (*Turner and Bajwa 1999*) and the books by Lazzeretti et al. (*Lazzeretti et al. 1995*) and Mathioudakis and Sieverding (*Mathioudakis and Sieverding 2003*). Gas-turbine analysis is the subject of many theses, research papers, and technical presentations. The Department of Propulsion and Power in the School of Mechanical Engineering at Cranfield University has sponsored many MSc and PhD theses on this subject. Giovanni Bechini's thesis (*Bechini 2007*) investigates the shortcomings and limitations of current techniques and proposes a pattern-matching technique (which brings together Bayesian statistics and fuzzy logic) supported by nonlinear observability analysis for sensor selection. The author's PhD thesis (*Provost 1994*) discusses the "concentrator" enhancement to the Kalman Filter as applied to aeroengine gas-path analysis in some detail: this was also the subject of two patent applications (*Provost and Nevell 1992, 1994*). *Li* (*2002*) reviews many of the methods available at that time, while Riti Singh's paper (*Singh 2003*) reviews the impact of such methods on the business and operation of aeroengines as well as some of the methodologies. Ranjan Ganguli's book (*Ganguli 2013*) describes many of the latest gas-path analysis techniques in some detail. Dave Doel's presentation (*Doel 2008*) makes the point that even the most sophisticated gas-path analysis techniques may not be enough to properly diagnose the state of a gas turbine, so a more holistic approach incorporating oil analysis, vibration analysis, and analysis of fault messages is needed. The book by Stephen Johnson and others (*Johnson et al. 2011*) is an extremely comprehensive overview of system health management for military and space applications. The MSc thesis by Gonçalo Matos dos Santos Marques (*dos Santos Marques 2010*) describes the use of the Rolls-Royce COMPASS™ condition monitoring system to monitor an aircraft fleet. Gas Path Analysis Ltd. (www.gpal.co.uk) offers performance monitoring systems for gas turbines, process compressors and combined heat and power (CHP) systems.

References

* denotes source that the author found particularly helpful and/or inspiring.

Ariduru, S., "Fatigue Life Calculation by Rainflow Cycle Counting Method," MSc thesis, Graduate School of Natural and Applied Sciences, Middle East Technical University, Ankara, Turkey, 2004. Pg. 95

Batson, J., "Implementing Predictive Analytics," *3rd Annual Predictive Asset Management Seminar*, Dexter House, London, UK, December 4-5, 2013 (London, UK: IBC Energy/Informa UK Ltd). Pg. 95

Bechini, G., "Performance Diagnostics and Measurement Selection for On-Line Monitoring of Gas Turbine Engines," Ph.D. thesis, Department of Propulsion and Power, School of Engineering, Cranfield University, Cranfield, UK, 2007. Pg. 96

Bozic, S., *Digital and Kalman Filtering* (London, UK: Edward Arnold, 1979). Pg. 96

Doel, D., "Is Gas Path Analysis Enough?," *Conference on the Industrial Application of Gas Turbines*, Ottawa, Ontario, October 21 2008 (Ottawa, Ontario: National Research Council Canada, Institute for Aerospace Research). Pg. 96

dos Santos Marques, G., "Engine Condition Monitoring as a Route to Savings: PGA – Portugália Airlines as a Case Study," Master's thesis, Universidade Técnica de Lisboa, Lisbon, Portugal, 2010. Pg. 96

Faragher, R., "Understanding the Basis of the Kalman Filter via a Simple and Intuitive Description," *IEEE Signal Processing Magazine* (September 2012): 128-132. Pg. 95

Ganguli, R., *Gas Turbine Diagnostics: Signal Processing and Fault Isolation* (Boca Raton, FL: CRC Press/Taylor & Francis, 2013). Pg. 96

Gelb, A. eds., *Applied Optimal Estimation* (Cambridge, MA: MIT Press, 1974). Pg. 96

Grewal, M. and Andrews, A., *Kalman Filtering: Theory and Practice Using MatLab®*, 3rd ed. (Hoboken, NJ: John Wiley & Sons Inc., 2008). Pg. 96

Johnson, S., Gormley, T., Kessler, S., Mott, C. et al. eds., *System Health Management, with Aerospace Applications* (Hoboken, NJ: John Wiley & Sons Inc., 2011). Pg. 96

Kalman, R., "A New Approach to Linear Filtering and Prediction Problems," *Transactions of the ASME, Journal of Basic Engineering* 82, no. Series D (1960): 35-45. Pg. 93

Kumar, T., "Analytics Driven Asset Lifecycle Management," *2nd Annual Predictive Asset Management Seminar*, Royal Garden Hotel, Kensington, London, UK, September 11-13, 2012 (London, UK: IBC Energy/Informa UK Ltd). Pg. 95

Lazzeretti, R., Andrenucci, M., Barry, B., Chevalier, J. et al., eds., "The Role of Engine Health Monitoring in Aero-Engine Design, Manufacture and Operation," *Proceedings of the AIAA/AAAF/DGLR/RAeS 5th European Propulsion Forum*, University of Pisa, Italy, 1995 (Pisa, Italy: Associazione Italiana di Aeronautica e Astronautica/Association Aéronautique et Astronautique de France/Deutsche Gesellschaft für Luft- und Raumfahrt/Royal Aeronautical Society). Pg. 96

Li, Y., "Performance Analysis Based Gas Turbine Diagnostics: A Review," *Proceedings of the Institution of Mechanical Engineers Part A: Journal of Power and Energy* 216, no. 5 (2002): 363-377. Pg. 96

Mathioudakis, K. and Sieverding, C. eds., *Gas Turbine Condition Monitoring & Fault Diagnosis* (Rhode Saint Genèse, Belgium: Von Karman Institute for Fluid Dynamics, 2003). Pgs. 95, 96

Muthuraman, S., "Condition Monitoring at SSE," *Condition Monitoring and Analysis Seminar*, Dexter House, London, UK, February 5-6 2014 (London, UK: IBC Energy/Informa UK Ltd). Pg. 95

Parlikad, A., "Predictive Analytics in Engineering Asset Management," *2nd Annual Predictive Asset Management Seminar*, Royal Garden Hotel, Kensington, London, UK, September 11-13 2012 (London, UK: IBC Energy/Informa UK Ltd). Pg. 95

* Provost, M., "The Use of Optimal Estimation Techniques in the Analysis of Gas Turbines," Ph.D. thesis, Department of Propulsion and Power, School of Engineering, Cranfield University, Cranfield, UK, 1994. Pgs. 87, 95, 95, 96

Provost, M., "A Simple Guide to Kalman Filtering," Unpublished presentation, Nottingham, UK, 2008. Pg. 94, 96

Provost, M. and Nevell, D., Data processing system using a Kalman filter. US Patent 5,105,372, 1992. Filed January 30, 1991, granted April 14, 1992. Pgs. 95, 96

Provost, M. and Nevell, D., Performance data processing system. European Patent 0 315 307 B1, 1994. Filed May 10, 1989, granted June 22, 1994. Pgs. 95, 96

Simon, D., "Kalman Filtering," *Embedded Systems Programming* 14, no. 6 (2001): 72-79. Pg. 95

Simon, D., *Optimal State Estimation* (Hoboken, NJ: Wiley-Interscience, 2006). Pg. 96

Singh, R., "Advances and Opportunities in Gas Path Diagnostics," *16th International Symposium on Air Breathing Engines*, Renaissance Cleveland Hotel, Cleveland, OH, August 31-September 5 2003 (Reston, VA: American Institute of Aeronautics and Astronautics). Pg. 96

Smith, S., *The Scientist and Engineer's Guide to Digital Signal Processing*, 2nd ed. (San Diego, CA: California Technical Publishing, 1999). Pg. 95

Turner, I. and Bajwa, A., "A Survey of Aircraft Engine Health Monitoring Systems," *35th AIAA/ASME/SAE/ASEE Joint Propulsion Conference and Exhibit*, Los Angeles, CA, June 20-24 1999 (Reston, VA: American Institute of Aeronautics and Astronautics). Pg. 96

Welch, G. and Bishop, G., *SIGGRAPH 2001 Course 8: An Introduction to the Kalman Filter* (Chapel Hill, NC: Department of Computer Science, University of North Carolina at Chapel Hill, 2001a). Pg. 96

Welch, G. and Bishop, G., *SIGGRAPH 2001 Course 8: An Introduction to the Kalman Filter*, Presentation (Chapel Hill, NC: Department of Computer Science, University of North Carolina at Chapel Hill, 2001b). Pg. 96

Welch, G. and Bishop, G., *An Introduction to the Kalman Filter* (Chapel Hill, NC: Department of Computer Science, University of North Carolina at Chapel Hill, 2006). Pg. 96

Websites

Gas Path Analysis Ltd. Pg. 96	www.gpal.co.uk
Scientist and Engineer's Guide to Digital Signal Processing Pg. 95	www.dspguide.com

Additional Material

References

* denotes source that the author found particularly helpful and/or inspiring.

* Bower, D., *Rail Vehicle Testing* (Nottingham, UK: Salient Books, 2018).

Visser, W, *Generic Analysis Methods for Gas Turbine Engine Performance: The Development of the Gas Turbine Simulation Program GSP* (Enschede, the Netherlands: Ipskamp Drukkers, 2014).

CHAPTER

5

Observability

Introduction

This chapter (based on the author's PhD thesis (*Provost 1994*) and the author's papers in *Lazzeretti et al. 1995* and *Mathioudakis and Sieverding 2003*) discusses a method developed for determining the ability of different measurement sets to provide the information needed to perform analysis of component changes from nominal, as well as determining measurement sensor biases. This area seems to be neglected in the literature but is one of the important requirements before any analysis is attempted, no matter how sophisticated it may be.

Setting Up an Analysis

As explained in Chapter 3, the purpose of analysis is to use measurements taken while an asset is operating to understand the causes of any changes in the overall performance of the asset, by identifying the changes in performance of the individual components that make up that asset.

The performance analyst needs to decide:

- The hardware changes that are to be looked for which could be the cause of changes in the asset's overall performance. This is usually based on engineering judgement, backed up as necessary by a failure modes and effects analysis (FMEA; see Chapter 10).
- How those hardware changes would be modelled as component performance changes. For example, pipe fouling would be modelled as an increase in pressure loss, leaks would be modelled as fluid losses from the asset, compressor

blade fouling would be modelled as compressor efficiency loss, etc. It must be recognized, though, that there will often be more possible hardware changes than available component performance parameters, so the resulting analyses will inevitably contain some ambiguities.

- How the measurements taken from sensors fitted to the asset vary as the performance of the components that make up the asset change.

The available measurements will usually vary, depending on the asset being analyzed. Assets in service usually have a comparatively limited set of sensors fitted, whereas development assets running on company premises can have anything from a limited sensor fit tailored to monitoring of endurance running to a full set of sensors carefully placed in many positions on the asset specifically for detailed determination of the performance of all the asset's components.

In addition, measurements from sensors fitted to determine environmental conditions (such as ambient temperature, ambient pressure, ambient relative humidity, forward speed, etc.) and asset operation setting(s) (such as power output, fluid throughput, control settings, etc.) must always be available before any form of analysis can proceed, because these measurements define the basic conditions in which the asset is operating. They are needed to define a datum against which any subsequent analysis can be compared.

Generation of Exchange Rate Tables

Having generated two lists, one containing possible component performance parameter changes and the other containing available diagnostic measurements, the analyst then needs to generate a component exchange rate table showing the percentage change in each diagnostic measurement for a 1% change in each component performance parameter (or unit change if the datum value is zero) in turn, at the environmental conditions and asset operation setting at which the analysis is to be done. Note that parameters normally thought of in terms of percentages (e.g., efficiency, cooling bleed flow) usually have their exchange rates calculated by adding 1% (e.g., 88% => 89%, 3% => 4%) while other parameters have their exchange rates calculated by factoring by 1.01 (e.g., 100 => 101). Figure 5.1 shows the typical output from such a set of calculations carried out using the fuel cell model discussed in Appendix B.

If sensor-bias determination is to be included in the analysis, it is usually necessary to determine the effects of 1% biases in some, or all, of the environmental condition and operation setting measurements, which are fixed when the exchange rate table in Figure 5.1 is generated. This is done by changing the environmental condition and/or operating setting measurements by −1% and calculating the resulting percentage changes in the other measurements. The presence of the negative sign is to account for the fact that if the operating setting measurement sensor (say) reads high (i.e., a positive

FIGURE 5.1 Component exchange rate table for a notional fuel cell system.

	R Datum	ηconv	ΔPfan	Ndesign	*E0 Datum*	*ηfan*	*Wfandesign*
Istack	0.33%	-1.75%	0.03%	0.00%	*-2.47%*	*-0.04%*	*-0.05%*
Vstack	-0.30%	0.53%	-0.01%	0.00%	*2.18%*	*0.01%*	*0.02%*
Ifan	2.26%	-7.52%	1.28%	0.00%	*-17.07%*	*-1.85%*	*-2.22%*
Nfan	0.75%	-2.49%	0.09%	1.00%	*-5.56%*	*-0.06%*	*-1.07%*

bias), then all the other measurements will appear to be reading low. Figure 5.2 shows these exchange rates; for simplicity, only operation setting sensor biases are considered.

In Figures 5.1 and 5.2, the following nomenclature is used:

FIGURE 5.2 Operation setting exchange rate table for a notional fuel cell system.

	PWnet	Tstack
Istack	-1.57%	0.68%
Vstack	0.47%	-0.54%
Ifan	-6.72%	9.05%
Nfan	-2.23%	2.99%

Measurements

Measurements used for diagnosis of component performance changes

- Istack Fuel cell stack current (A)
- Vstack Fuel cell stack voltage (V)
- Ifan Fuel cell system fan current (A)
- Nfan Fuel cell system fan rpm

Operation setting measurements

- PWnet Fuel cell system net output power (kW)
- Tstack Fuel cell stack temperature (°C)

Component performance parameters

- R Datum Datum value of fuel cell stack resistance (mΩcm^2)
- ηconv DC to DC converter efficiency (%)
- ΔPfan Fan back pressure (kPa)
- Ndesign Fan design rpm
- *E0 Datum* *Datum value of fuel cell voltage at zero load* (mV)
- *ηfan* *Fan efficiency* (%)
- *Wfandesign* *Fan design volume flow rate* (m^3/s)

For reasons that will become apparent as the discussion progresses, two sets of component performance parameters have been considered:

- The full set of seven items, including those in italics.
- The reduced set of four items, excluding those in italics.

Correlations Between Measurements

The first thing that should be checked is whether any of the diagnostic measurements chosen respond in similar ways to other measurements to all (or nearly all) of the component performance parameter changes. This test indicates whether there is any redundancy in the measurement set. This may be beneficial if the sensors are present for reasons other than asset performance analysis, but may be an indication of overspecification of required sensors if the analyst is trying to determine an optimum set of sensors for analysis in the future. Note, however, that with a small number of diagnostic measurements, the presence of large amounts of redundancy between measurements may indicate that there is insufficient information to do the required analysis because the diagnostic sensor set is so small that all the component performance parameter changes are indistinguishable.

FIGURE 5.3 Two-dimensional "component change space": five measurement vectors (*Provost 1994*).

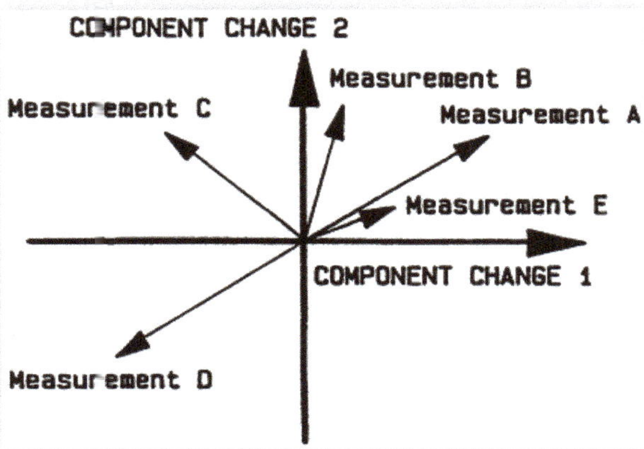

The method of checking whether the chosen measurements respond in similar ways to the component changes being sought is a general one, which is best understood by imagining a multidimensional "component change space" coordinate system in which each component performance parameter is represented by an axis that is perpendicular to all the other component performance parameter axes. While human experience is limited to three dimensions, matrix algebra recognizes no such limitations, so the above "component change space" coordinate system is quite valid even if more than three component performance parameters are considered. Examples given here will, of course, be limited to two dimensions: it is hoped that the reader will be able to grasp the concepts, even if imagining a 10-dimensional space can be a little tricky!

The "component change space" described above will contain measurement vectors for each measurement being considered: the coordinates of the endpoints of the vectors will be defined by the elements in each measurement row in the component exchange rate table, as shown in Figure 5.3.

Some of the measurement vectors will be nearly parallel to others. For example, in Figure 5.3, measurement vectors A, D, and E are nearly parallel (the fact that D "points" in the opposite direction does not affect the argument) whereas measurement vectors B and C are not, either with each other or with A, D, and E. Parallel measurement vectors indicate similar response by the measurements to the component changes. For example, in Figure 5.3, it is difficult to distinguish component change 1 from component change 2 using only measurements A, D, and/or E.

It is obvious that the direction, not the length, of the vectors is important and that we should evaluate the angles between pairs of vectors to see if they are nearly parallel or not. This is done as follows:

- Divide each element in each row of the component exchange rate table by the square root of the sum of squares of the elements in that row: this normalizes the rows defining the measurement vectors by converting them to unit length.
- Multiply this matrix by its transpose: this effectively works out the dot product of every pair of normalized measurement vectors, which produces the cosine of the angle between them.
- The resulting symmetric matrix of cosines can then be inspected for possible correlations between the measurements, that is, measurements which respond in similar ways to the component changes being sought. The values will vary between +1 and −1 with:

  +1 indicating perfect positive correlation: the pair of measurements change in the same direction, but not necessarily by the same amount, for all the component changes being sought.

  0 indicating no correlation: the pair of measurements do not respond in the same way.

  −1 indicating perfect negative correlation: the pair of measurements change in the opposite direction, but not necessarily by opposite amounts, for all the component changes being sought.

Cosines with magnitudes greater than 0.7 are worthy of note: levels above this may indicate that the pair of measurements only respond differently to a small subset of the component changes. The following simple classifications can therefore be made:

- 0.7 => 0.8 indicates some correlation.
- 0.8 => 0.9 indicates a high degree of correlation.
- 0.9 => 1.0 indicates a very high degree of correlation: the measurements are redundant.

Figures 5.4 and 5.5 show the results of the above calculations for both the full (seven component performance parameters) and reduced (four component performance parameters) sets, respectively, in the component exchange rate table in Figure 5.1. The figures show that, for both the full and reduced sets of component performance parameters, all the diagnostic sensors are redundant. Unlike the gas turbine example in the references quoted in the chapter introduction, the sensor set for these cases is not sufficient to carry out unambiguous analyses.

FIGURE 5.4 Measurement correlation matrix (component changes only)—full set.

	Istack	Vstack	Ifan	Nfan
Istack	1.0000			
Vstack	-0.9303	1.0000		
Ifan	0.9704	-0.9724	1.0000	
Nfan	0.9573	-0.9575	0.9807	1.0000

FIGURE 5.5 Measurement correlation matrix (component changes only)—reduced set.

	Istack	Vstack	Ifan	Nfan
Istack	1.0000			
Vstack	-0.9482	1.0000		
Ifan	0.9843	-0.9652	1.0000	
Nfan	0.9282	-0.9109	0.9261	1.0000

Inclusion of Sensor Bias

The effects of the inclusion of sensor bias in the analysis on the correlation between measurements can be done in a similar manner to that described above; however, a matrix that includes the effects of sensor bias must first be created.

An analysis that includes the effects of sensor bias is looking for three distinct sets of items:

- Component changes.
- Environmental and operation setting measurement sensor biases.
- Other measurement sensor biases.

The information about the way that the measurements change when the abovementioned component changes and sensor biases occur is given by:

- The component exchange rate table (Figure 5.1).
- The environmental and operation setting exchange rate table (also known as the slope table: Figure 5.2).
- An identity matrix, which merely states that a 1% bias in a diagnostic sensor produces a change of 1% in that measurement, while leaving the other measurements unaffected.

A combined matrix, subsequently called the system matrix, is created by concatenating the columns of the environmental and operation setting exchange rate table and the identity matrix to the end of the component exchange rate table to extend it. Figure 5.6 shows the process diagrammatically.

FIGURE 5.6 Generation of system matrix, showing dimensions (*Provost 1994*).

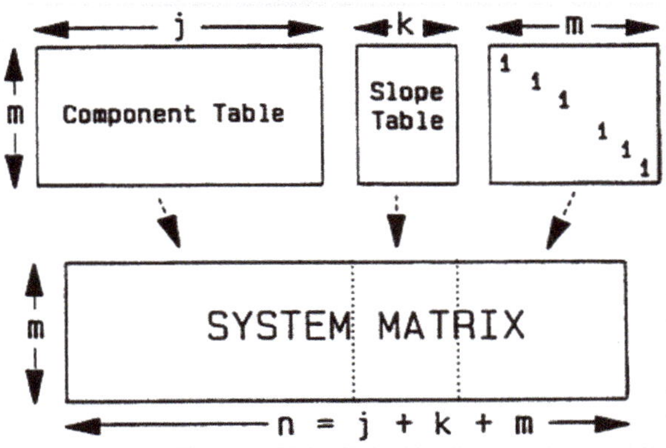

FIGURE 5.7 Measurement correlation matrix (component changes and sensor biases)—full set.

	Istack	Vstack	Ifan	Nfan
Istack	1.0000			
Vstack	-0.8038	1.0000		
Ifan	0.9071	-0.8755	1.0000	
Nfan	0.8989	-0.8652	0.9847	1.0000

FIGURE 5.8 Measurement correlation matrix (component changes and sensor biases)—reduced set.

	Istack	Vstack	Ifan	Nfan
Istack	1.0000			
Vstack	-0.5839	1.0000		
Ifan	0.8323	-0.6693	1.0000	
Nfan	0.8182	-0.6579	0.9716	1.0000

Note that if some diagnostic sensor biases are not considered possible (e.g., speeds), then the appropriate columns are removed from the identity matrix at the end of the system matrix.

Having done this, the system matrix can be analyzed in exactly the same way as the component exchange rate table was previously to determine which pairs of diagnostic measurements respond in a similar manner to both component changes and sensor biases. Generally, there will be fewer correlations due to the diagonal nature of the identity matrix. Figures 5.7 and 5.8 show the results of the above calculations for the system matrix made up of the tables in Figures 5.1 and 5.2 for both the full (seven component performance parameters) and reduced (four component performance parameters) sets, respectively, in the component exchange rate table in Figure 5.1.

Correlations Between Component Performance Parameter Changes and/or Sensor Biases

The next thing the analyst needs to check is whether any of the component performance parameter changes and/or sensor biases produce similar changes to any other change or bias in all (or nearly all) of the measurements. This test indicates whether any component performance parameter changes or sensor biases cannot be distinguished from one another, which may be unacceptable.

The method of checking whether component performance parameter changes and diagnostic sensor biases are distinguishable from each other is similar to the methods for checking diagnostic measurement redundancy. A multi-dimensional "measurement space" coordinate system, in which each diagnostic measurement is represented by an axis that is perpendicular to all the other measurement axes, is considered, which contains component/bias vectors for each component performance parameter change and sensor bias being considered. The coordinates of the endpoints of the vectors will be defined by the elements in each component performance parameter change or sensor bias column in the system matrix; see Figure 5.9.

FIGURE 5.9 Two-dimensional "measurement space": five component performance parameter change vectors and three sensor bias vectors (*Provost 1994*).

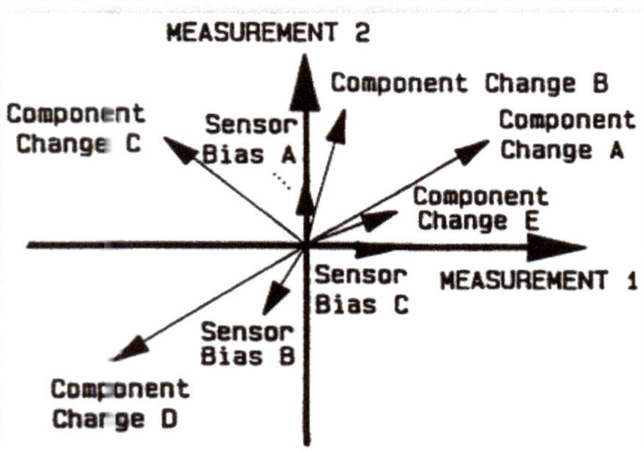

Some of the component performance parameter change or sensor bias vectors will be nearly parallel to others. For example, in Figure 5.9, component change vectors A, D, and E and sensor bias vector B are nearly parallel (the fact that

D and B "point" in the opposite direction does not affect the argument), whereas the others are not, either with each other or the group stated above. Parallel component performance parameter change and/or sensor bias vectors indicate that they have similar effects on the measurements. For example, in Figure 5.9, it will be difficult to distinguish 2% changes in component E or −1% changes in component D from 1% changes in component A. As before, the direction, not the length, of the vectors is important. Again, we need to evaluate the angles between pairs of vectors to see if they are nearly parallel or not. This is done as follows:

- Divide each element in each column of the system matrix by the square root of the sum of squares of the elements in that column: this normalizes the columns defining the component performance parameter change and sensor bias vectors by converting them to unit length.
- Multiply the transpose of this matrix by the matrix: this effectively works out the dot product of every pair of normalized component performance parameter change and sensor bias vectors, which produces the cosine of the angle between them.
- The resulting symmetric matrix of cosines can then be inspected for possible correlations between the component performance parameter changes and/or sensor biases, that is, component performance parameter changes and/or sensor biases that will be difficult to distinguish from one another. The values will vary between +1 and −1 with:

  +1 indicating perfect positive correlation: the pair of component performance parameter changes and/or sensor biases affect the measurements in the same sense and are therefore indistinguishable.

  0 indicating no correlation: the pair of component performance parameter changes and/or sensor biases are distinguishable.

  −1 indicating perfect negative correlation: the pair of component performance parameter changes and/or sensor biases affect the measurements in the opposite sense and are therefore indistinguishable.

Cosines with magnitudes greater than 0.7 are worthy of note: levels above this may indicate that the pair of measurements only respond differently to a small subset of the component changes and/or sensor biases. The following simple classifications can therefore be made:

- 0.7 => 0.8 indicates some correlation.
- 0.8 => 0.9 indicates a high degree of correlation.
- 0.9 => 1.0 indicates a very high degree of correlation: the component performance parameter changes and/or sensor biases are effectively indistinguishable.

Figure 5.10 shows the results of the above calculations for the system matrix made up of the tables in Figures 5.1 and 5.2. Again, it appears that unambiguous analyses are not possible.

Superficially, this appears to be a very disappointing situation. However, by recognizing that:

- The fan back pressure (ΔPfan) cannot be analyzed and therefore should be taken out of the analysis and measured directly, possibly using a differential pressure sensor for higher accuracy.
- The datum value of fuel cell resistance (R Datum) is best analyzed using data from a range of fuel cell operating conditions (rather than data only taken at a single point) and therefore can be taken out of the analysis.

FIGURE 5.10 System correlation matrix.

	R Datum	ηconv	ΔPfan	Ndesign	*E0 Datum*	*ηfan*	*Wfandesign*	PWnet	Tstack	Istack	Vstack	Ifan
R Datum	**1.0000**											
ηconv	-0.9954	**1.0000**										
ΔPfan	0.9564	-0.9502	**1.0000**									
Ndesign	0.3106	-0.3064	0.0732	**1.0000**								
E0 Datum	*-1.0000*	*0.9953*	*-0.9587*	*-0.3045*	***1.0000***							
ηfan	*-0.9458*	*0.9397*	*-0.9992*	*-0.0326*	*0.9483*	***1.0000***						
Wfandesign	*-0.9781*	*0.9710*	*-0.9299*	*-0.4349*	*0.9779*	*0.9142*	***1.0000***					
PWnet	-0.9953	1.0000	-0.9498	-0.3071	0.9953	0.9393	0.9709	**1.0000**				
Tstack	0.9955	-0.9895	0.9676	0.3126	-0.9959	-0.9570	-0.9890	-0.9894	**1.0000**			
Istack	0.1380	-0.2147	0.0231	0.0000	-0.1354	-0.0229	-0.0210	-0.2154	0.0714	**1.0000**		
Vstack	-0.1227	0.0650	-0.0070	0.0000	0.1195	0.0069	0.0064	0.0652	-0.0561	0.0000	**1.0000**	
Ifan	0.9324	-0.9251	0.9970	0.0000	-0.9352	-0.9992	-0.9002	-0.9247	0.9455	0.0000	0.0000	**1.0000**

FIGURE 5.11 System matrix—final set.

	ηconv	Ndesign	Istack	Vstack
Istack	-1.75%	0.00%	1.00%	0.00%
Vstack	0.53%	0.00%	0.00%	1.00%
Ifan	-7.52%	0.00%	0.00%	0.00%
Nfan	-2.49%	1.00%	0.00%	0.00%

FIGURE 5.12 Measurement correlation matrix (component changes and sensor biases)—final set.

	Istack	Vstack	Ifan	Nfan
Istack	**1.0000**			
Vstack	-0.4051	**1.0000**		
Ifan	**0.8677**	-0.4669	**1.0000**	
Nfan	0.8052	-0.4332	**0.9280**	**1.0000**

FIGURE 5.13 System correlation matrix—final set.

	ηconv	Ndesign	Istack	Vstack
ηconv	**1.0000**			
Ndesign	-0.3064	**1.0000**		
Istack	-0.2147	0.0000	**1.0000**	
Vstack	0.0650	0.0000	0.0000	**1.0000**

- Sensor errors in fan current (Ifan) and fan rpm (Nfan) are likely to be small or have a small influence on the analysis and therefore can be disregarded.
- Fuel cell net power (PWnet) and stack temperature (Tstack) are key control parameters, so are likely to be measured accurately by the fuel cell control system.

It is possible to construct a more suitable analysis scheme where the system matrix has only four columns; see Figure 5.11.

Figures 5.12 and 5.13 show the measurement correlation matrix (with all the component performance parameter changes and sensor biases that are in the analysis included) and the system correlation matrix (which is a subset of Figure 5.10), respectively.

The lack of large correlations in Figures 5.12 and 5.13 (apart from that between Istack and Ifan in Figure 5.12, which is not large enough to cause concern) indicates that this is a much more viable analysis scheme than those discussed previously.

The reader should now be able to appreciate the importance of the methods discussed in this chapter in ensuring that the set of sensors fitted to an asset is optimized, the results from any analysis that uses those sensor readings are as meaningful and unambiguous as possible, and an appropriate balance is struck between analysis of single operating points and operating point ranges.

References

* denotes source that the author found particularly helpful and/or inspiring.

Lazzeretti, R., Andrenucci, M., Barry, B., Chevalier, J. et al. eds., "The Role of Engine Health Monitoring in Aero-Engine Design, Manufacture and Operation," *Proceedings of the AIAA/AAAF/DGLR/RAeS 5th European Propulsion Forum*, University of Pisa, Italy, 1995 (Pisa, Italy: Associazione Italiana di Aeronautica e Astronautica/Association Aéronautique et Astronautique de France/Deutsche Gesellschaft für Luft- und Raumfahrt/Royal Aeronautical Society). Pg. 99

Mathioudakis, K. and Sieverding, C. eds., *Gas Turbine Condition Monitoring & Fault Diagnosis* (Rhode Saint Genèse, Belgium: Von Karman Institute for Fluid Dynamics, 2003). Pg. 99

* Provost, M., "The Use of Optimal Estimation Techniques in the Analysis of Gas Turbines," Ph.D. thesis, Department of Propulsion and Power, School of Engineering, Cranfield University, Cranfield, UK, 1994. Pg. 99, 102, 103, 104

CHAPTER

6

Time Series Analysis

Introduction

This chapter (based on *Provost 2003* and reproduced with permission from the IET) gives an overview of the analysis of time series and demonstrates the advantages of applying Kalman Filtering theory and a related technique, the Optimal Tracker, to time series analysis. Examples of time series analysis using the Optimal Tracker are given in Appendix B.

Time Series Basics

A time series is any set of observations of a single parameter that varies with time. Examples of typical time series are observations of fuel flow, pressures, temperatures, or shaft speeds in a gas turbine, fuel consumption of a car, wind speed or direction at a weather station, share prices, foreign exchange rates, etc., all of which rise and fall as time passes. Each observation of the time series has a time value associated with it (or a pseudo-time value such as cycle number or fuel tank fill-up number, which always increases as the observations are gathered) which determines the order in which the observations are displayed and processed.

A typical time series for an arbitrary parameter is sketched in Figure 6.1.

In this case, the observations appear to be gradually rising as time passes. This is often referred to as a ramp (which could, of course, be either up or down and could, in theory, change its slope or rate of change as time passes). Another important pattern that often occurs is a step, in which the observations suddenly appear to settle at a new level (either higher or lower) after a particular time. A typical step, again for an arbitrary parameter, is sketched in Figure 6.2.

FIGURE 6.1 Typical time series—ramp (*Provost 2003*).

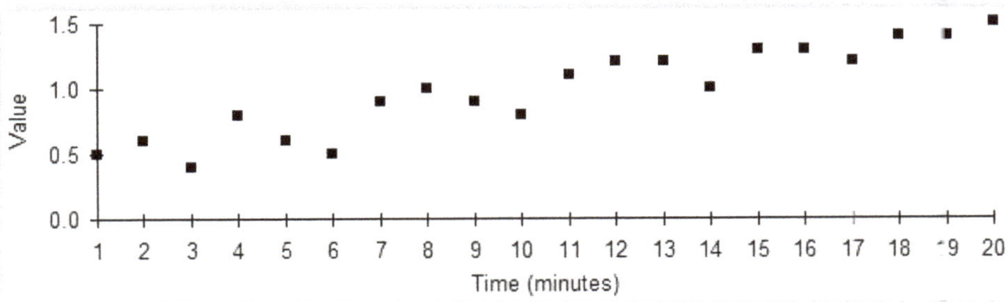

FIGURE 6.2 Typical time series—step (*Provost 2003*).

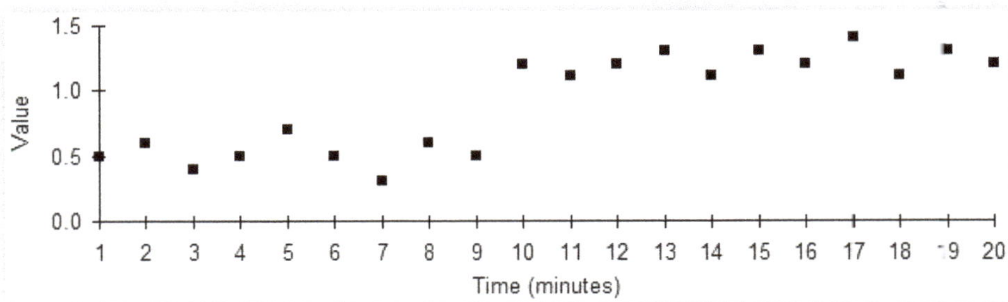

A final effect often seen in real time series is an outlier, an observation that is much higher (or lower) than its neighbors in the time series. A typical outlier, again for an arbitrary parameter, is sketched in Figure 6.3: the observation taken at 10 minutes appears to be odd.

The three sketches also show a common problem with time series observations: each observation appears to be arbitrarily shifted up or down relative to its near neighbors, giving the plots a "lumpy" appearance. The observations have been corrupted by random noise, which is the name given to the sum of all the random errors in the processes (measurement, data transmission and reduction, etc.) that generate each of the observed values. Random noise can be a major problem when trying to understand time series, because it is unclear whether changes in the observations over time are genuine effects that should be investigated or spurious noise effects that should be ignored. As the ratio of the genuine effects being looked for to the average noise level (the signal-to-noise ratio) becomes smaller, the difficulties become greater.

FIGURE 6.3 Typical time series—outlier (*Provost 2003*).

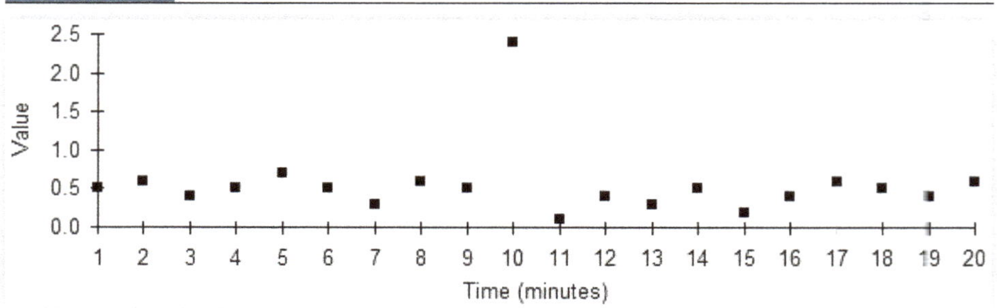

Moving Average

A very common method of reducing the effects of random noise in order to get an idea of the "true" value of the parameter generating the observations (a process called smoothing) is to calculate what is known as a moving average. At a point in time **t**, a moving average of order **n** (or n-point moving average) is calculated as follows:

$$\textbf{moving average}_t = (\textbf{observation}_t + \textbf{observation}_{t-1} + \ldots + \textbf{observation}_{t-n+1}) \div \textbf{n}$$

For example, a 3-point moving average, calculated at time = 10 minutes, is:

$$\textbf{moving average}_{10} = (\textbf{observation}_{10} + \textbf{observation}_9 + \textbf{observation}_8) \div 3$$

At time = 11 minutes, $\textbf{observation}_{11}$ is included in the numerator and $\textbf{observation}_8$ is discarded, thus generating $\textbf{moving average}_{11}$.

The choice of **n** is crucial to the performance of the moving average technique:

- If **n** is large, then the noise is smoothed out (because the average of all the noise terms is zero) but genuine changes in the series do not affect the value of the moving average for some time after the change appears in the observations.
- If **n** is small, genuine changes in the series affect the value of the moving average soon after they appear in the observations, but noise may not be sufficiently smoothed out.

Taking the technique to extremes, a 1-point moving average merely repeats the observations (including the noise) while a moving average with an order equal to the total number of points in the entire series merely gives an average of all the observations, thus obscuring features in the time series data.

A value of **n** should be chosen that provides the best compromise between smoothing out the noise in the observations and allowing the underlying features of the time series to show up. This choice depends on the experience and judgement of the user, taking into account knowledge of the characteristics of the time series and what the results of the time series analysis will be used for. Often two or more moving averages, with different values of **n**, are calculated for a single time series in order to provide an assessment of whether slopes are present in the series.

The moving average technique has a number of problems which make it unsuitable for modern condition monitoring techniques:

- It is a surprisingly cumbersome calculation, particularly when **n** is large. Historical observations have to be remembered, so they can be discarded from the moving average calculation when new observations are gathered. If a large number of parameters are being analyzed, this is a serious drawback.
- If there is a ramp in the series, the moving average lags the observations, as demonstrated in Figure 6.4.
- The moving average technique does not give a direct estimate of the slope of any ramps in the time series. In many cases the slope is as important as the absolute level, and a numerical estimate of the slope is necessary to generate alerts for the user. If two or more moving averages are calculated, their relative levels can indicate the beginning of a slope but give no clue about its magnitude.
- The moving average technique does not take proper account of variable time periods between observations.

FIGURE 6.4 Lag in moving average calculation (*Provost 2003*).

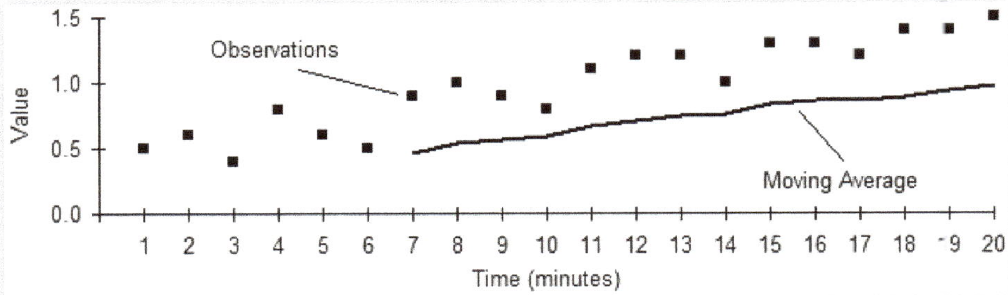

Exponential Smoothing

Exponential smoothing is another very common method of smoothing out the random noise in a time series. The calculation process is somewhat simpler than for the moving average technique, since it only depends on information from the current and immediately previous observations in the time series (a property known as recursion). The formula used is as follows:

$$\textbf{smoothed level}_t = \textbf{smoothed level}_{t-1} + \alpha \times (\textbf{observation}_t - \textbf{smoothed level}_{t-1})$$

For example, at time = 10 minutes:

$$\textbf{smoothed level}_{10} = \textbf{smoothed level}_9 + \alpha \times (\textbf{observation}_{10} - \textbf{smoothed level}_9)$$

α is known as the exponential smoothing constant, which is set by the user between 0 and 1. A value of 0 means that the smoothed level stays at the value set at the start of the time series analysis, while a value of 1 gives a smoothed level equal to the observation. Like the choice of **n** in the moving average technique, the choice of **α** depends on the user's view of the characteristics of the time series:

- If **α** is small, then the noise is smoothed out, but genuine changes in the series do not affect the value of the smoothed level for some time after the change appears in the observations.
- If **α** is large, genuine changes in the series affect the value of the smoothed level soon after they appear in the observations, but noise may not be sufficiently smoothed out.

Again, two or more smoothed levels with different values of **α** can be calculated for a single time series, in order to provide an assessment of whether slopes are present in the series.

Exponential smoothing suffers from most of the problems that affect the moving average technique, except that it is a more efficient calculation:

- No direct numerical estimates of the slopes of any ramps in the time series are calculated.
- No account is taken of variable time periods between observations.
- The smoothed level lags behind the observations when a ramp is present, as demonstrated in Figure 6.5.

FIGURE 6.5 Lag in exponential smoothing calculation (*Provost 2003*).

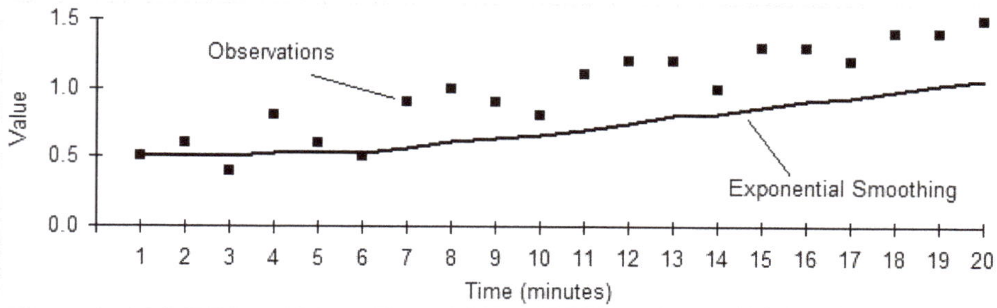

Kalman Filtering

The Kalman Filtering technique is similar in principle to exponential smoothing, except that it includes a slope term in the equation which is generated using an equation similar to the one used for exponential smoothing. The main equations are as follows:

- $\mathbf{delta}_t = \mathbf{observation}_t - (\mathbf{smoothed\ level}_{t-1} + \Delta\mathbf{t} \times \mathbf{smoothed\ slope}_{t-1})$

  where $\Delta\mathbf{t}$ is the difference between the time $\mathbf{observation}_t$ was taken and the time $\mathbf{observation}_{t-1}$ was taken.
- $\mathbf{smoothed\ level}_t = (\mathbf{smoothed\ level}_{t-1} + \Delta\mathbf{t} \times \mathbf{smoothed\ slope}_{t-1}) + \alpha_1 \times \mathbf{delta}_t$
- $\mathbf{smoothed\ slope}_t = \mathbf{smoothed\ slope}_{t-1} + \alpha_2 \times \mathbf{delta}_t$

For example, at time = 10 minutes (1 minute after the last observation):

$$\mathbf{delta}_{10} = \mathbf{observation}_{10} - (\mathbf{smoothed\ level}_9 + 1 \times \mathbf{smoothed\ slope}_9)$$

$$\mathbf{smoothed\ level}_{10} = (\mathbf{smoothed\ level}_9 + 1 \times \mathbf{smoothed\ slope}_9) + \alpha_1 \times \mathbf{delta}_{10}$$

$$\mathbf{smoothed\ slope}_{10} = \mathbf{smoothed\ slope}_9 + \alpha_2 \times \mathbf{delta}_{10}$$

α_1 and α_2 are constants similar in function to the exponential smoothing constant: they are derived from statistical inputs describing the uncertainties in the observations, levels, and slopes, taking proper account of the time intervals between each successive observation. The derivation of α_1 and α_2 from the required statistical inputs is discussed in detail in the author's PhD thesis (*Provost 1994)* and the author's paper on time series analysis (*Mathioudakis and Sieverding 2003)*. These statistical inputs can be set up to enable the Kalman Filter to calculate values for α_1 and α_2 that provide the best compromise between smoothing out the noise and quick response to genuine changes in the time series.

There are several features of the Kalman Filter technique that make it superior to both the moving average and exponential smoothing techniques:

- The function is recursive, relying only on information generated from the current and immediately previous observations in the time series.
- A numerical estimate of the slopes of any ramps in the time series is calculated as each observation is gathered, enabling alerts to be generated if slopes exceed predetermined limits (see Figure 6.6).
- Proper account is taken of any changes in the time periods between observations.

FIGURE 6.6 Smoothed slope output from Kalman Filter calculation (*Provost 2003*).

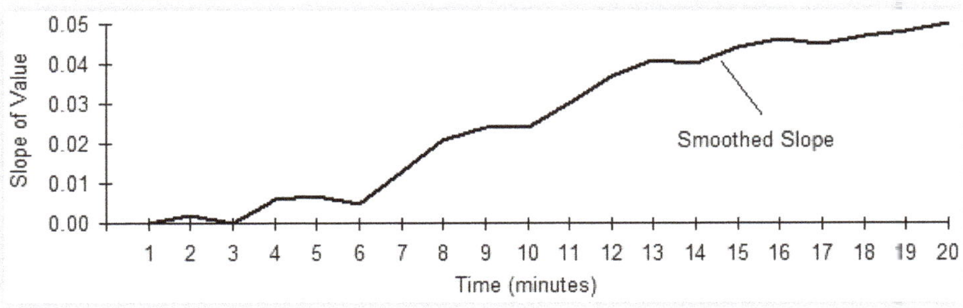

FIGURE 6.7 Absence of lag in Kalman Filter calculation (*Provost 2003*).

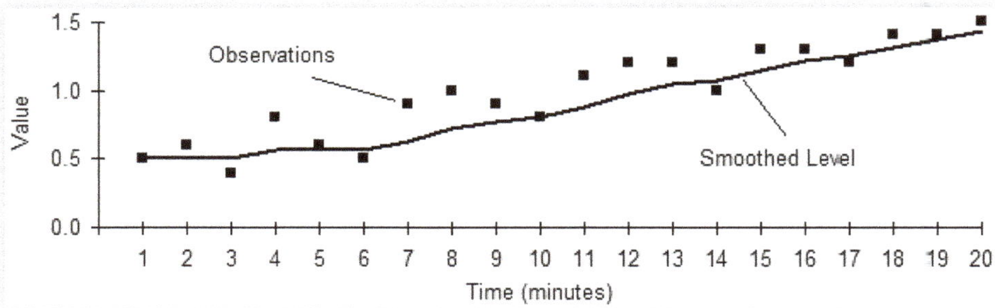

- The delta term in the equations can be thought of as the difference between the actual observation and the expected observation. This difference is useful for detecting steps and outliers in the time series.
- The presence of the smoothed slope term in the equations removes the lag between the smoothed level and the observations when a ramp is present, as shown in Figure 6.7.

When combined with alerting functions, the Kalman Filter technique is a powerful tool that alerts the user to events of interest without needing to view each series manually. Time is therefore spent on investigation of real events, rather than reviewing run-of-the-mill data.

The Optimal Tracker

The paper by Benedict and Bordner (*Benedict and Bordner 1962*) shows that the optimal form of the above Kalman Filter equations (the Optimal Tracker) is given by the very simple formula:

$$\alpha_2 = \alpha_1{}^2 \div \left((2 - \alpha_1) \times \Delta t\right)$$

where the terms have the meanings described above. This means that $\boldsymbol{\alpha_1}$ is the only constant that has to be chosen by the user; like the exponential smoothing constant $\boldsymbol{\alpha}$, it can take any value between 0 and 1, with lower values providing more smoothing out of noise in the observations and higher values producing a greater response to features in the time series.

Experience has shown that better results can sometimes be obtained by a further simplification, particularly if there are large gaps in the time series or **Δt** is consistently less than unity:

$$\alpha_2 = \alpha_1^{\ 2} \div (2 - \alpha_1)$$

Improving the Optimal Tracker

There are two situations which the basic Optimal Tracker does not handle well:

- Large outliers in the data caused by zero readings or other sensor failures.
- Changes in the character of the series caused by maintenance intervention, etc.

It is desirable that the Optimal Tracker responds appropriately to these situations by detecting their occurrence using an appropriate test, then responding correctly by ignoring large outliers and/or resetting the calculations when the series has significantly changed its character. A presentation by the author (*Provost 2007*) describes an extension to the Optimal Tracker to enable it to better cope with outliers and discontinuities in time series data. The basic outlier detection and calculation reset logic is shown in Figure 6.8.

It may also be appropriate in some situations to smooth out the effects of operational variability by averaging observations taken over a suitable time period prior to them being fed into the Optimal Tracker.

The Optimal Tracker can fail because of excessive measurement noise. If the requirement to analyze data as soon as it is gathered is relaxed, data can be averaged first either over suitable time periods or once a defined number of data points has been gathered. The averaging process will significantly reduce the measurement noise, thus giving the Optimal Tracker more chance to pick out the relevant features in the data.

It is also advantageous to track the differences between data measurements and the values that would have been observed had the asset been behaving as expected, rather

FIGURE 6.8 Logic for Optimal Tracker outlier detection and calculation reset (*Provost 2007*).

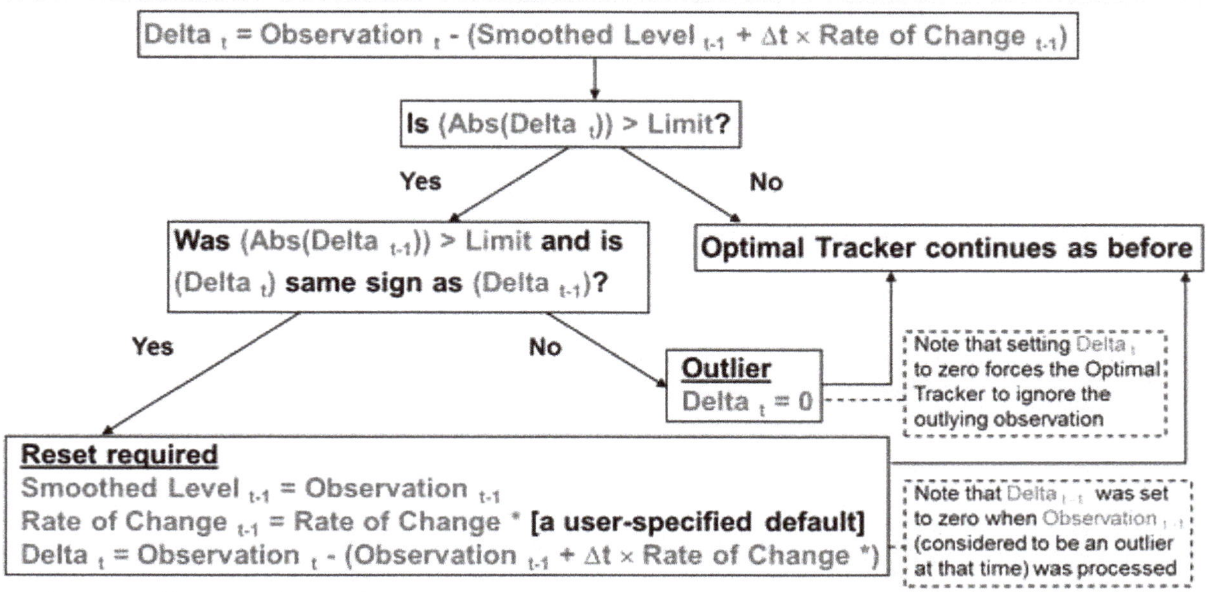

than the raw data measurements themselves, particularly if the performance of the asset is sensitive to environmental conditions and asset operation set point.

A combination of averaging, comparison with a datum, use of the Optimal Tracker, and outlier detection and reset can be extremely effective at turning a mass of seemingly incomprehensible time series data into meaningful information (*Provost 2010*). The process can be summarized as follows:

- Select appropriate measurements from a data stream, based on operating conditions and stability of operation.
- Compare measurements with an appropriate datum model to produce differences that have had the effects of variations over time of operating conditions, environment, etc. removed.
- Average the calculated differences over an appropriate time period, to produce observations for input into the Optimal Tracker.
- Choose a suitable time base (calendar time, up time, start/stop cycles, etc.).
- Run the observations through the Optimal Tracker calculation, as described above.
- Process the output from the Optimal Tracker to remove outliers and initiate restarts where the series has changed its character.
- Use the smoothed level and smoothed slope estimates from the Optimal Tracker to predict the amount of time that will elapse before any alert limits are breached, filtering out spurious warnings as necessary.

Books and Papers

A presentation by the author (*Provost 2006*) gives a simple introduction to time series analysis.

The author's 2003 IEE paper (*Provost 2003*) reproduced here and simple guide (*Provost 2008*) give an easy explanation of the advantages of the above approach over more common methods of time series analysis such as moving average and exponential smoothing.

Previously, the author had used Kalman Filter theory for time series analysis (see *Provost 1994*, and the author's paper on time series analysis in *Mathioudakis and Sieverding 2003*) but the simpler equations based on the paper by Benedict and Bordner (*Benedict and Bordner 1962*) proved easier to understand and manipulate and form the basis of all the author's subsequent work on this subject.

Paul Goodwin describes a similar approach (Holt-Winters) for producing forecasts in the presence of trends and seasonality (*Goodwin 2010*) while the books by Warren Gilchrist (*Gilchrist 1976*), Chris Chatfield (*Chatfield 2009*), and Douglas Montgomery and others (*Montgomery et al. 2008*) cover a range of both simple and complex time series analysis techniques that the reader is likely to encounter. The classic text by George Box and others (*Box et al. 2016*) is a rigorous treatment of the subject, but the mathematics is complex and not really suitable for efficient real-time processing.

Steve Morgan and the author produced a patent (*Morgan and Provost 1988*) for a version of this theory embedded in the control system of an aircraft engine for the detection of steady-state operation. A patent by the author, Hugh Torry and others (*Provost et al. 2012*) describes a method and system for predicting flashovers in brushed electrical machines which makes reference to time series analysis using Kalman Filtering.

The paper by Victoria Hodge and Jim Austin (*Hodge and Austin 2004*) presents a comprehensive survey of outlier detection methods and discusses their relative advantages and disadvantages.

References

* denotes source that the author found particularly helpful and/or inspiring.

Benedict, T. and Bordner, G., "Synthesis of an Optimal Set of Radar Track-While-Scan Smoothing Equations," *IRE Transactions on Automatic Control* AC-7 (July 1962): 27-32. Pg. 114

Box, G., Jenkins, G., Reinsel, G., and Ljung, G., *Time Series Analysis: Forecasting and Control*, 5th ed. (Hoboken, NJ: John Wiley & Sons Inc, 2016). Pg. 116

Chatfield, C., *The Analysis of Time Series: An Introduction*, 6th ed (Boca Raton, FL: CRC Press/Taylor & Francis, 2009). Pg. 116

Gilchrist, W., *Statistical Forecasting*, (Chichester, UK: John Wiley & Sons Ltd., 1976). Pg. 116

Goodwin, P., "The Holt-Winters Approach to Exponential Smoothing: 50 Years Old and Going Strong," *Foresight: The International Journal of Applied Forecasting* no. 19 (2010): 30-33. Pg. 116

Hodge, V. and Austin, J., "A Survey of Outlier Detection Methodologies," *Artificial Intelligence Review* 22, no. 2 (2004): 85-126. Pg. 117

Mathioudakis, K. and Sieverding, C. eds. *Gas Turbine Condition Monitoring & Fault Diagnosis*, (Rhode Saint Genèse, Belgium: Von Karman Institute for Fluid Dynamics, 2003). Pg. 113, 116

Montgomery, D., Jennings, C. and Kulahci, M., *Introduction to Time Series Analysis and Forecasting*, (Hoboken, NJ: Wiley-Interscience, 2008). Pg. 116

Morgan, S. and Provost, M., Steady state sensor. European Patent 0 691 631 B1, 1988. Filed July 3, 1995, granted September 2, 1998. Pg. 116

* Provost, M., "The Use of Optimal Estimation Techniques in the Analysis of Gas Turbines," Ph.D. thesis, Department of Propulsion and Power, School of Engineering, Cranfield University, Cranfield, UK, 1994. Pgs. 113, 116

Provost, M., "Kalman Filtering Applied to Time Series Analysis," *Aircraft Airborne Condition Monitoring, British Energy*, Gloucester, UK, May 14, 2003, (Stevenage, UK: Institution of Electrical Engineers Aerospace Professional Network). Pgs. 109, 110, 110, 110, 112, 113, 114, 114, 116

Provost, M., "Introduction to Time Series Analysis," Unpublished presentation, Nottingham, UK, 2006. Pg. 116

Provost, M., "Further Developments of the Optimal Tracker," Unpublished presentation, Nottingham, UK, 2007. Pgs. 115, 115

Provost, M., "A Simple Guide to Kalman Filtering," Unpublished presentation, Nottingham, UK, 2008. Pg. 116

Provost, M., "From Data to Information: An Example," Internal presentation, Loughborough, UK, Intelligent Energy Ltd., 2010. Pg. 116

Provost, M., Torry, H., and Chan, S., Method and system for predicting flashovers in brushed electrical machines. UK Patent Application GB 2481657 A, Published January 4, 2012. Pg. 116

Additional Material

References

Hyndman, R., *Forecasting: Principles and Practice*, (Perth, Australia: University of Western Australia, 2014).

Hyndman, R. and Athanasopoulos, G., *Forecasting: Principles and Practice*, (Perth, Australia: OTexts, 2014).

Website

OTexts www.otexts.org

CHAPTER 7

Data Visualization

Introduction

One of the most important aspects of asset management is the visualization of the data and information produced by the various methods and processes discussed in this book. It is important for the reader to realize that the aim of the whole exercise is to influence key decision-makers at every level (from boardroom to shop floor, both internally and externally) to take action; otherwise all efforts are in vain. A good visualization of data gathered from an asset and/or the results of an analysis can electrify an audience and create real enthusiasm, whereas a poor one can severely hamper progress. Goodness is in the eye of the beholder, not the analyst, so the creators of visualizations must always be careful to tailor them to user needs.

Note that the use of color is important in data visualization, but examples in this book have had to be presented in black and white to reduce printing costs.

Fundamental Information Requirements

Typical users of asset management information will come from all parts of an organization (finance, engineering, operations, services, etc.) as well as external parties such as customers and suppliers. Each information consumer needs an information solution built for them, which is made up of one or more linked, user-customizable information building blocks. Generally, information users will be looking for:

- Visibility of current asset status (via dashboards and maps).
- Details of which assets need attention (via alerts and weighted lists).

- Standard information displays and reports about asset operations and health.
- The ability to answer questions about what assets are doing and how they are behaving (via x-y plots, bar and column charts, fault/event data cubes, summary statistics, time series, and bespoke analyses).
- The capability to derive other parameters from sensor readings (e.g., power from voltage and current measurements), which could extend to sophisticated online simulation and modelling.
- Views of what assets may do in the future (via predictive analytics, based on many of the methods mentioned in this book).
- The ability to export data to external software tools for off-line processing and methods development.

Typical Information Solutions

The following are typical examples of what information users may ask for:

- Management and finance will require regular reports, weighted lists of Key Performance Indicators (KPIs), and alerts, in order to answer questions such as the length of time assets have been running, what they have delivered to the customer (electrical energy, passenger-km or ton-km, thrust, etc.), lists of "rogue" assets that are particularly costly to keep running, and lists of fleet-wide issues affecting many assets, as well as information that can be used for billing customers. Regularity, consistency, ease of understanding, and presentation quality are critical. Much of this information will almost certainly be reviewed regularly at board level.
- Operations and services will require more real-time information about the assets in their care, including high-level dashboards showing relevant information about both single and multiple assets, maps showing where assets are, time series plots, KPIs (which may be different to those required by finance), analyses that produce alerts from both failed or failing assets and regular reports, probably at a higher frequency and a more detailed level than those required by finance. Speed and ease of understanding are critical.
- Engineering will require access to very large quantities of current and historical asset data, together with the means to carry out both regular and *ad hoc* analyses using a range of techniques in order to truly understand how assets are behaving in service. Capabilities to carry out calculations, display data and information in a range of ways, and export data are essential.
- Customers and suppliers will require subsets of the information discussed above. Adherence to contractual obligations (in terms of frequency, format, and content of information delivered) and secure data and information delivery to external systems are essential.

Typical Information Building Blocks

The following section describes typical building blocks that, when put together in appropriate ways, will deliver the various information needs discussed above.

Alerts

Alerts automatically highlight situations which require action to be taken, such as assets being operated incorrectly, actual or impending failures of assets or their component parts, or any other situations that may be of interest to the user. The logic that generates alerts can either be simple (such as a comparison of a single analogue parameter with a warning or action threshold, or checking for the presence of single fault or event codes) or complex (such as comparing various analogue parameters with appropriate thresholds within a given time period or checking for the simultaneous occurrence of multiple fault or event codes). In the case of connected assets (e.g., vehicles making up a train, engines in an aircraft), one or more alerts from one or more of these could generate an alert for that entire asset group. Alerts need to be easily and quickly created and modified by users, since the logic behind them may need to be changed rapidly to meet operational or customer requirements. Alerts need to be transmitted quickly to those responsible for taking action, using email, text message, or highlighting on a web-based dashboard. Some form of acknowledgment of receipt and action taken also needs to be put in place with escalation mechanisms provided if alerts are particularly critical and responses are not forthcoming.

Bar and Column Charts

Bar (horizontal) and column (vertical) charts are frequently used to compare values of one parameter that have been gathered from many assets (usually during the same time period) in order to easily compare their values and highlight discrepancies. They are also used to compare similar measurements taken on a single asset (e.g., individual fuel cell voltages in a fuel cell stack) and to show fault and event-code counts gathered during specific time periods on one or more assets or their components.

Figures 7.1 and 7.2 show examples of typical column charts.

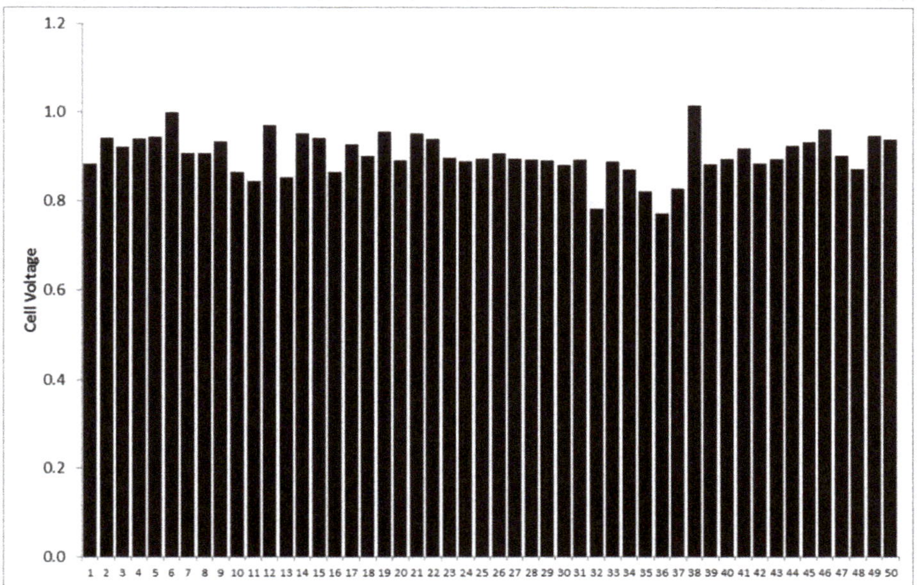

FIGURE 7.1 Typical column chart comparing fuel cell voltages in a fuel cell stack.

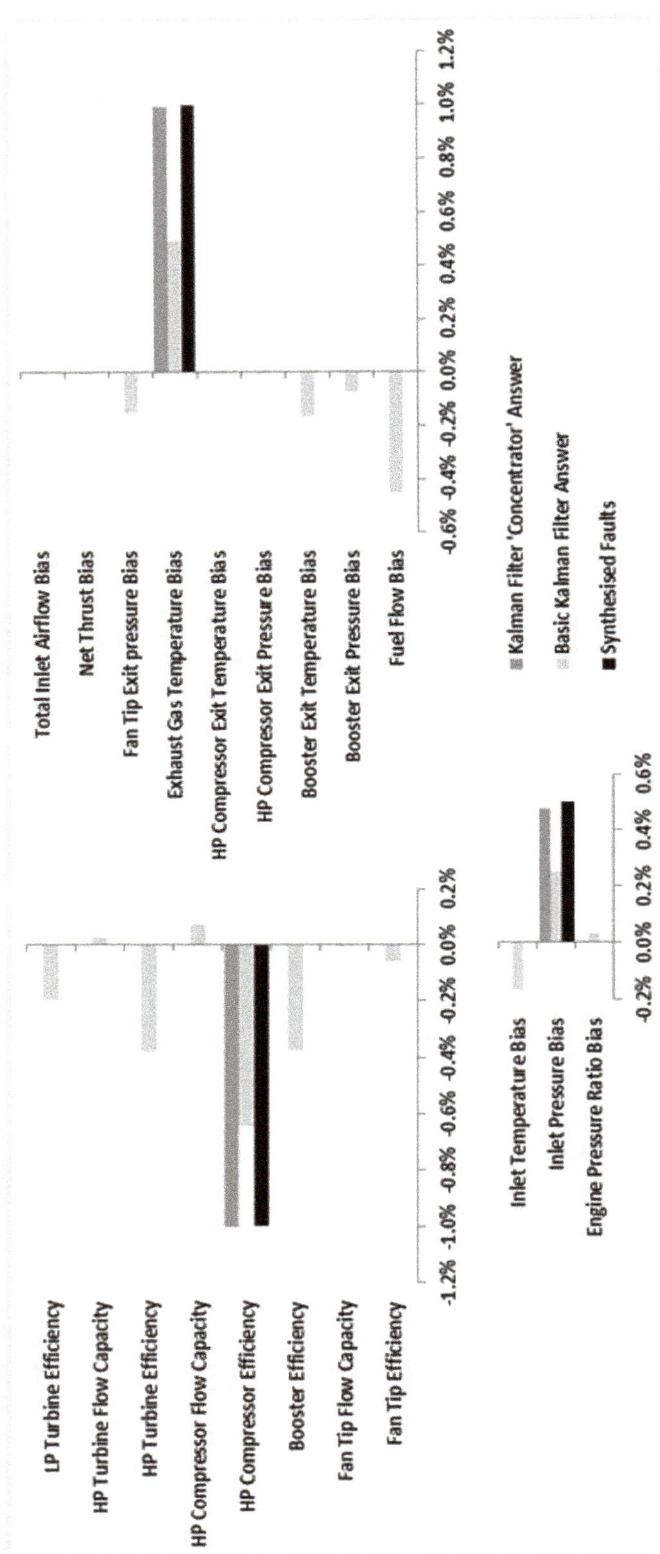

FIGURE 7.2 Typical bar charts comparing analysis outputs.

Dashboards

Dashboards are displays of critical asset information (from either a single asset or an entire fleet) that convey a quickly and easily assimilated assessment of asset state(s), either currently or at a particular point in history. Users generally have very particular views of what they want to see, how it should be displayed, and how they can "drill down" into the details of individual asset behavior if needed. Some dashboards are based around pictorial representations of the assets being monitored, with relevant parameters displayed either numerically or graphically near their physical locations on the asset.

Figures 7.3 through 7.5 show typical dashboards.

Note the combination of several elements in Figure 7.3: maps, tables, color-coded symbols, numbers, and time series plots, as well as hyperlinks to other displays, an external source of weather forecasts, and a diagram of the system being monitored.

Data Export Capabilities

A data export capability needs to be provided that can deliver data in a format and layout that can be input into the system receiving the information. Common formats include comma-separated values (csv), Microsoft Excel® spreadsheet formats (xls and xlsx), portable document format (pdf), and extensible markup language (xml), but other proprietary formats (e.g., MatLab®) should also be available, as agreed with recipients.

Fault/Event Cubes

Fault/event cubes (also known as data or OLAP cubes) are dynamic column charts of counts of fault and/or event codes for user-specified assets (and their subcomponents, if relevant) over a user-specified time period, presented by asset, system, or time as required. They provide good visibility of patterns in the data and are particularly helpful when combined with a drill-down facility that enables problem assets or components to be quickly isolated.

Figure 7.6 shows the principle of fault/event cubes, with example column charts.

Mapping

The display of asset position is particularly useful and easy to interpret, particularly when the assets move. Color coding the asset position markers to indicate operational outputs, ownership, faults, events, or levels of critical parameters can be particularly informative. If the assets are stationary and reasonably close together, contour plots of analogue data can be drawn that provide information about what could be happening over a wide area.

Figures 7.7 and 7.8 show typical mapping displays.

Maps need not necessarily use geographic coordinates: Figure 7.9 shows a contour map of ambient temperature versus time of day (vertical axis) and date (horizontal axis).

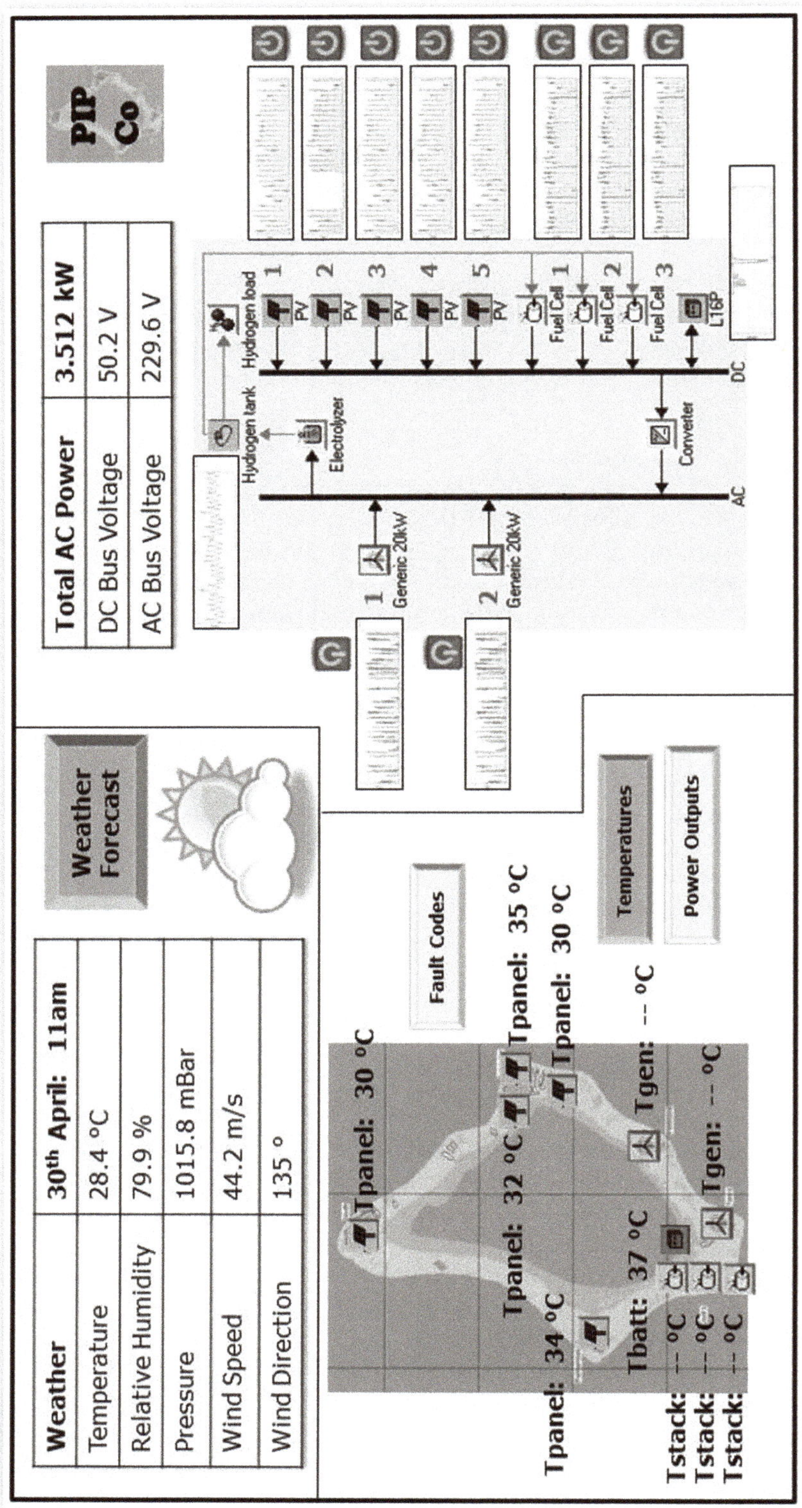

FIGURE 7.3 Notional dashboard for monitoring the fictitious power system in Appendix B (Palmerston Island map: Flickr/EVS-Islands).

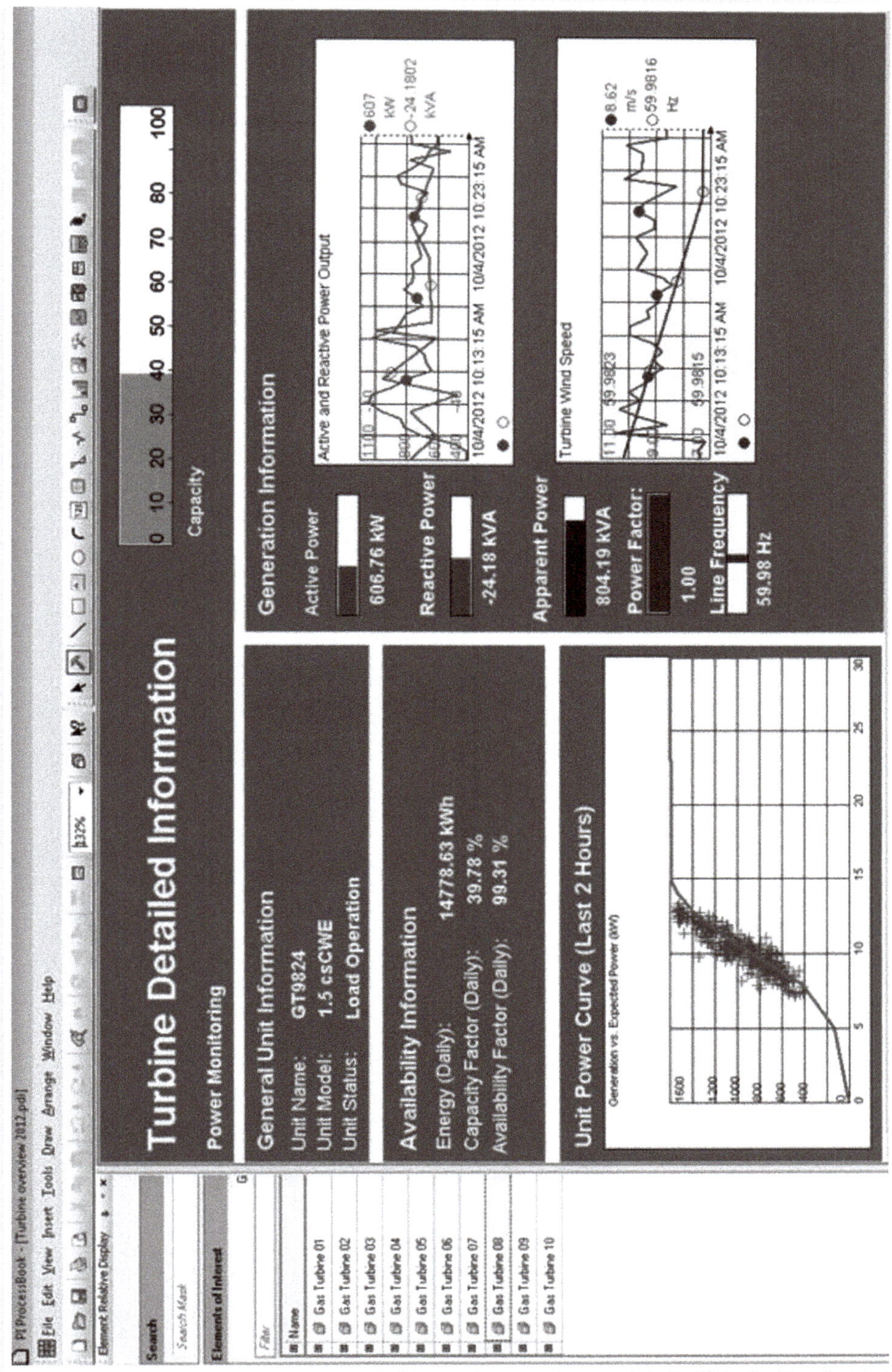

FIGURE 7.4 Typical dashboard produced using PI ProcessBook (OSIsoft (www.osisoft.com): reproduced with permission).

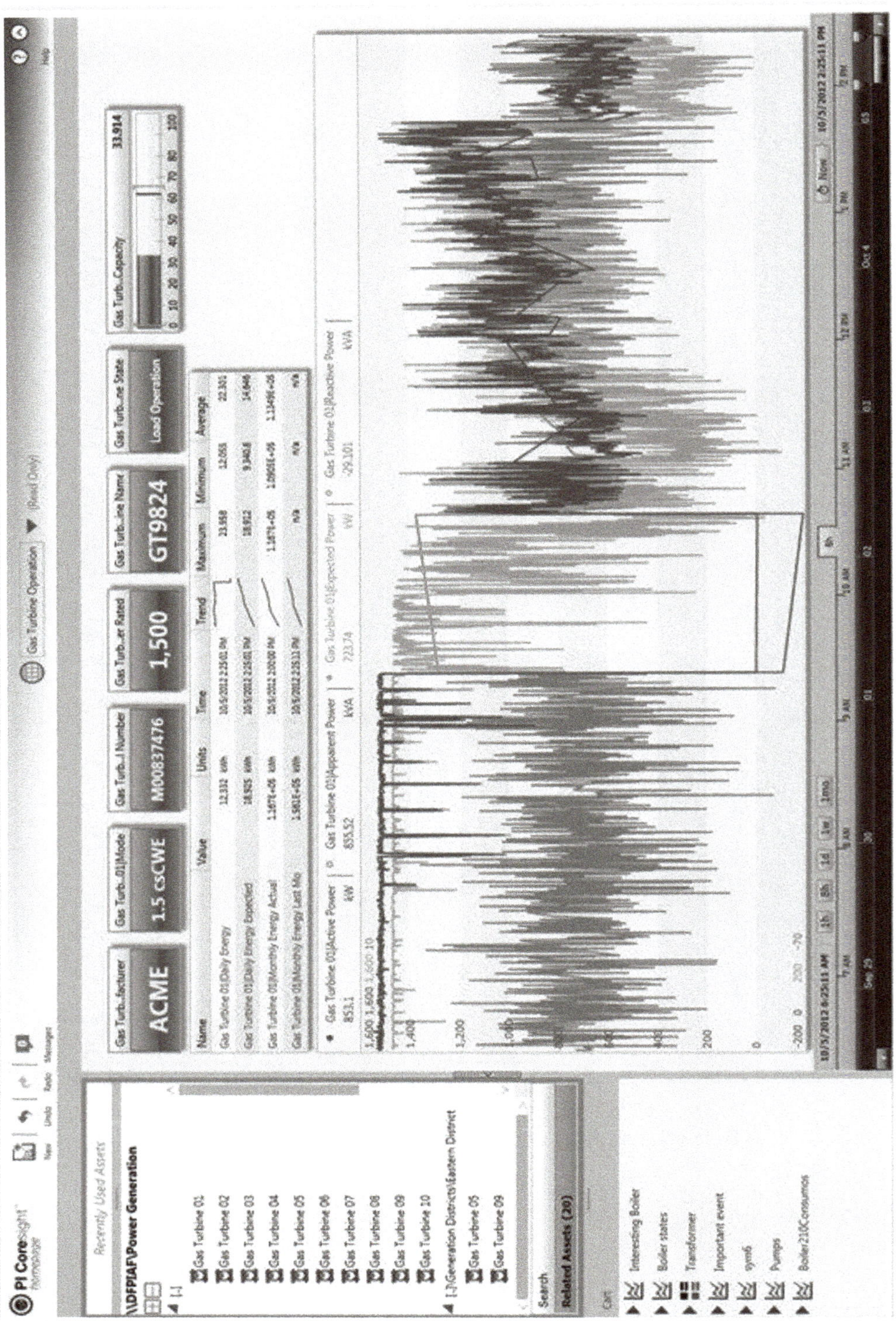

FIGURE 7.5 Typical dashboard produced using PI Coresight (OSIsoft: reproduced with permission).

FIGURE 7.6 Example of fault/event cube.

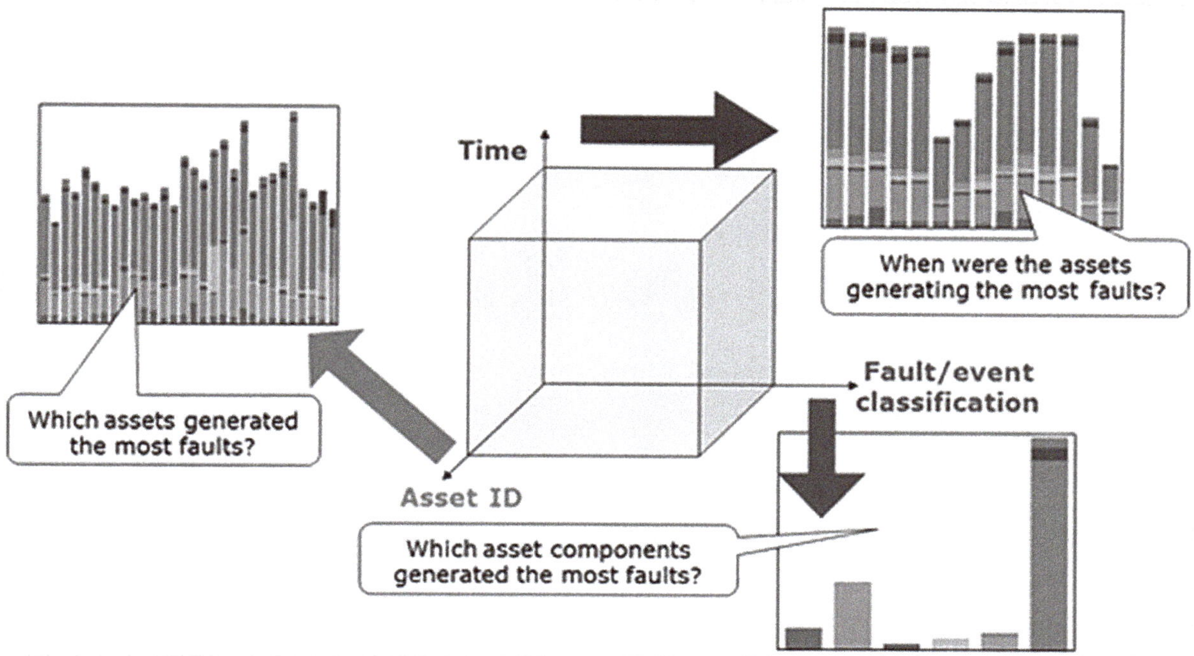

FIGURE 7.7 Typical location map, with drill-down superimposed (Shoothill Ltd. (www.shoothill.com), reproduced with permission).

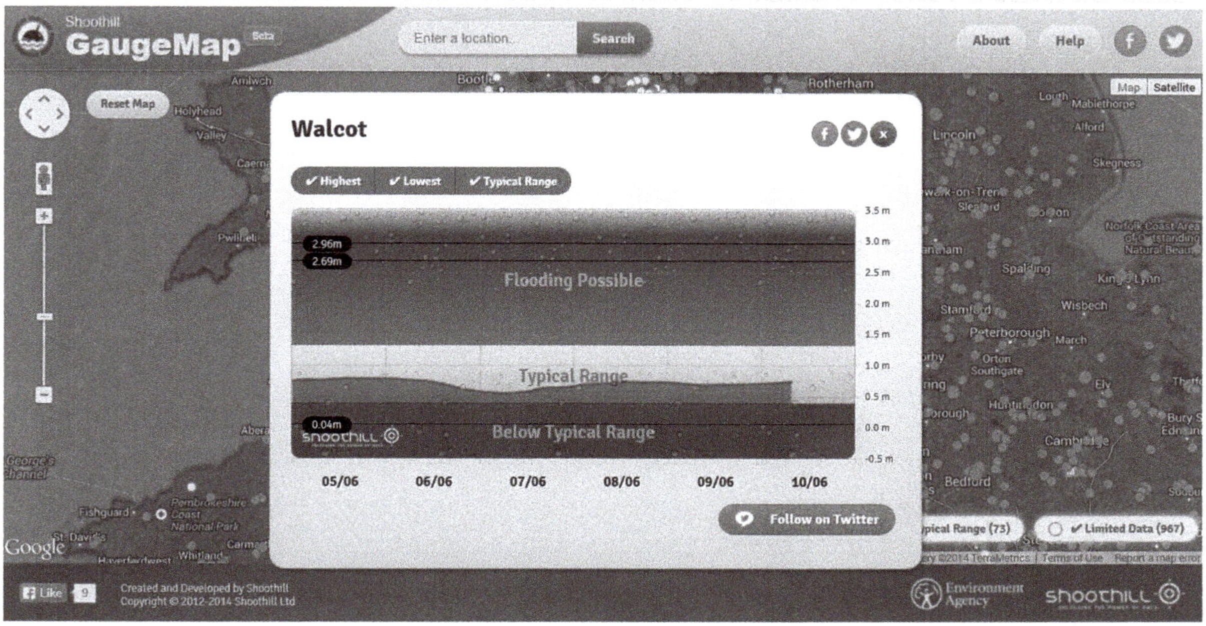

FIGURE 7.8 Typical contour map (Shoothill Ltd.: reproduced with permission).

FIGURE 7.9 Contour plot of ambient temperature, using data from the fictitious power system in Appendix E.

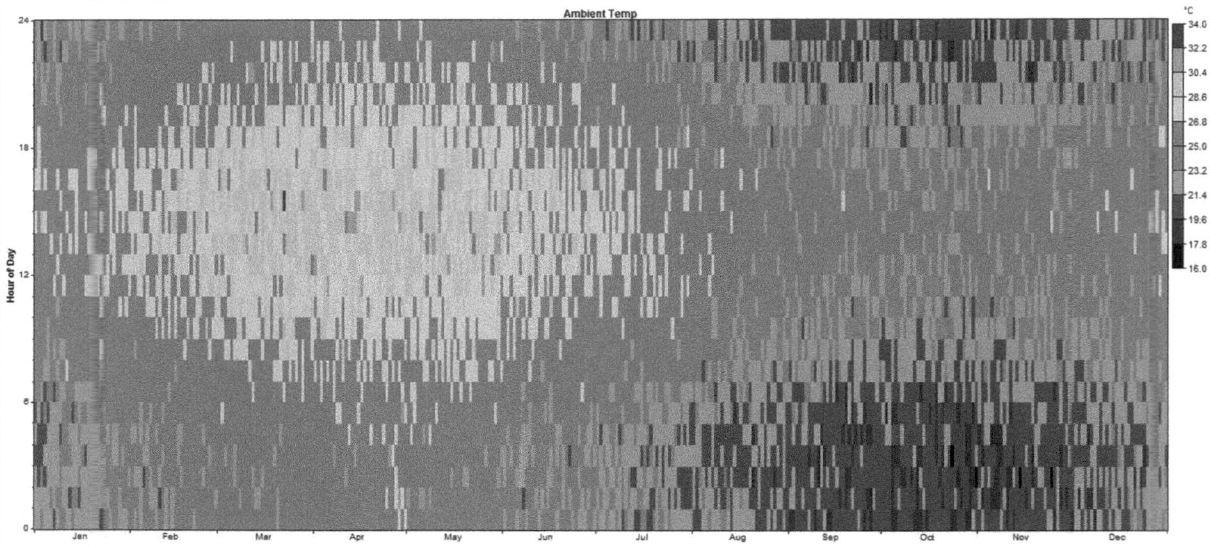

Parallel Coordinates Plots

This specialized display consists of multiple analogue parameters recorded on an asset plotted on different vertical axes, with the values recorded at each given time joined by straight lines. Over time, a picture is built up of the ranges covered by each parameter. Color coding the lines can also show how the parameters vary relative to each other.

Figure 7.10 shows a typical parallel coordinates plot, in this case highlighting in light gray the operating conditions when broadband (BB) vibration (the third axis from the right) on one of the wind turbines included in the fictitious power system in Appendix B was higher than normal.

FIGURE 7.10 Parallel coordinates plot, using data from the fictitious power system in Appendix B.

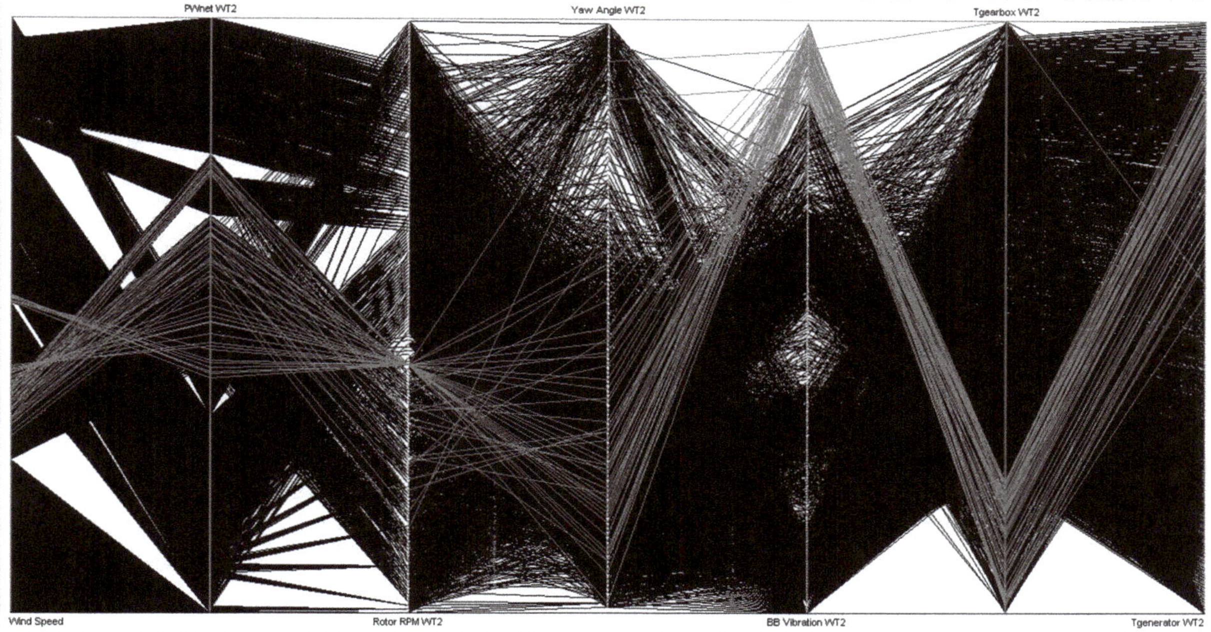

Reporting

The output of standard reports in a consistent format is crucial to provide organizations and their customers with traceable and repeatable records that provide both short-term and long-term understanding of what is happening to the assets used by a business and the value that these assets generate.

"Smart" Analytics

Proprietary and nonproprietary analysis and visualization techniques that are used by asset and business experts to derive new knowledge and create new business understanding and value are an important addition to generic visualization toolsets. Initial testing of these to prove their usefulness will probably be done using stand-alone

FIGURE 7.11 "Smart" analytics plots, using data from the fictitious power system in Appendix B.

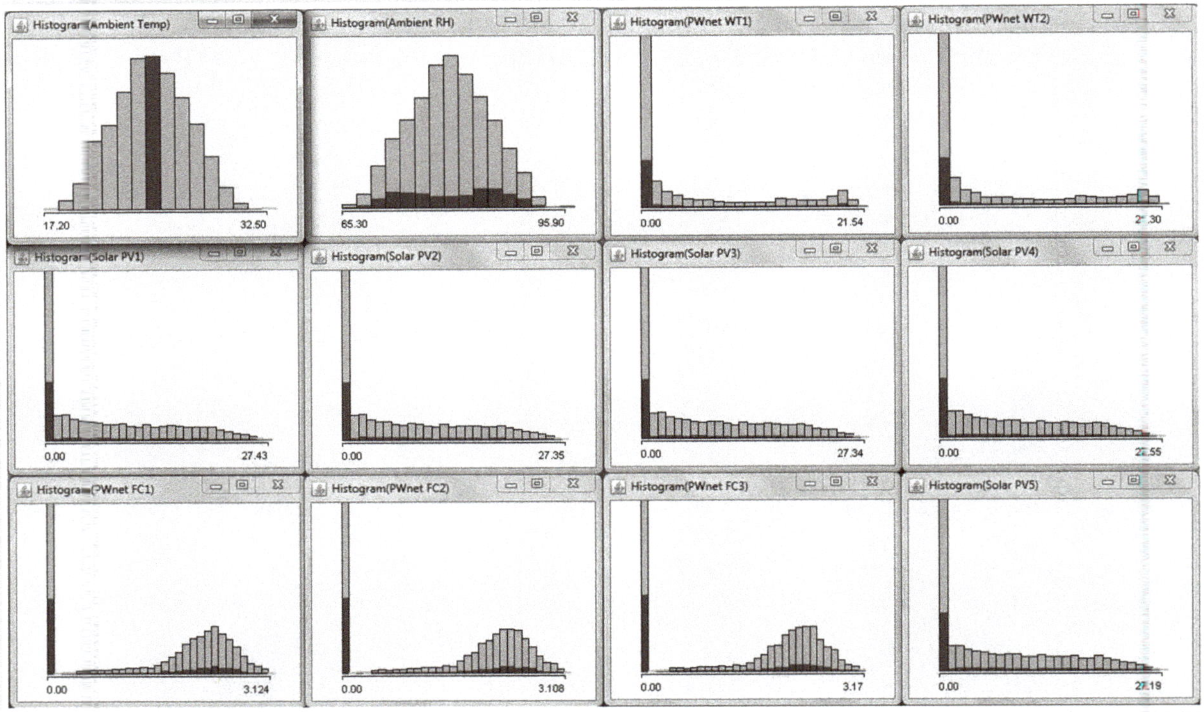

off-line software, but the need to incorporate them into production systems eventually must be recognized, particularly if their application delivers value and competitive advantage to an organization.

Figure 7.11 shows the distribution of power outputs over a year from all the power units in the fictitious power system in Appendix B, together with the distributions of ambient temperature and ambient relative humidity. Power outputs and relative humidity when the ambient temperature was in the dark highlighted range are indicated.

Summary Statistics

Values that summarize the central tendency (mean, median, etc.) and spread (minimum, maximum, standard deviation, quartiles, deciles, etc.) of values gathered across time periods or fleets of assets are important measures that need to be calculated and displayed, either in specialized ways, such as box-and-whisker plots and probability distributions, or in conjunction with other displays, because they generate expectations and/or limits.

Figures 7.12 and 7.13 show a typical probability distribution and box-and-whisker plot, respectively.

Time Series

Plotting analogue parameters or fault/event occurrences against calendar time (or a time-related parameter, such as start counts or operating time) is one of the most important and best-understood data visualization techniques. Users must be able to select one or

FIGURE 7.12 Probability distribution of annual AC primary load, using data from the fictitious power system in Appendix B.

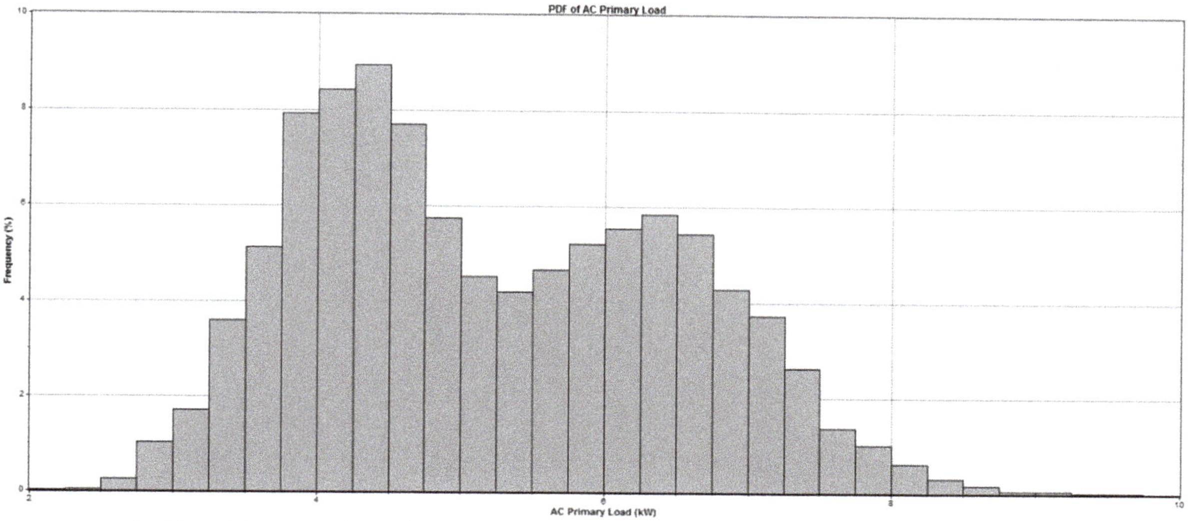

FIGURE 7.13 Box-and-whisker plot of monthly values of AC primary load, using data from the fictitious power system in Appendix B.

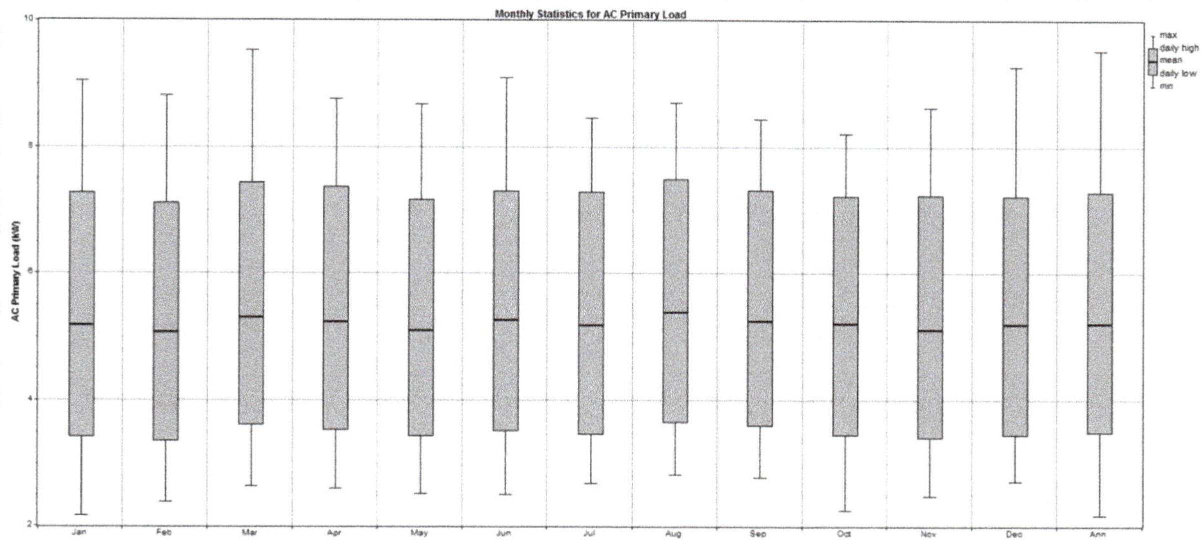

FIGURE 7.14 Time series plots of ambient relative humidity (top) and ambient temperature (bottom), using data from the fictitious power system in Appendix B.

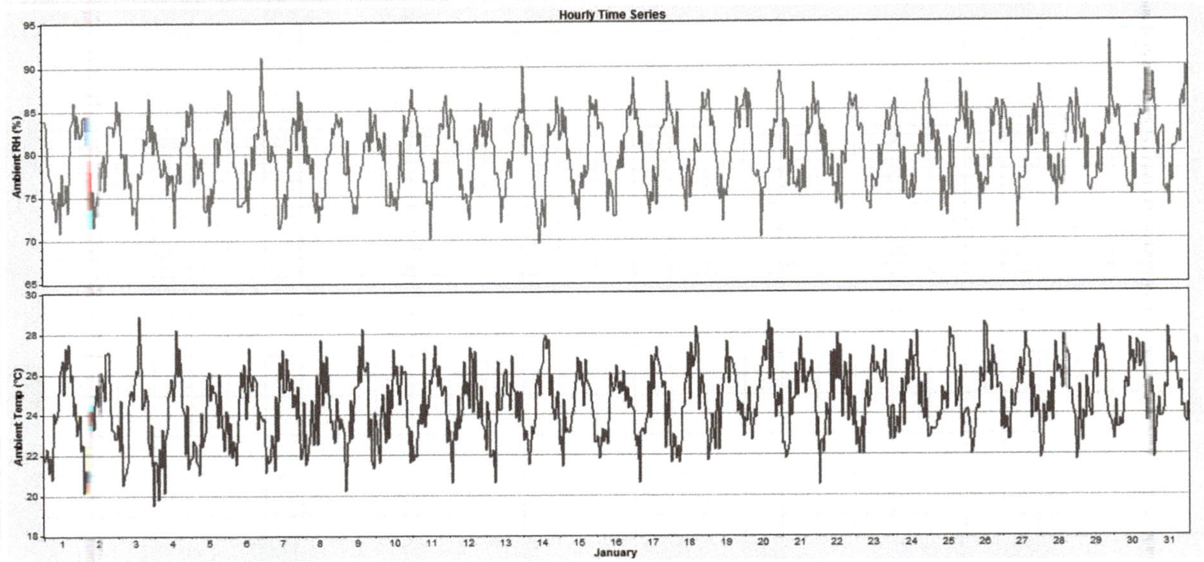

more assets and define the time period displayed. Often several different parameters are displayed against a common time base, either on the same plot using multiple scales or on different synchronized plots: care needs to be taken to avoid confusion if the number of parameters is large. Figure 7.14 shows a pair of typical time series plots.

User-Defined Calculations

Users need the ability to derive new parameters from measurements taken on an asset, since these will provide additional detailed and focused information about asset and component health.

Weighted Lists

A list of assets ranked by values of a user-selected KPI provides an easily understood view of the relative behavior of each asset in a fleet.

X-Y Plots

Users need the ability to visualize the variation of one or more y parameters against a base parameter x, since this provides an easily understood display of how parameters relate to each other as asset operational conditions vary and assets deteriorate over time.

Figure 7.15 shows a typical x-y plot, clearly showing the three operating regimes of one of the wind turbines included in the fictitious power system in Appendix B: normal (the z-shaped relationship between zero and 24 m/s wind speed), pitch failure (rotor rpm proportional to wind speed), and feathered (wind speed greater than 24 m/s).

FIGURE 7.15 Typical x-y plot of wind turbine rotor rpm versus wind speed, using data from the fictitious power system in Appendix B.

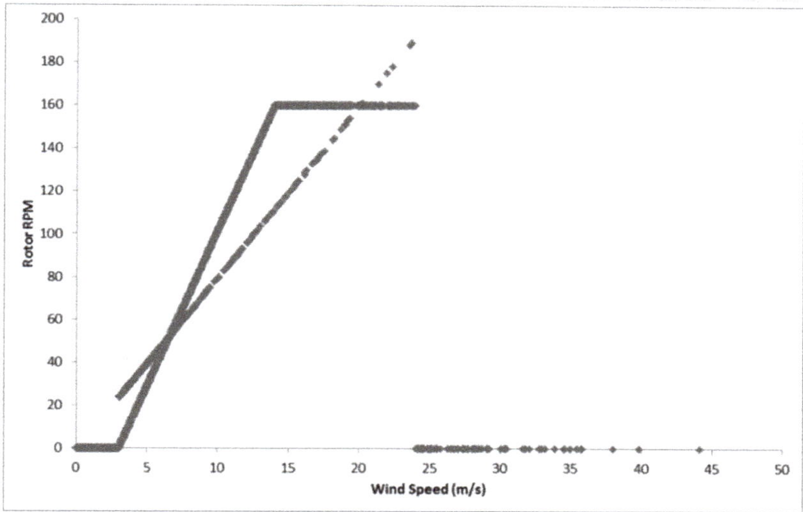

Drill-Down

One very important capability that any data visualization system needs is the ability to quickly move from high-level summary displays into the detail that makes up that summary. For example, a map showing where assets are located should allow the user to get time series information about a particular asset merely by clicking on that asset's marked location, rather than typing a query into another screen or system. Similarly, dashboards should be configured to allow users to easily "drill" into the detail in order to quickly determine the reasons behind any anomalies displayed. As far as possible selection of assets, display formats and time ranges should be via buttons, hyperlinks, or drop-down menus, particularly for operations and services which will usually be under considerable time pressure to use all available information to produce quick and effective solutions to problems. If operators can evolve or specify consistent, repeatable ways of working with data and information, then those working practices should be automated as far as possible without removing the ability to answer "out of the box" queries.

Business Intelligence

Business intelligence (BI) is becoming an important skillset for companies who wish to make complex information easily accessible to management and employees. Cindi Howson's book (*Howson 2013*) discusses many of the issues involved in successfully setting up and using a BI system. Her company's website (www.biscorecard.com) makes available paid-for evaluations of many of the better-known BI packages. There is also a presentation (*BI Scorecard 2012*) which usefully lines up the needs of various potential users with some of the available packages. MicroStrategy (www.microstrategy.com) has provided a thorough overview of the BI process (*MicroStrategy 2010a*) and some good examples of the dashboards that can be produced (*MicroStrategy 2010b*), as well as a detailed white paper (*MicroStrategy 2010c*) that provides a framework for understanding the diverse range of BI functionality that has evolved in the market over the past 15 years. The paper by Matillion (*Matillion 2014*) also discusses how to create effective BI displays. The yearly Gartner BI magic quadrant (*Sallam et al. 2014*: www.gartner.com) gives a useful high-level comparison of most of the BI market offerings. Daniel Murray's book (*Murray 2013*) provides a comprehensive guide to the setup and use of Tableau® (www.tableausoftware.com). Papers by suppliers such as Birst (*Birst 2013*: www.birst.com), Tableau® (*Tableau 2014*), Targit® (*Targit 2013*: www.targit.com), Qualitin (*Qualitin 2015*: www.qualitin.com), and Matillion (*Matillion 2013a, 2013b*: www.matillion.com), despite being sales and marketing platforms, are useful sources of information. The Spotfire® BI application from TIBCO™ software (www.tibco.com) is also worthy of note.

The excellent book on KPIs by Bernard Marr (*Marr 2015*) discusses the correct design, implementation, and use of KPIs for maximum business impact, while the article by Lance Jakob and Anthony Honaker (*Jakob and Honaker 2015*) states that KPIs, unlike metrics, must clearly identify where a business wants to go, when it is expected to get there, and how it is currently performing, all while providing actionable opportunities to ensure it reaches the required destination. A metric can evolve into a KPI when it clearly indicates either progress, problems, or opportunities, the relationship of the measure to key objectives is identified, a target or threshold is decided, and the actions and planned reactions necessary to drive performance are established. The article by Jeff Smith (*Smith 2009*) argues that it is important to realize that numbers are not everything and that KPIs, while offering guidance, can be misleading if the user does not go deeper than the numbers to find the ingredients and drivers for success. *The Goal*, Eli Goldratt's best-selling novel on the theory of constraints (*Goldratt and Cox 2013*), is essentially about how misleading KPIs can destroy a business: there is also a very telling example of individual KPIs producing a suboptimal overall business solution in Chris Lloyd's book (*Lloyd 2010*).

Books and Papers

The books by Edward Tufte (*Tufte 1990, 1997, 2001, 2006*) are widely regarded as classics. They demonstrate the essential principles of data visualization in a clear and beautiful way, drawing on many famous examples, such as Napoleon's advance and retreat in Russia, John Snow's visualization of the 1854 cholera epidemic in London (which led to an understanding and eventual eradication of the disease by providing proper sanitation), and the visualization that could have prevented the Challenger space shuttle accident. Tufte also clearly demonstrates how not to visualize data. The book by Jason Lankow, Josh Ritchie, and others (*Lankow et al. 2012*) extends Tufte's ideas, looking at the history and practical use of the new art and science of infographics. Andy Cotgreave's paper (*Cotgreave 2012*) gives some examples of influential visualizations.

Stephen Few has written several excellent books on effective visual communication and the design of graphical displays (*Few 2004, 2006a, 2009*). The books by Nathan Yau (*Yau 2011, 2013*), Judie Steele and Noah Iliinsky (*Steele and Iliinsky 2010*), and Ben Fry (*Fry 2008*) go into more detail, with examples and code in several computer languages.

The Mondrian general-purpose statistical data visualization system (which can be downloaded from www.theusrus.de/Mondrian) is a very capable free package, described in a paper by Martin Theus (*Theus 2002*) and comprehensively written up in a book Martin has co-authored with Simon Urbanek (*Theus and Urbanek 2009*), which includes many examples. A companion website to the book (www.interactivegraphics.org) covers each of the features in Mondrian in a series of presentations, including some examples.

The JMP® statistical analysis software package from SAS® Institute Inc. (www.jmp.com) enables users to visualize data in many ways, as described in the booklet by Curt Hinrichs and Chuck Boiler (*Hinrichs and Boiler 2010*). Other very powerful data visualization and analysis packages have been produced by SAS® Institute Inc. (www.sas.com), LIONsolver Inc. (www.lionsolver.com, www.grapheur.com), RapidMiner (www.rapidminer.com), and Omniscope (www.visokio.com).

The presentation by Alan Mahoney and others (*Mahoney et al. 2012*) describes a commercial system (geometric process control (GPC), available from Process Plant Computing Ltd.: www.ppcl.com) that visualizes many parameters simultaneously and enables the user to simply define operating envelopes of both normal and abnormal operation of an asset visually, which can then form the basis for alarm generation. Stephen Few has also written an article on this subject (*Few 2006b*).

MatLab® can also be used to create both simple and complex data visualizations. The book by Wendy and Angel Martinez (*Martinez and Martinez 2005*) presents the theory behind several complex data visualization methods, together with the MatLab® code used to create them.

Microsoft Excel® and Microsoft PowerPoint®, when used correctly, can produce powerful and effective data visualizations. Nancy Duarte has produced two books (*Duarte 2008, 2010*) that discuss how to create presentations that connect with an audience. However, PowerPoint® has to be used with care: Edward Tufte points out in his short critique of PowerPoint® (*Tufte 2004*) that it is very easy to become lazy and imprecise when creating slides, particularly when using built-in templates.

The presentation by Cyient (*Cyient 2014*: www.cyient.com) discusses a range of advanced analysis, data visualization, and maintenance management techniques used in the medical, heavy equipment, aircraft engine, transportation, utility, and other sectors.

Two presentations by Tim Sharpe of Sabisu (*Sharpe 2015a, 2015b*: www.sabisu.co) describe the Sabisu visualization and decision support platform and describe an application in the energy industry.

The presentation by Caxton Okoh (*Okoh 2015*) discusses a proposal to generate visibility of aeroengine historical and current health events on a timeline to help assess through-life performance and display concise and clear information to help decision-making.

References

* denotes sources that the author found particularly helpful and/or inspiring.

BI Scorecard, *Positioning Users and BI Tool Modules*, (Sparta, NJ: ASK LLC/BI Scorecard, 2012). Pg. 133

Birst, *The Top 10 Business Questions that Drive Your BI Technology Requirements*, (San Francisco, CA: Birst Inc., 2013). Pg. 134

Cotgreave, A., *The 5 Most Influential Data Visualizations of All Time*, (London, UK: Tableau® Software, 2012). Pg. 134

Cyient, "Cyient Insights: Advanced Analytics for Actionable Business Insights," Unpublished presentation, Hyderabad, India: Cyient Ltd., 2014. Pg. 135

Duarte, N., *Slide:Ology: The Art and Science of Creating Great Presentations*, (Sebastopol, CA: O'Reilly Media, 2008). Pg. 135

Duarte, N., *Resonate: Present Visual Stories that Transform Audiences*, (Sebastopol, CA: O'Reilly Media, 2010). Pg. 135

Few, S., *Show Me the Numbers*, (Oakland, CA: Analytics Press, 2004). Pg. 134

Few, S., *Information Dashboard Design*, (Beijing: O'Reilly Media, 2006a). [Pg. 134]

Few, S., *Multivariate Analysis Using Parallel Coordinates*, (Berkeley, CA: Perceptual Edge, 2006b). [Pg. 135]

Few, S., *Now You See It*, (Oakland, CA: Analytics Press, 2009). [Pg. 134]

Fry, B., *Visualizing Data*, (Sebastopol, CA: O'Reilly Media, 2008). [Pg. 134]

Goldratt, E. and Cox, J., *The Goal: A Process of Ongoing Improvement*, 3rd ed. (Farnham, UK: Gower Publishing Ltd., 2013). [Pg. 134]

Hinrichs, C. and Boiler, C.. *JMP® Essentials: An Illustrated Step-by-Step Guide for New Users*, (Cary, NC: SAS® Institute, 2010). [Pg. 135]

* Howson, C., *Successful Business Intelligence: Unlock the Value of BI & Big Data*, 2nd ed (New York, NY: McGraw-Hill Osborne, 2013). [Pg. 133]

Jakob, G. and Honaker, A., "Metrics vs KPIs: Lessons from 'Alice in Wonderland'," *Uptime* 10, no. 68 (December/January 2015): 18-20. [Pg. 134]

Lankow, J., Ritchie, J., and Crooks, R., *Infographics: The Power of Visual Storytelling*, (Hoboken, NJ: John Wiley & Sons Inc., 2012). [Pg. 134]

* Lloyd, C. eds., *Asset Management: Whole-Life Management of Physical Assets*, (London, UK: Thomas Telford, 2010). [Pg. 134]

Mahoney, A., Brooks, R., and Wilson, J., "Predictive Condition Monitoring with Geometric Process Control," *2nd Annual Predictive Asset Management Seminar*, Royal Garden Hotel, Kensington, London, UK, September 11-13, 2012. (London, UK: IBC Energy/Informa UK Ltd). [Pg. 135]

* Marr, B., *Key Performance Indicators for Dummies*, (Chichester, UK: John Wiley & Sons Ltd., 2015). [Pg. 134]

Martinez, W. and Martinez, A., *Exploratory Data Analysis with MatLab®*, (Boca Raton, FL: Chapman & Hall/CRC, 2005). [Pg. 135]

Matillion, *How to Create Compelling Business Dashboards*, (Knutsford, UK: Matillion Ltd., 2013a). [Pg. 134]

Matillion, *Successfully Implementing Business Intelligence*, (Knutsford, UK: Matillion Ltd., 2013b). [Pg. 134]

Matillion, *The Bumper Book of Business Intelligence*, (Knutsford, UK: Matillion Ltd., 2014). [Pg. 134]

MicroStrategy, *Enterprise Business Intelligence: Improving Corporate Performance through Integration*, (Tysons Corner, VA: MicroStrategy, Inc., 2010a). [Pg. 133]

MicroStrategy, *MicroStrategy Dynamic Enterprise Dashboards*. (Tysons Corner, VA: MicroStrategy, Inc., 2010b). [Pg. 133]

MicroStrategy, *The 5 Styles of Business Intelligence: Industrial-Strength Business Intelligence*, (Tysons Corner, VA: MicroStrategy, Inc., 2010c). [Pg. 133]

* Murray, D., *Tableau Your Data! Fast and Easy Visual Analysis with Tableau® Software*, (Indianapolis, IN: John Wiley & Sons Inc., 2013). [Pg. 134]

Okoh, C., "Data Fusion for Timeline of Engine Events in Through-Life Engineering Services," Unpublished Ph.D. proposal presentation, Cranfield, UK: Cranfield University, 2015. [Pg. 135]

Qualitin, *ICG (Strategy Execution System)*, (London, UK: Qualitin, 2015). [Pg. 134]

Sallam, R., Tapadinhas, J., Parenteau, J., Yuen, D. et al., *Magic Quadrant for Business Intelligence and Analytics Platforms*, (Stamford, CT: Gartner Inc., 2014). [Pg. 134]

Sharpe, T., *Aggregating Energy Management KPIs from Distributed Sites for SABIC KSA*, (Sale, UK: Sabisu Ltd, 2015a). Pg. 135

Sharpe, T., *Decision Support with Sabisu*, (Sale, UK: Sabisu Ltd, 2015b). Pg. 135

Smith, J., "Looking beyond the Numbers," *Uptime* 4, no. 35 (June/July 2009): 8-17. Pg. 134

Steele, J. and Iliinsky, N. eds., *Beautiful Visualization: Looking at Data through the Eyes of Experts*, (Sebastopol, CA: O'Reilly Media, 2010). Pg. 134

Tableau, *Visual Analysis Best Practices: Simple Techniques for Making Every Visualisation Useful and Beautiful*, (Seattle, WA: Tableau® Software, 2014). Pg. 134

Targit, *Five Steps to Becoming a Data-Driven Services Organization*, (Hjørring, Denmark: Targit A/S, 2013). Pg. 134

Theus, M., "Interactive Data Visualization Using Mondrian," *Journal of Statistical Software* 7, no. 11 (2002). Pg. 134

* Theus, M. and Urbanek, S., *Interactive Graphics for Data Analysis: Principles and Examples*, (Boca Raton, FL: CRC Press/Taylor & Francis, 2009). Pg. 134

* Tufte, E., *Envisioning Information*, (Cheshire, CT: Graphics Press, 1990). Pg. 134

* Tufte, E., *Visual Explanations*, (Cheshire, CT: Graphics Press, 1997). Pg. 134

* Tufte, E., *The Visual Display of Quantitative Information*, 2nd ed. (Cheshire, CT: Graphics Press, 2001). Pg. 134

Tufte, E., *The Cognitive Style of PowerPoint®*, (Cheshire, CT: Graphics Press, 2004). Pg. 135

Tufte, E, *Beautiful Evidence*, (Cheshire, CT: Graphics Press, 2006). Pg. 134

Yau, N., *Data Points: Visualization that Means Something*, (Indianapolis, IN: John Wiley & Sons Inc., 2013). Pg. 134

Yau, N., *Visualize This: The Flowingdata Guide to Design, Visualization and Statistics*, (Indianapolis, IN: John Wiley & Sons Inc., 2011). Pg. 134

Websites

BI Scorecard Pg. 133	www.biscorecard.com
Birst Inc. Pg. 134	www.birst.com
Cyient Pg. 135	www.cyient.com
Gartner Inc. Pg. 134	www.gartner.com
Interactive Graphics for Data Analysis Pg. 134	www.interactivegraphics.org
JMP® Software Pg. 135	www.jmp.com
LIONsolver Inc. Pg. 135	www.lionsolver.com, www.grapheur.com
Matillion Ltd. Pg. 134	www.matillion.com
MicroStrategy Inc. Pg. 133	www.microstrategy.com
Mondrian Pg. 135	www.theusrus.de/Mondrian
OSIsoft LLC Pg. 125	www.osisoft.com
Process Plant Computing Ltd. (PPCL) Pg. 135	www.ppcl.com
Qualitin Pg. 134	www.qualitin.com

RapidMiner Pg. 135	www.rapidminer.com
Sabisu Pg. 135	www.sabisu.co
SAS Institute Inc. Pg. 135	www.sas.com
Shoothill Ltd. Pg. 127	www.shoothill.com
Tableau® Software Pg. 134	www.tableausoftware.com
Targit® Pg. 134	www.targit.com
TIBCO™ Software Pg. 134	www.tibco.com
Visokio Pg. 135	www.visokio.com

Additional Material

References

* denotes sources that the author found particularly helpful and/or inspiring.

Board International, *Board BEAM: Overview*, (Chiasso, Switzerland: Board International, 2016).

* Camões, J., *Data at Work: Best Practices for Creating Effective Charts and Information Graphics in Microsoft Excel®*, (San Francisco, CA: New Riders/Peachpit/Pearson Education, 2016).

* Cohen, G. and Kotorov, R., *Organizational Intelligence*, (New York, NY: Information Builders, 2016).

Dean, A, *Mathematics in the Engineering Workplace: An Application*, (Derby, UK: Bombardier Transportation, 2009).

Evergreen, S., *Effective Data Visualization: The Right Chart for the Right Data*, (Thousand Oaks, CA: SAGE Publications, Inc., 2016).

* Nori, S. and Scott, J., *The Definitive Guide to BI and Analytics on a Data Lake*, (Santa Clara, CA: MapR Technologies, Inc., 2017).

Nussbaumer Knaflic, C., *Storytelling with Data: A Data Visualization Guide for Business Professionals*, (Hoboken, NJ: John Wiley & Sons Inc., 2015).

Websites

AtScale Inc.	http://atscale.com
Board International	www.board.com
Excel® Charts	www.excelcharts.com, www.dataatworkbook.com
Grafana	http://grafana.org
Information Builders	www.informationbuilders.com
Initial State Technologies, Inc.	www.initialstate.com
MapR Technologies, Inc.	http://mapr.com
Wia Technologies Ltd.	www.wia.io

CHAPTER 8

More Complex Techniques

Introduction

This chapter introduces some of the more complex methods that can be used for the analysis of many types of asset data, with a particular emphasis on diagnosis. Inevitably, the subject matter is wide-ranging and can be quite difficult to understand, so a full treatment of all the techniques presented can (and has...) filled many books! In the interests of brevity, this chapter can only provide a list of the main ideas. Good explanations and details are in the references given to the excellent and extensive work done by others that the author has found to be useful. It goes without saying that the field is developing all the time, so what is written here should be seen as a snapshot of some of the main methods that are currently available.

Appendix B presents applications of some of these methods to data from a fictitious power system for a Pacific island, which should help the reader put some of these ideas in context.

Figure 8.1 gives a high-level overview of the methods available to the analyst, while Figures 8.2 and 8.3 show several commonly used diagnosis techniques that can be used in many common asset monitoring situations, summarize the knowledge required to make the methods work, and show the strong and weak points of each technique. The analysis techniques discussed in Chapter 3 that are based on mathematical models of assets fall broadly into the boxes in the bottom left of Figure 8.1 (depending on the level of detail in the models), the third column of Figure 8.2, and the third row of Figure 8.3. This chapter looks at some of the other methods mentioned in the three figures.

FIGURE 8.1 General approach to the choice of a diagnostic model (*BS ISO 13379-1:2012*: reproduced with permission).

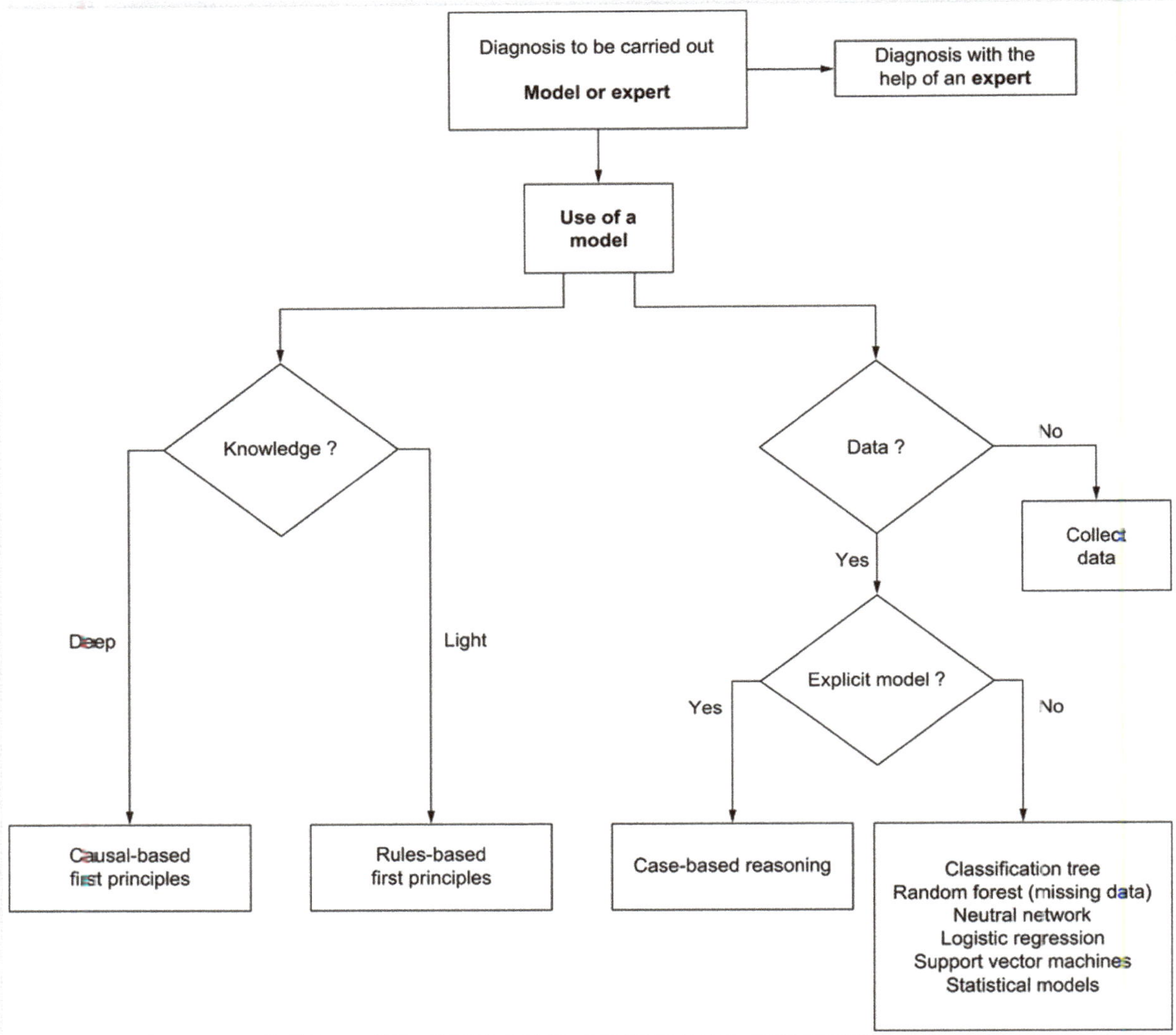

Statistics

Many people find the subject of statistics quite daunting. The following references are ones that the author has found to be particularly helpful and useful.

Derek Rowntree's book (*Rowntree 1981*) is a good introduction to the basics, with a minimum of mathematics. The book by Eddie Martin and John Firth (*Martin and Firth 1983*) is a good short introduction to statistics, while the book by Thomas and Ronald Wonnacott (*Wonnacott and Wonnacott 1984*) covers most of the key points of the subject in a very readable way, using the simplest possible mathematics consistent with sound presentation. The books by Dougal Swinscow (*Swinscow 1983*) and Richard Boddy and Gordon Smith (*Boddy and Smith 2009*) contain short chapters on many of the main methods; the latter book has an associated website with a Microsoft Excel® spreadsheet programmed with several of the tools. The book by Murray Spiegel

FIGURE 8.2 Most commonly used diagnostic models by monitoring technique (*BS ISO 13379-1:2012*: reproduced with permission).

Diagnostic model/ monitoring technique	Knowledge-based			Data-driven						
	Rule-based	Causal fault	First principle	Statistical methods	Case-based reasoning	Neural network	Classification trees	Random forest	Logistic regression	Support vector machines
Vibration	M	D	P	M	D	D	—	D	—	—
Thermography	M	—	—	M	—	D	—	P	—	—
Oil analysis	M	P	—	M	D	D	—	D	D	D
Process parameters	M	—	D	M	M	M	M	M	M	M
Performance	M	—	D	M	M	M	M	M	M	M
Acoustic emission	M	—	—	M	—	D	P	D	—	—
Acoustic monitoring	M	—	—	M	—	D	—	D	—	—
Electrical monitoring	M	—	—	M	—	D	—	—	—	—

M: Mature and commonly applied in industrial applications.

D: Under development and some initial applications.

P: Promising and potential.

(*Spiegel 1972*) also gives an introduction to general statistical principles, with many worked examples to reinforce the points made. More detail behind many of the statistical methods in these books can be found in the books by John Kennedy and Adam Neville (*Kennedy and Neville 1986*) and Susan Milton and Jesse Arnold (*Milton and Arnold 1990*). The course by the author (*Provost 1991*) used material from several of these books: the use of the word "statistics" was deliberately kept back until the very end to make the subject approachable.

There are also many general reference books, ranging from Roger Porkess' statistical dictionary (*Porkess 1988*) to a very detailed work by Thomas Hill and Pawel Lewicki (*Hill and Lewicki 2006*), which is also available online (www.statsoft.com/textbook). Another very comprehensive online resource is the *NIST/SEMATECH e-Handbook of Statistical Methods*, edited by Carroll Croarkin and Paul Tobias (*Croarkin and Tobias 2012*: www.itl.nist.gov/div898/handbook). Allen Downey's book (*Downey 2011*) emphasizes the use of statistics to explore large datasets, taking a computational approach to the problem.

Curve fitting is a valuable tool for modelling the relationships between variables: all the books mentioned above cover this topic. A detailed discussion of best practices can be found in a series of guides from ID Business Solutions (*ID Business Solutions 2008*: www.idbs.com) which helps the reader identify the causes of poor curve fits and use that information to produce more robust results with greater meaning. William Jacoby's paper on LOESS (LOcalized rEgreSSion: *Jacoby 2000*) presents a curve-fitting technique in which more weight is given to points nearer the region of each local curve fit than those further away, creating curve fits that more closely align with data that exhibits complex relationships.

Finally, the book by Darrell Huff (*Huff 1988*) discusses, in a humorous way, the serious subject of how statistical methods can be used to mislead.

FIGURE 8.3 Comparative analysis of diagnostic models (*BS ISO 13379-1:2012*: reproduced with permission).

Diagnostic method	Knowledge used	Strong points	Weak points	Typical applications and references
Rule-based	Human expertise	— Relatively simple to implement	— Incompleteness — Difficulty in explaining multiple faults — Poor explicative capacity — Brittleness to system changes	— Rotating machinery diagnosis — Medical diagnosis
Causal fault	Description of fault mechanism and propagation	— Explicative diagnosis — Handling of multiple independent faults	— Requires good knowledge of possible faults (tested equipment) — Incomplete	— Rotating machinery diagnosis — Medical diagnosis
First principles	Decomposition and transfer function of equipment	— Does not require knowledge of faults (new equipment) — Handles multiple faults well — Gives flexibility to system modification, FMEA, test generation, diagnosis analysis	— Non-explicative diagnosis — Possible aberrant diagnosis — Model complexity in certain domains	— Electronic or fluid circuit diagnosis — Automotive engines and control systems
Statistical **Case-based reasoning**	Samples of significant past diagnosis cases	— Approach well understood — Does not require in-depth knowledge of dysfunctions	— Difficulty in obtaining a sufficient number of significant, well-described cases	— Aeroplane engine diagnosis
Classification trees **Random forests (RFs)** **Logistic regression (LR)** **Neural networks** **Support vector machines (SVMs)**	Samples of significant past diagnosis cases and associated data	— Does not require in-depth knowledge of dysfunctions — RF can accommodate missing data	— Non-explicative diagnosis — Difficulty in obtaining a sufficient number of significant, well-described cases	— Any application

Data Science

An excellent introduction to data science (defined as the collection, preparation, analysis, visualization, management, and preservation of large collections of information) is provided by Jeffrey Stanton's e-book (*Stanton 2013*), while the book by Foster Provost and Tom Fawcett (*Provost and Fawcett 2013*) provides a broad but not overly technical guide to the fundamental principles of data science. Cosma Shalizi's course notes (*Shalizi 2013*) set out to provide an understanding of the range of modern methods of data analysis and the considerations which go into choosing the right method for the job at hand (rather than distorting the problem to fit the methods the reader happens to know): statistical theory is kept to a minimum and mainly introduced as needed. The excellent book and websites curated by Vincent Granville (*Granville 2014*: www.analyticbridge.com, www.datasciencecentral.com) provide rich sources of advice, source code, recipes, rules of thumb, and datasets for the aspiring data scientist. The report by Harlan Harris and others (*Harris et al. 2013*) looks at the psychological characteristics and skillsets of various types of data scientist.

The books by Roberto Battiti and Mauro Brunato (*Battiti and Brunato 2011, 2017*) discuss many simple and complex methods for data modelling, visualization, and analysis in an easy-to-read style. Christian Borgelt's course notes (*Borgelt 2008*) explain the difference between data and knowledge and go into further detail on a number of statistical techniques. The book by Michael Berthold and others (*Berthold et al. 2010*) provides hands-on instruction for many data analysis techniques, including application of the R (see below) and Knime (a free, user-friendly graphical workbench for the entire analysis process, including data access, data transformation, initial investigation, powerful predictive analytics, visualization, and reporting, available from www.knime.org) software packages. The book by Robert Nisbet and others (*Nisbet et al. 2009*) also guides the reader through the stages of data analysis, including discerning the technical and business issues and understanding the strengths and weaknesses of current algorithms. The tutorials in this book use several leading commercially available tools, including SAS® (www.sas.com), Statistica (www.statsoft.com), and SPSS® (www-01.ibm.com/software/uk/analytics/spss). The book by James Wu and Stephen Coggeshall (*Wu and Coggeshall 2012*) presents the fundamental background required for analyzing data and building models for many practical applications. The books by Anasse Bari and others (*Bari et al. 2014*) and Michael Wessler (*Wessler 2014b*) discuss predictive analytics in some detail. Finally, the book by Warren Gilchrist (*Gilchrist 1984*) presents the statistical ideas that can be used to model systems dominated by chance events.

Kevin Roebuck's compendium of articles from the internet (*Roebuck 2012*) brings together a collection of miscellaneous modern analysis techniques and a list of some of the major business intelligence (BI) system vendors, while the book by Michael Wessler (*Wessler 2014a*: www.alteryx.com) discusses techniques for blending data from disparate sources in order to produce meaningful information.

Bayesian Statistics

Bayesian statistics (named after the Reverend Thomas Bayes, who lived from 1701 to 1761) is a set of techniques that use new data to alter a prior belief about a situation, rather than analyze the data without reference to previous knowledge.

Bayes' theorem is summarized in Figure 8.4.

A very useful diagram explaining the derivation of Bayes' theorem is reproduced in Figure 8.5. Figure 8.6 shows an example in the field of cancer diagnosis (where it is seen that a test result is not all it seems), while Figure 8.7 summarizes some other common terms in this field, using the data in Figure 8.6 as an example. Kalman Filtering (Chapter 4) and the Optimal Tracker (Chapter 6) should be thought of as extensions of Bayes' theorem: both methods essentially update prior beliefs when given new evidence.

The book by James Stone (*Stone 2013*) and course notes by Floyd Bullard's (*Bullard 2001*) give good introductions to the subject, as do many of the books mentioned above. Allen Downey's book (*Downey 2013*) goes into more detail, while the books by David Barber (*Barber 2012*), Cameron Davidson-Pilon (*Davidson-Pilon 2013*), and Andrew Gelman and others (*Gelman et al. 2004*) are very comprehensive discussions.

FIGURE 8.4 Bayes' theorem.

$$Pr(H|O) = Pr(O|H) \times Pr(H)/Pr(O)$$

- **Pr(H|O)** = Probability of Hypothesis **H** given Observation **O** (also known as the Posterior Probability of **H**), which represents the updated degree of belief in Hypothesis **H** after Observation **O** is made
- **Pr(O|H)** = Probability of Observation **O** given Hypothesis **H** (also known as the Likelihood of Hypothesis **H**)
- **Pr(H)** = Probability of Hypothesis **H** before Observation **O** is made (also known as the Prior Probability of Hypothesis **H**)
- **Pr(O)** = Probability of Observation **O**, irrespective of any hypotheses. This can be rewritten as **Pr(O|H) × Pr(H) + Pr(O|~H) × Pr(~H)**, where **~H** is 'not Hypothesis **H**'

FIGURE 8.5 Bayes' theorem in pictures (*Cousins 2009*: reproduced with permission).

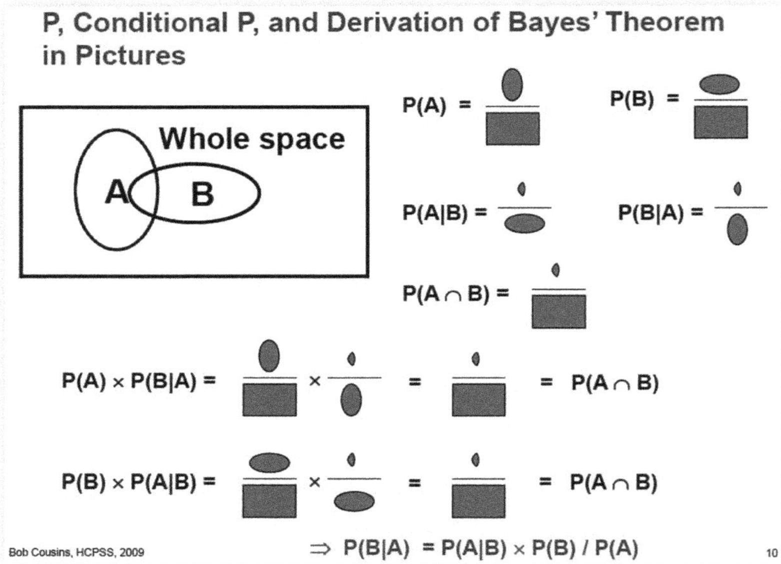

Finally, the article by Joe Marasco and others (*Marasco et al. 2011*) looks at the use of Bayesian statistics in medical diagnosis, showing how medical test results should be treated with care.

R and Microsoft Excel®

The R package for statistical computing and graphics, which is available free of charge from the R Project for Statistical Computing (www.r-project.org), is a very widely used and respected environment for carrying out both simple and complex statistical calculations on both small and large datasets. Many books and online guides are available, such as the documentation from the R Development Core Team (*Venables and Smith 2008*, *R Development Core Team 2009*), an introductory guide by John Verzani (*Verzani 2002*), and the comprehensive book by Michael Crawley (*Crawley 2007*).

FIGURE 8.6 Bayes' theorem and breast cancer.

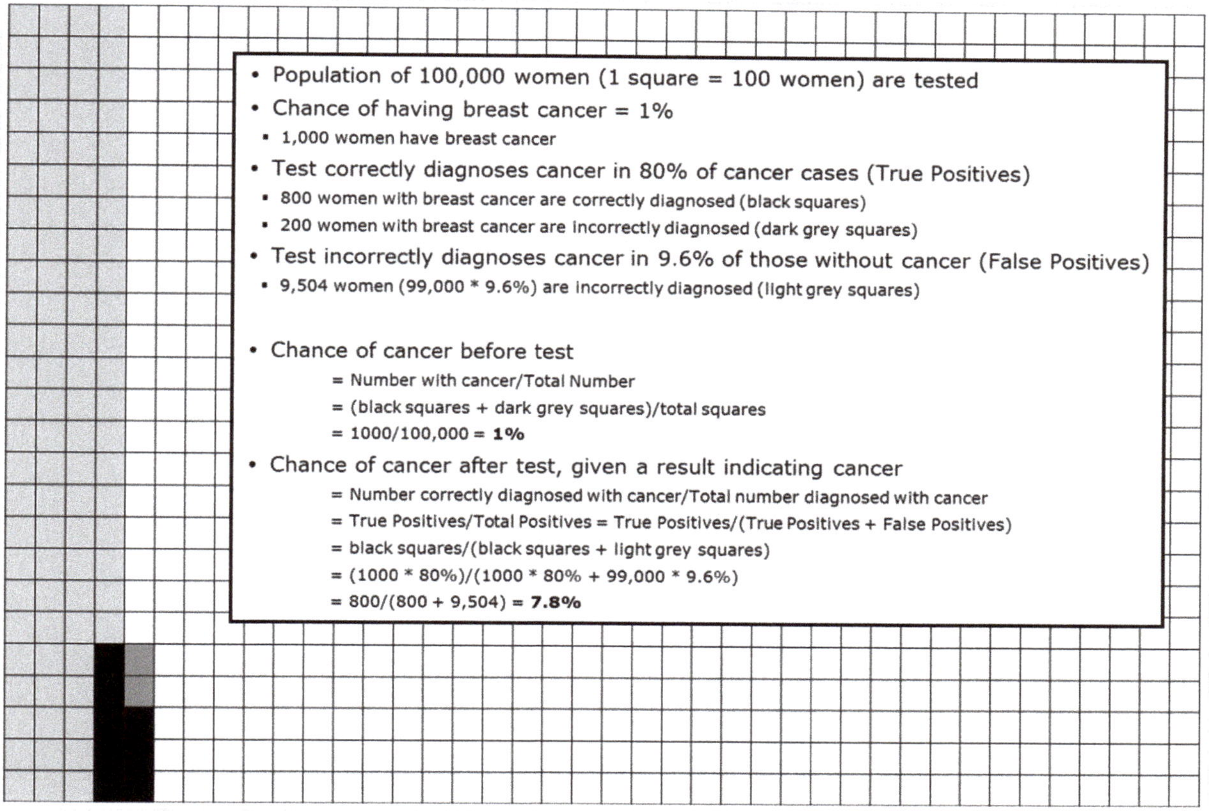

The Microsoft Excel® spreadsheet program is also widely used for a host of statistical analysis tasks. Many add-ins are available to augment Excel®'s native capabilities, for example, DataMinerXL (www.dataminerxl.com).

Principal Component Analysis

Lindsay Smith's tutorial (*Smith 2002*) is an excellent introduction to principal component analysis, a useful statistical technique that has found application in fields such as face recognition and image compression. It is a standard technique for finding patterns in data of high dimension and is related to the concept of observability discussed in Chapter 5. Anahì Balbi's presentation (*Balbi 2008a*) also provides a good overview of this method.

Artificial Intelligence

An *Economist* article (*Economist 2015*) provides a concise summary of the technology, achievements, and current developments in artificial intelligence. The book by Michael Negnevitsky (*Negnevitsky 2002*) provides a good overview of the various techniques, such

FIGURE 8.7 Bayes' theorem summary (condition/test outcome source: Wikipedia).

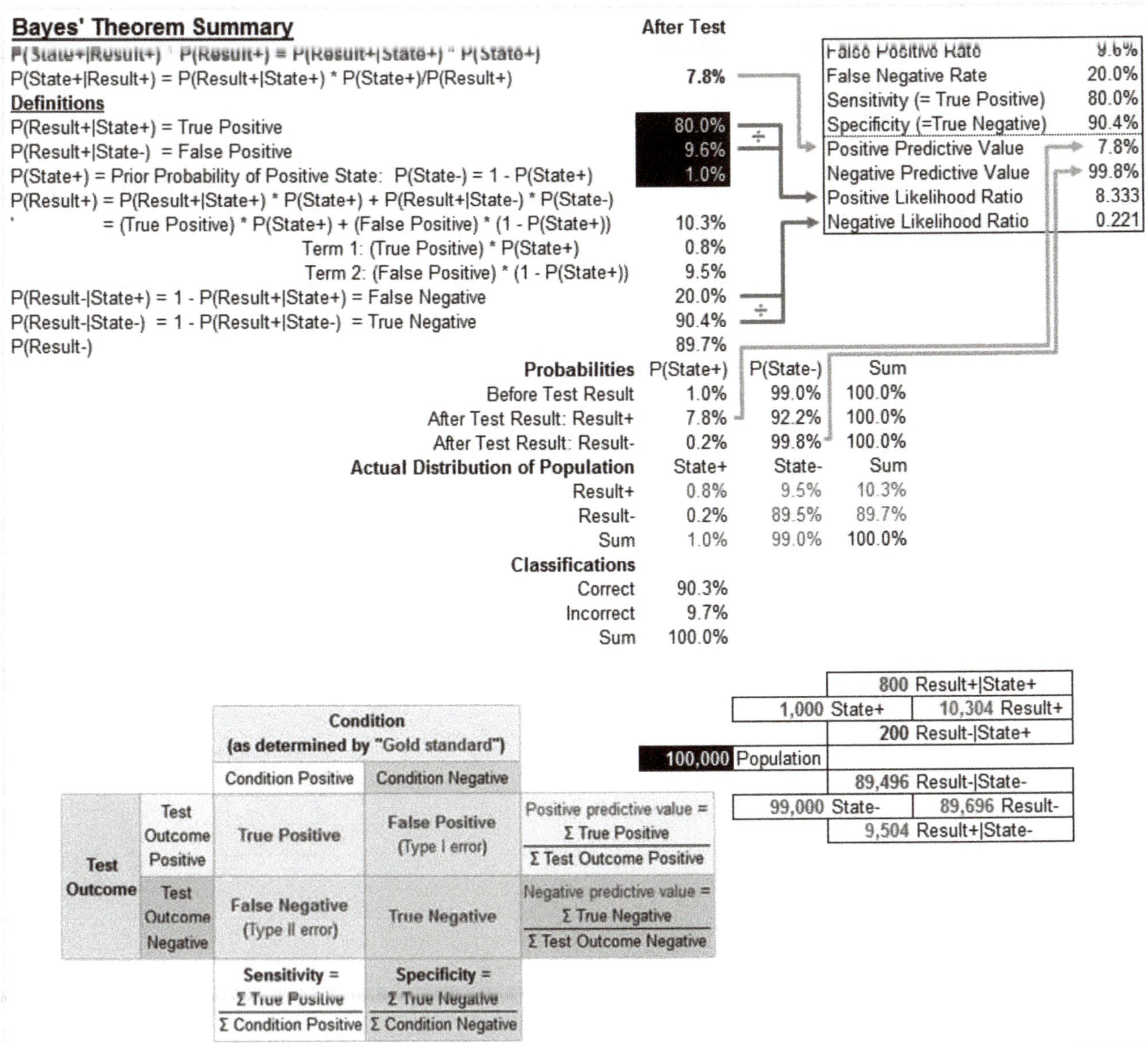

as rule-based expert systems, Bayesian reasoning, fuzzy logic, neural networks, genetic algorithms, and hybrid systems that fall under the generic classification of artificial intelligence, as well as discussing data mining and knowledge engineering.

Fuzzy Logic

Fuzzy logic is a method that enables processing of variables that have a degree of uncertainty associated with them, rather than the precise values more commonly used in computation. Jan Jantzen has written a tutorial on the subject (*Jantzen 1998*) while Timothy Ross' book (*Ross 2010*) provides a complete description of the topic.

Neural Networks

Neural networks mimic the functioning of the way the cells in the human brain connect and are activated, hence the name. Lionel Tarassenko's book (*Tarassenko 1998*) provides an excellent overview of the different types of neural network and their mathematical background, discusses the issues around application of this technology, and includes some case studies. Kevin Gurney's book (*Gurney 1997*) delves more deeply into the mathematics, while Christopher Bishop's book (*Bishop 1995*) is widely regarded as a classic and goes into much more theoretical detail.

Neural network software packages are available from Alyuda Research LLC (www.alyuda.com), NeuroDimension Inc. (www.neurosolutions.com), OLSOFT LLC (who produce Microsoft Excel® add-ins, available from www.neuroxl.com), Palisade Corp. (www.palisade.com), and Neural Planner Software (www.easynn.com, www.justnn.com).

Case-Based Reasoning

Case-based reasoning (CBR) solves new problems by adapting solutions that have been used to solve past problems. The method remembers a former situation similar to the current one and uses that to solve the new problem. The article by Janet Kolodner (*Kolodner 1992*) and presentation by Ralph Bergmann (*Bergmann 2000*) introduce the subject well. More detail can be found in the book by Ian Watson (*Watson 1998*), his course notes (*Watson 2012, 2013a, 2013b*), and a tutorial by Julie Main and others (*Main et al. 2000*).

KEEL®

Knowledge Enhanced Electronic Logic (KEEL®) is a methodology developed by Compsim LLC (www.compsim.com) that allows an analyst to capture complex situations with a dynamic graphical language that can be then packaged into conventional source code. The method is clearly described in an article by Compsim (*Compsim 2003a*) that starts with the reasons a person might choose to smoke, or not to smoke, and uses these to calculate a metric indicating the propensity to continue smoking. The articles by Compsim (*Compsim 2003b*) and Tom Keeley (*Keeley 2004*) discuss the history and logic behind KEEL® in more detail.

Pattern Matching

The paper by Cybula Ltd. (*Cybula 2004*: www.cybula.com) provides an overview of the use of Cybula's patented AURA technology in a wide variety of pattern matching tasks, highlighting the universal nature of the technology. It explains how the AURA process allows simple fusion of information resulting from pattern matching on different data sources. The papers by Jim Austin and others (*Austin et al. 2003, 2005*) and Martyn Fletcher and others (*Fletcher et al. 2004*) describe the Distributed Aircraft Maintenance Environment (DAME) which uses AURA, CBR, and other proprietary tools to provide a grid-based collaborative and interactive workbench that can support remote analysis of vibration and performance data by multiple geographically dispersed users (maintenance engineers, maintenance analysts, and other domain experts). The DAME environment was built around a workflow system and an extensive set of data analysis tools which can provide automated diagnosis for known conditions. Where automated diagnosis is not possible, DAME provides remote experts with a collaborative and interactive diagnosis and analysis environment. The paper by Jim Austin and others (*Austin et al. 2010*) and the presentation by Cybula Ltd. (*Cybula 2011*) describe a further extension of the AURA technology to process multiple parameters gathered simultaneously from the same asset.

As described in the article by Steve Coppock and others (*Coppock et al. 2003*), the SmartSignal similarity-based modelling software (called eCM) uses actual process variable measurements to construct empirical models of equipment operation that capture normal operational behavior. In real-time operation, the model uses actual measurements to generate estimates of expected values for normal operation and then generates residuals (differences between actual and estimated values). Statistically significant residuals imply abnormal deviations that can be linked to expected failure modes. The paper by Stephan Wegerich (*Wegerich 2013*) describes the mathematics behind similarity-based modelling and gives a performance comparison between this and other condition monitoring approaches. The paper by James Herzog and others (*Herzog et al. 2005*) discusses its use to monitor aeroengines, while the paper from SmartSignal (*SmartSignal 2009*) looks at the benefits that result from using the method to monitor power plants. SmartSignal are now part of GE Intelligent Platforms (www.ge-ip.com/products/proficy-smartsignal/p3704).

As described in the article by Aaron Hussey and others (*Hussey et al. 2003*), the SureSense™ software produced by Expert Microsystems (http://expmicrosys.com) produces a parameter estimate through utilization of an advanced pattern recognition technique called the multivariate state estimation technique (MSET) that was developed by the U.S. Argonne National Laboratory. Patterns or relationships among the signals included in a model are learned to define an expected operating space. During online operation, a diagnostic procedure determines whether the signal agrees with the learned model or the equipment is operating outside the learned operating space because of a process anomaly, instrumentation, or equipment problem. The technique has been used to validate sensor readings and monitor equipment operation in aeroengines and power plants, as described in papers by Randolph Bickford, Eddie Davis, Aaron

Hussey, and others (*Bickford and Malloy 2002*; *Bickford et al. 2001*; *Davis et al. 2002*; *Hussey et al. 2010*).

Prognostics

Prognostics is the process of predicting the time at which a component will no longer perform a particular function. This is usually expressed in terms of remaining useful life, which can be either time- or usage-based. The presentation by Fatih Camci (*Camci 2013*) explains the basics of prognostics very well, while the presentation by Lodovico Menozzi (*Menozzi 2013*) goes into more detail. The book by George Vachtsevanos and others (*Vachtsevanos et al. 2006*) is a comprehensive review of diagnostic and prognostic techniques, taking a systems approach, including discussion of sensors and logistics and some case studies. Victoria Catterson's presentation (*Catterson 2012*) discusses the use of linear regression with upper and lower bounds to predict when parameters reach thresholds. The paper by Allan Volponi and others (*Volponi et al. 2004*) discusses the fusion of data from disparate sources to obtain more accurate and comprehensive diagnostic and prognostic information regarding the health of aircraft engines. The article by Moritz von Plate and others (*von Plate et al. 2015*) discusses seven basic steps that should be taken to successfully apply prognostics.

Humaware (www.humaware.com) has developed CFAR-Autotrend, a specialized tool for the analysis of Health and Usage Monitoring System (HUMS) data gathered from helicopters that has a very low false alarm rate. The technology is also applicable to rail, marine, petrochemical, and other assets. The papers by Kenneth Pipe (*Pipe 2006, 2008a, 2008b, 2008c, 2009, 2013*) describe the methodology, features, and functionality of this and related maintenance management software. Two papers by the UK Civil Aviation Authority (CAA) (*CAA 2012a, 2012b*) demonstrate the application of Advanced Anomaly Detection (AAD) methods successfully developed and applied to helicopter HUMS data by General Electric (GE): they show that the techniques used are both effective and practical.

The paper by Andrew Jardine and others (*Jardine et al. 2006*) summarizes and reviews recent research and developments in diagnostics and prognostics of mechanical systems implementing Condition-Based Maintenance (CBM), with emphasis on models, algorithms, and technologies for data processing and maintenance decision-making. The authors also discuss different techniques for multiple sensor data fusion and conclude with a brief discussion on current practices and possible future CBM trends.

The paper by Joanna Sikorska and others (*Sikorska et al. 2011*) discusses business issues that need to be considered when selecting an appropriate Remaining Useful Life (RUL) modelling approach and presents classification tables and process flow diagrams to help select appropriate prognostic models for predicting the RUL of assets within a specific business environment. The paper also explores the strengths and weaknesses of the main prognostics model classes to establish what makes them better suited to certain applications than to others and summarizes how each has been applied.

The Prognostics and Health Management Society (PHM Society: www.phmsociety.org) is a nonprofit organization dedicated to the advancement of PHM as an engineering discipline. As well as producing an online journal (the *International Journal of Prognostics and Health Management* (IJPHM)), the society also runs an annual conference. The society's conference proceedings (*Bregon and Saxena 2012*; *Celaya et al. 2011*; *Roychoudhury et al. 2012*;

Sankararaman and Roychoudhury 2013; Daigle and Bregon 2014; Bregon and Daigle 2014) contain much useful information.

Data Mining

Data mining is the process of discovering insightful, interesting, and novel patterns from large-scale data, as well as deriving descriptive, understandable, and predictive models from it. The report by Graham Williams (*Williams 2005*) provides a concise overview of data mining, including basic concepts and descriptions of applications, techniques, and open-source and commercially available software tools. The book and paper by Xindong Wu and others (*Wu and Kumar 2009*; *Wu et al. 2008*) describe the ten most popular algorithms, with the book devoting a chapter on each one written by an expert in the field. The book by Ian Witten and Eibe Frank (*Witten and Frank 2000*: www.cs.waikato.ac.nz/ml/weka) is a good introduction to the subject, while the document by Pete Chapman and others (*Chapman et al. 2000*) and neatly summarized on a single page by Nicole Leaper (*Leaper 2009*) describes the cross-industry standard process for data mining (CRISP-DM) process model. The paper and presentation by Klaus ten Hagen (*ten Hagen 2011, 2013*) describes an application of data mining to predictive maintenance, while the paper by Rosaria Silipo and Phil Winters (*Silipo and Winters 2013*) describes how the Knime software package was used to mine data on electricity usage. Finally, the book by Mohammed Zaki and Wagner Meira (*Zaki and Meira 2014*) delves into the theoretical detail of the subject.

Python is a general-purpose, high-level programming language available free of charge from the Python Software Foundation (www.python.org/psf) that is widely used in this field. The book by Allen Downey (*Downey 2012*) describes Python and uses it to teach readers how to think like data scientists, while the book by Wes McKinney (*McKinney 2012*) discusses the use of Python for data analysis.

Machine Learning

The article by Dorian Pyle and Cristina San Jose (*Pyle and San Jose 2015*) is an executive's guide to machine learning. Peter Flach's book (*Flach 2012*) is a very accessible but comprehensive discussion of the major machine learning techniques, while the books by Max Welling (*Welling 2010*) and Alex Smola and Vishy Vishwanathan (*Smola and Vishwanathan 2008*) provide more detailed introductions. The book by Trevor Hastie and others (*Hastie et al. 2013*) brings together many of the important new ideas in machine learning and explains them in a statistical framework. While some mathematical details are included, the book emphasizes the methods and their conceptual underpinnings rather than their theoretical properties. A companion volume by Gareth James and others (*James et al. 2013*) takes a broader and less technical view and provides many examples written in R. The book by Ashok Srivastava and Jiawei Han (*Srivastava and Han 2012*) presents state-of-the-art tools for automatically detecting, diagnosing, and predicting adverse events in an engineered system, bringing together the two areas of machine learning and systems health management.

The presentation and thesis by Anahì Balbi (*Balbi 2008b, 2009*) and the paper by her and others (*Balbi et al. 2010*) considers the problem of detecting anomalies in time series data obtained from measuring diesel engine turbocharger exit pressure gathered from a fleet of in-service passenger trains: an automated methodology for labelling time series samples as normal, abnormal, or noisy, then training supervised classifiers with labelled historical data, and finally combining classifiers to filter new data are described. Evidence that this methodology yields error rates comparable to those of equivalent manual processes is presented.

David Mackay's book (*MacKay 2003*) covers many topics in information theory and inference, including data mining, machine learning, pattern recognition, and Bayesian statistics. The book by Arieh Ben-Naim (*Ben-Naim 2012*) is a simple discussion of information theory as a prelude to understanding the concept of entropy.

The research paper by Mercer (*Mercer 2003*) reviews current methods for clustering large quantitative datasets, while the book by Ted Dunning and Ellen Friedman (*Dunning and Friedman 2014*) discusses some new techniques in anomaly detection and classification.

The book by Carl Rasmussen and Christopher Williams (*Rasmussen and Williams 2006*: www.gaussianprocess.org) is an advanced treatise on the application of Gaussian methods to machine learning, while the book by James Stone (*Stone 2015*) is a useful introduction to information theory.

References

* denotes sources that the author found particularly helpful and/or inspiring.

Austin, J., Brewer, G., Jackson, T., and Hodge, V., "AURA-Alert - The Use of Binary Associative Memories for Condition Monitoring Applications," *Proceeding of the 7th International Conference on Condition Monitoring and Machinery Failure Prevention Technologies*, Ettington Chase, Stratford-upon-Avon, UK, June 22-24, 2010 (Red Hook, NY: Curran Associates Inc), 699-711. Pg. 148

Austin, J., Davis, R., Fletcher, M., Jackson, T. et al., "DAME: Searching Large Data Sets within a Grid-Enabled Engineering Application," *Proceedings of the Institute of Electrical and Electronics Engineers* 93, no. 3 (2005): 496-509. Pg. 148

Austin, J., Jackson, T., Fletcher, M., Jessop, M. et al., "Predictive Maintenance: Distributed Aircraft Engine Diagnostics," in: Foster, I. and Kesselman, C. eds., *The GRID 2: Blueprint for a New Computing Infrastructure*, 2nd ed. (San Francisco, CA: Morgan Kaufmann, 2003), 69-80. Pg. 148

* Balbi, A., "Principal Component Analysis," Unpublished presentation, Genoa, Italy: Department of Informatics, Systems and Telematics, University of Genoa, 2008a. Pg. 145

Balbi, A., "Processing Noisy and Irregular Time Series Using Machine Learning," Unpublished presentation, Genoa, Italy: Department of Informatics, Systems and Telematics, University of Genoa, 2008b. Pg. 150

Balbi, A., "Detecting Abnormalities in Noisy and Irregular Time Series: The 'Turbodiesel Charging Pressure' Case Study," Ph.D. thesis, Department of Informatics, Systems and Telematics, University of Genoa, Genoa, Italy, 2009. Pg. 150

Balbi, A., Provost, M., and Tacchella, A., "Anomaly Detection in Noisy and Irregular Time Series: The Turbodiesel Charging Pressure Case Study," *Trends in Applied Intelligent Systems: Lecture Notes in Computer Science*, 6096 (Berlin, Germany: Springer-Verlag, 2010), 123-132. Pg. 150

Barber, D., *Bayesian Reasoning and Machine Learning* (Cambridge, UK: Cambridge University Press, 2012). Pg. 143

Bari, A., Chaouchi, M., and Jung, T., *Predictive Analytics for Dummies* (Hoboken, NJ: John Wiley & Sons Inc, 2014). Pg. 143

* Battiti, R. and Brunato, M., "Reactive Business Intelligence," Trento, Italy: Reactive Search Srl, 2011. Pg. 143

* Battiti, R. and Brunato, M., *The LION Way: Machine Learning Plus Intelligent Optimization – Version 3.0* (Seattle WA: CreateSpace/Amazon.com Inc, 2017). Pg. 143

Ben-Naim, A., *Entropy and the Second Law: Interpretation and Misss-Interpretations* (Singapore: World Scientific Publication Co. Pte. Ltd, 2012). Pg. 151

Bergmann, R., "Introduction to Case-Based Reasoning," Kaiserslautern, Germany: Kaiserslautern University, 2000. Pg. 147

Berthold, M., Borgelt, C., Höppner, F., and Klawonn, F., *Guide to Intelligent Data Analysis: How to Intelligently Make Sense of Real Data* (London, UK: Springer-Verlag London Ltd, 2010). Pg. 143

Bickford, R. and Malloy, D., "Development of a Real-Time Turbine Engine Diagnostic System," *38th AIAA/ASME/SAE/ASEE Joint Propulsion Conference*, Indiana Convention Center & RCA Dome, Indianapolis, IN, July 7-10, 2002 (Reston, VA: American Institute of Aeronautics and Astronautics). Pg. 149

Bickford, R., Meyer, C., and Lee, V., "Online Signal Validation for Assured Data Quality," *47th International Instrumentation Symposium*, Holiday Inn, Denver, CO, May 6-10, 2001 (Research Triangle Park, NC: The Instrument Society of America), 107-110. Pg. 149

Bishop, C., *Neural Networks for Pattern Recognition* (Oxford, UK: Oxford University Press, 1995. Pg. 147

* Boddy, R. and Smith, G., *Statistical Methods in Practice* (Chichester, UK: John Wiley & Sons Ltd, 2009). Pg. 140

Borgelt, C., "Intelligent Data Analysis: Course Notes," Magdeburg, Germany: Magdeburg University, 2008. Pg. 143

Bregon, A. and Daigle, M. eds., *PHME 2014: Proceedings of the European Conference of the Prognostics and Health Management Society 2014*, La Cité, Nantes Events Center, Nantes, France, July 8-10, 2014 (New York: Prognostics and Health Management Society). Pg. 150

Bregon, A. and Saxena, A. eds., *Proceedings of the 1st European Conference of the Prognostics and Health Management Society 2012*, MARITIM Hotel & International Congress Center, Dresden, Germany, July 3-5, 2012 (New York: Prognostics and Health Management Society). Pg. 149

BS ISO 13379-1:2012, "Condition Monitoring and Diagnostics of Machines - Data Interpretation and Diagnostics Techniques - Part 1: General Guidelines," London, UK: British Standards Institution, 2012. Pg. 140, 141, 142

Bullard, F., *A Brief Introduction to Bayesian Statistics* (Durham, NC: The North Carolina School of Science and Mathematics, 2001). Pg. 143

CAA, "CAA Paper 2011/01: Intelligent Management of Helicopter Vibration Health Monitoring Data," London, UK: Civil Aviation Authority, 2012a. Pg. 149

CAA, "CAA Paper 2012/01: The Application of Advanced Anomaly Detection to Tail Rotor HUMS Data," London, UK: Civil Aviation Authority, 2012b. Pg. 149

Camci, F., *Fundamentals of Prognostics* (Cranfield, UK: Cranfield University, 2013). Pg. 149

Catterson, V., "Predicting Asset Behaviour to Plan Maintenance Strategies," *2nd Annual Predictive Asset Management Seminar*, Royal Garden Hotel, Kensington, London, UK, September 11-13, 2012 (London, UK: IBC Energy/Informa UK Ltd). Pg. 149

Celaya, J., Saha, S., and Saxena, A. eds., *Proceedings of the Annual Conference of the Prognostics and Health Management Society 2011*, Hotel Hilton Montréal Bonaventure, Montréal, Québec, September 25-29, 2011 (New York: Prognostics and Health Management Society). Pg. 149

Chapman, P., Clinton, J., Kerber, R., Khabaza, T. et al., *CRISP-DM 1.0 Step-by-Step Data Mining Guide* (Chicago, IL: SPSS® Inc, 2000). Pg. 150

Compsim, "Decisions and Actions in a Knowledge Enhanced Electronic Logic (KEEL®) Domain," Brookfield, WI: Compsim LLC, 2003a. Pg. 147

Compsim, "KEEL® (Knowledge Enhanced Electronic Logic) Explained Using Smoking as an Example," Brookfield, WI: Compsim LLC, 2003b. Pg. 147

Coppock, S., Maner, R., Cassidy, K., and Holtan, T., "Improved Equipment Condition Monitoring Software," Lisle, IL: SmartSignal Corp., 2008. Pg. 148

Cousins, B., "Advanced Statistics for High-Energy Physics," *Fourth CERN-Fermilab Hadron Collider Physics Summer School*, CERN, Geneva, Switzerland, June 8, 2009 (Geneva, Switzerland: CERN (Conseil Européen pour la Recherche Nucléaire)). Pg. 144

Crawley, M., *The R Book* (Chichester, UK: John Wiley & Sons Ltd, 2007). Pg. 144

Croarkin, C. and Tobias, P. eds., *NIST/SEMATECH e-Handbook of Statistical Methods* (Washington DC: Technology Administration, US Department of Commerce, 2012). Pg. 141

Cybula, "AURA Universal Pattern Matching Technology for Multiple Data Types – White Paper, 2nd ed. (York, UK: Cybula Ltd, 2004). Pg. 148

Cybula, "AURA-Alert: The Use of Binary Associative Memories for Condition Monitoring Applications," Unpublished presentation, York, UK: Cybula Ltd, 2011. Pg. 148

Daigle, M. and Bregon, A. eds., *PHM 2014: Proceedings of the Annual Conference of the Prognostics and Health Management Society 2014*, Hilton Fort Worth, Fort Worth, TX, September 28-October 2, 2014 (New York: Prognostics and Health Management Society). Pg. 150

Davidson-Pilon, C., "Probabilistic Programming and Bayesian Methods for Hackers," http://github.com/camdavidsonpilon/probabilistic-programming-and-bayesian-methods-for-hackers/blob/master/pdf/june82013.pdf, 2013. Pg. 143

Davis, E., Bickford, R., Colgan, P., Nesmith, K. et al., "On-Line Monitoring at Nuclear Power Plants – Results from the EPRI On-Line Monitoring Implementation Project," *45th Annual ISA Power Industry (POWID)/12th Annual Joint ISA POWID/EPRI Conference*, Catamaran Resort Hotel, San Diego, CA, June 2-7, 2002 (Research Triangle Park, NC: International Society of Automation). Pg. 148

Downey, A., *Think Stats: Probability and Statistics for Programmers* (Sebastopol, CA: O'Reilly Media, 2011). Pg. 141

Downey, A. *Think Python: How to Think Like a Computer Scientist* (Sebastopol, CA: O'Reilly Media, 2012). Pg. 150

Downey, A., *Think Bayes: Bayesian Statistics in Python* (Sebastopol, CA: O'Reilly Media, 2013). Pg. 143

Dunning, T. and Friedman, E., *Practical Machine Learning - A New Look at Anomaly Detection* (Sebastopol, CA: O'Reilly Media, 2014). Pg. 151

Economist, "Artificial Intelligence: Rise of the Machines," *The Economist*, May 9, 2015. Pg. 145

* Flach, P., *Machine Learning: The Art and Science of Algorithms That Make Sense of Data* (Cambridge, UK: Cambridge University Press, 2012). Pg. 150

Fletcher, M., Austin, J., and Jackson, T., "Distributed Aero-Engine Condition Monitoring and Diagnosis on the Grid: DAME," *17th International Congress & Exhibition on Condition Monitoring and Diagnostic Engineering Management (COMADEM 2004 International)*, Robinson College, Cambridge, UK, August 23-25, 2004 (Birmingham, UK: COMADEM 2004 Secretariat). Pg. 148

Gelman, A., Carlin, J., Stern, H., and Rubin, D., *Bayesian Data Analysis*, 2nd ed. (Boca Raton, FL: Chapman & Hall/CRC, 2004). Pg. 143

Gilchrist, W., *Statistical Modelling* (Chichester, UK: John Wiley & Sons Ltd, 1984). Pg. 143

* Granville, V., *Developing Analytic Talent: Becoming a Data Scientist* (Indianapolis, IN: John Wiley & Sons Inc., 2014). Pg. 143

Gurney, K., *An Introduction to Neural Networks* (London, UK: UCL Press, 1997. Pg. 147

Harris, H., Murphy, S., and Vaisman, M., *Analyzing the Analyzers: An Introspective Survey of Data Scientists and Their Work* (Sebastopol, CA: O'Reilly Media, 2013). Pg. 143

Hastie, T., Tibshirani, R., and Friedman, J., *The Elements of Statistical Learning: Data Mining, Inference and Prediction*, 2nd ed. (New York: Springer-Verlag, 2009). Pg. 150

Herzog, J., Hanlin, J., Wegerich, S., and Wilks, A., "High Performance Condition Monitoring of Aircraft Engines," *ASME Turbo Expo 2005: Power for Land, Sea and Air*, Reno-Tahoe, NV, June 6-9, 2005 (New York: American Society of Mechanical Engineers). Pg. 148

Hill, T. and Lewicki, P., *Statistics: Methods and Applications* (Tulsa, OK: StatSoft Ltd, 2006). Pg. 141

Huff, D., *How to Lie with Statistics* (London, UK: Pelican Books, 1988). Pg. 141

Hussey, A., Hesler, S., and Bickford, R., "Automation of Troubleshooting and Diagnostics of Power Plant Equipment Faults," *53rd Annual ISA POWID Controls and Instrumentation Conference*, JW Marriott, Summerlin, NV, June 6-11, 2010 (Research Triangle Park, NC: International Society of Automation). Pg. 149

Hussey, A., Shankar, R., Davis, E., and Bickford, R., "Automated Equipment Condition Monitoring," *46th Annual POWID Symposium/13th Joint EPRI-ISA Conference*, Woodlands Hotel, Williamsburg, VA, June 15-20, 2003 (Research Triangle Park, NC: International Society of Automation). Pg. 148

ID Business Solutions, "Curve Fitting Best Practice," Guilford, UK: ID Business Solutions Ltd, 2008. Pg. 141

Jacoby, W., "LOESS: A Non-Parametric, Graphical Tool for Depicting Relationships between Variables," *Electoral Studies* 19, no. 4 (2000): 577-613. Pg. 141

* James, G., Witten, D., Hastie, T., and Tibshirani, R., *An Introduction to Statistical Learning: With Applications in R* (New York: Springer-Verlag, 2013). Pg. 150

Jantzen, J., "Tutorial on Fuzzy Logic," Lyngby, Denmark: Department of Automation, Technical University of Denmark, 1998. Pg. 147

Jardine, A., Lin, D., and Banjevic, D., "A Review on Machinery Diagnostics and Prognostics Implementing Condition-Based Maintenance," *Mechanical Systems and Signal Processing* 20, no. 7 (October 2006): 1483-1510. Pg. 149

Keeley, T., "Application of KEEL® Technology in the Area of Computational Mathematics - Whitepaper," Brookfield, WI: Compsim LLC, 2004. Pg. 148

Kennedy, J. and Neville, A., *Basic Statistical Methods for Engineers and Scientists*, 3rd ed. (New York: Harper & Row, 1986). Pg. 141

Kolodner, J., "An Introduction to Case-Based Reasoning," *Artificial Intelligence Review* 6, no. 1 (1992): 3-34. [Pg. 147]

Leaper, N., "A Visual Guide to CRISP-DM Methodology," Portland, OR: www.nicoleleaper.com, 2009. [Pg. 150]

MacKay, D., *Information Theory, Inference and Learning Algorithms* (Cambridge, UK: Cambridge University Press, 2003). [Pg. 151]

Main, J., Dillon, T., and Shiu, S., "A Tutorial on Case-Based Reasoning, in: Pal, S., Dillon, T., and Yeung, D. eds., *Soft Computing in Case Based Reasoning* (London, UK: Springer-Verlag London Ltd, 2000), 1-28. [Pg. 147]

Marasco, J., Doerfler, R., and Roschier, L., "Doc, What Are My Chances?," *The UMAP Journal* 32, no. 4, (2011): 279-298. [Pg. 144]

Martin, E. and Firth, J., *Core Business Studies: Statistics* (London, UK: Mitchell Beasley, 1983). [Pg. 140]

McKinney, W., *Python for Data Analysis: Agile Tools for Real-World Data* (Sebastopol, CA: O'Reilly Media, 2012). [Pg. 150]

Menozzi, L., "The Future of Condition Monitoring: Adding Prognostics to Your Applications with NI Tools," Austin, TX: National Instruments Corp, 2013. [Pg. 149]

Mercer, D., *Clustering Large Datasets* (Oxford, UK: Linacre College, Oxford University, 2003). [Pg. 151]

Milton, J. and Arnold, J., *Introduction to Probability and Statistics*, 2nd ed. (Singapore: McGraw-Hill, 1990). [Pg. 141]

Negnevitsky, M., *Artificial Intelligence: A Guide to Intelligent Systems* (Harlow, UK: Addison-Wesley/Pearson, 2002). [Pg. 145]

Nisbet, R., Elder, J., and Miner, G., *Handbook of Statistical Analysis and Data Mining Applications* (Amsterdam, the Netherlands: Academic Press/Elsevier BV, 2009). [Pg. 143]

Pipe, K., "CFAR-Autotrend Technology Description," Petersfield, UK: Humaware, 2006. [Pg. 149]

Pipe, K., "CFAR-Autotrend Features & Capabilities," Petersfield, UK: Humaware, 2008a. [Pg. 149]

Pipe, K., "CFAR-Autotrend Functional Description," Petersfield, UK: Humaware, 2008b. [Pg. 149]

Pipe, K., "Maintenance Demand Forecaster Data Sheet," Petersfield, UK: Humaware, 2008c. [Pg. 149]

Pipe, K., "Robust Alert Generation for Non-Stationary PHM Data," *Annual Conference of the Prognostics and Health Management Society 2009*, Hilton San Diego Resort and Spa, San Diego, CA, September 27-October 1, 2009 (New York: Prognostics and Health Management Society). [Pg. 149]

Pipe, K., "Practical Prognostics: Maximising the Maintenance Benefits of CBM," *4th IMarEST Condition Based Maintenance Conference*, London, UK, September 25-26, 2013 (Petersfield, UK: Humaware). [Pg. 149]

Porkess, R., *Dictionary of Statistics* (Glasgow, UK: Collins Reference, 1988). [Pg. 141]

* Provost, F. and Fawcett, T., *Data Science for Business* (Sebastopol, CA: O'Reilly Media, 2013). [Pg. 142]

Provost, M., "Data Handling for Performance Engineers, or Ways of Coping with Variability and Uncertainty in Data," Internal presentation, Derby, UK: Rolls-Royce plc, 1991. [Pg. 141]

Pyle, D. and San Jose, C., *An Executive's Guide to Machine Learning* (New York: McKinsey & Company, 2015). Pg. 150

R Development Core Team, *R: A Language and Environment for Statistical Computing*, 10th ed. (Vienna, Austria: R Foundation for Statistical Computing, Vienna University of Economics and Business, 2009). Pg. 144

Rasmussen, C. and Williams, C., *Gaussian Processes for Machine Learning* (Cambridge, MA: MIT Press, 2006). Pg. 151

Roebuck, K. ed., *Predictive Analysis: High-Impact Emerging Technology - What You Need to Know: Definitions, Adoptions, Impact, Benefits, Maturity, Vendors* (Marston Gate, UK: Amazon.com, 2012). Pg. 143

Ross, T., *Fuzzy Logic with Engineering Applications*, 3rd ed. (Chichester, UK: John Wiley & Sons Ltd, 2010). Pg. 147

Rowntree, D., *Statistics without Tears: A Primer for Non-Mathematicians* (London, UK: Pelican Books, 1981). Pg. 140

Roychoudhury, I., Celaya, J., and Saxena, A. eds., *Proceedings of the Annual Conference of the Prognostics and Health Management Society 2012*, Hotel Regency Minneapolis, Minneapolis, MN, September 23-27, 2012 (New York: Prognostics and Health Management Society). Pg. 149

Sankararaman, S. and Roychoudhury, I. eds., *PHM 2013: Proceedings of the Annual Conference of the Prognostics and Health Management Society 2013*, New Orleans Downtown Convention Center, New Orleans, LA, October 14-17, 2013 (New York: Prognostics and Health Management Society). Pg. 150

Shalizi, C., *Advanced Data Analysis from an Elementary Point of View* (Pittsburgh, PA: Carnegie Mellon, 2013). Pg. 142

Sikorska, J., Hodkiewicz, M., and Ma, L., "Prognostic Modelling Options for Remaining Useful Life Estimation by Industry," *Mechanical Systems and Signal Processing* 25, no. 5 (July 2011): 1803-1836. Pg. 149

Silipo, R. and Winters, P., "Big Data, Smart Energy and Predictive Analytics: Time Series Prediction of Smart Energy Data," Zürich, Switzerland: KNIME.com AG, 2013. Pg. 150

SmartSignal, "The Value of Predictive Analytics for Fossil-Fuel Power Plants - White Paper," Lisle, IL: SmartSignal Corp, 2009. Pg. 148

Smith, L., "A Tutorial on Principal Component Analysis," Dunedin, New Zealand: Otago University, 2002. Pg. 145

Smola, A. and Vishwanathan, S., *Introduction to Machine Learning* (Cambridge, UK: Cambridge University Press, 2008). Pg. 150

Spiegel, M., *Schaum's Outline Series: Theory and Problems of Statistics*, SI edition (New York: McGraw-Hill, 1972). Pg. 141

Srivastava, A. and Han, J. eds., *Machine Learning and Knowledge Discovery for Engineering Systems Health Management* (Boca Raton, FL: CRC Press/Taylor & Francis, 2012). Pg. 150

Stanton, J., *Introduction to Data Science*, 2nd ed. (Syracuse, NY: Syracuse University, 2013). Pg. 142

Stone, J., *Bayes Rule: A Tutorial Introduction to Bayesian Analysis* (Berlin, Germany: Sebtel Press, 2013). Pg. 143

Stone, J., *Information Theory: A Tutorial Introduction* (Berlin, Germany: Sebtel Press, 2015). Pg. 151

Swinscow, T., *Statistics at Square One*, 8th ed. (London, UK: British Medical Association, 1983). Pg. 140

Tarassenko, L., *A Guide to Neural Computing Applications* (London, UK: Arnold, 1998). Pg. 147

ten Hagen, K., "Predictive Maintenance Using Data Mining," Dresden, Germany: Decision Optimisation GmbH, 2011. Pg. 150

ten Hagen, K., "Data Mining to Reduce Scheduled and Unscheduled Maintenance Visits," Dresden, Germany: Decision Optimisation GmbH, 2013. Pg. 150

Vachtsevanos, G., Lewis, F., Roemer, M., Hess, A. et al., *Intelligent Fault Diagnosis and Prognosis for Engineering Systems: Methods and Case Studies* (Hoboken, NJ: John Wiley & Sons Inc, 2006). Pg. 149

Venables, W. and Smith, D., *An Introduction to R* (Vienna, Austria: Vienna University of Economics, 2008). Pg. 144

Verzani, J., "SimpleR - Using R for Introductory Statistics," New York: Department of Mathematics, College of Staten Island, 2002. Pg. 144

Volponi, A., Brotherton, T., Luppold, R., and Simon, D., "Development of an Information Fusion System for Engine Diagnostics and Health Management," Cleveland, OH: Glenn Research Center, National Aeronautics & Space Administration, 2004. Pg. 149

von Plate, M., Kirschnick, F., and Heggemann, J., "Get Ready for Forecasting! The 7 Steps for Success with Prognostics," *Uptime* 10, no. 69 (February/March 2015): 48-50. Pg. 149

Watson, I., *Applying Case-Based Reasoning: Techniques for Enterprise Systems* (San Francisco, CA: Morgan Kaufmann Publishers Inc, 1998). Pg. 147

Watson, I., "How to Build a CBR System," Auckland, New Zealand: Auckland University, 2012. Pg. 147

Watson, I., "Case-Based Reasoning Course Notes," Auckland, New Zealand: Auckland University, 2013a. Pg. 147

Watson, I., "Case-Based Reasoning Gentle Intro," Auckland, New Zealand: Auckland University, 2013b. Pg. 147

Wegerich, S., "Condition-Based Monitoring Using Nonparametric Similarity-Based Modeling," Lisle, IL: SmartSignal Corp, 2013. Pg. 148

Welling, M., *A First Encounter with Machine Learning* (Irvine, CA: University of California, 2010). Pg. 150

Wessler, M., *Data Blending for Dummies*, Alteryx Special Edition (Hoboken, NJ: John Wiley & Sons Inc, 2014a). Pg. 143

Wessler, M., *Predictive Analytics for Dummies*, Alteryx Special Edition (Hoboken, NJ: John Wiley & Sons Inc, 2014b). Pg. 143

Williams, G., "Data Mining Desktop Survival Guide," Canberra, Australia: Togaware Pty Ltd, 2005. Pg. 150

Witten, I. and Frank, E., *Data Mining: Practical Machine Learning Tools and Techniques with Java Applications* (San Francisco, CA: Morgan Kaufmann, 2000). Pg. 150

* Wonnacott, T. and Wonnacott, R., *Introductory Statistics for Business and Economics*, 3rd ed. (New York: John Wiley & Sons Inc, 1984). Pg. 140

Wu, J. and Coggeshall, S., *Foundations of Predictive Analytics* (Boca Raton, FL: CRC Press/Taylor & Francis, 2012). Pg. 143

Wu, X. and Kumar, V. *The Top Ten Algorithms in Data Mining* (Boca Raton, FL: CRC Press, 2009). Pg. 150

Wu, X., Kumar, V., Quinian, J., Gosh, J. et al., "Top 10 Algorithms in Data Mining," *Knowledge and Information Systems* 14, no. 1 (2008): 1-37. Pg. 150

Zaki, M. and Meira, W., *Data Mining and Analysis: Fundamental Concepts and Algorithms* (Cambridge, UK: Cambridge University Press, 2014). Pg. 150

Websites

Alteryx Pg. 143	www.alteryx.com
Alyuda Research LLC Pg. 147	www.alyuda.com
Analytic Bridge Pg. 143	www.analyticbridge.com
Compsim LLC (KEEL®) Pg. 147	www.compsim.com
Cybula Ltd. Pg. 148	www.cybula.com
Data Science Central Pg. 143	www.datasciencecentral.com
DataMinerXL Pg. 145	www.dataminerxl.com
Expert Microsystems Pg. 148	http://expmicrosys.com
Gaussian Processes Pg. 151	www.gaussianprocess.org
GE SmartSignal Pg. 148	www.ge-ip.com/products/proficy-smartsignal/p3704
Humaware Pg. 149	www.humaware.com
IBM (SPSS®) Pg. 143	www-01.ibm.com/software/uk/analytics/spss
ID Business Solutions Ltd. Pg. 141	www.idbs.com
Konstanz Information Miner (Knime) Pg. 143	www.knime.org
LION Book Pg. 143	http://intelligent-optimization.org/lionbook
Neural Planner Software Pg. 147	www.easynn.com, www.justnn.com
NeuroDimension Inc. Pg. 147	www.neurosolutions.com
NIST/SEMATECH e-Handbook of Statistical Methods Pg. 141	www.itl.nist.gov/div898/handbook
OLSOFT LLC (NeuroXL) Pg. 147	www.neuroxl.com
Palisade Corp. (@Risk®, Decision Tools suite) Pg. 147	www.palisade.com
Prognostics and Health Management (PHM) Society Pg. 149	www.phmsociety.org
Python Software Foundation Pg. 150	www.python.org/psf
R Project for Statistical Computing Pg. 144	www.r-project.org
SAS Institute Inc. Pg. 143	www.sas.com
StatSoft Ltd. (Statistica) Pg. 141	www.statsoft.com
WEKA 3 Data Mining software Pg. 150	www.cs.waikato.ac.nz/ml/weka

Additional Material

References

* denotes sources that the author found particularly helpful and/or inspiring.

Bland, M., *An Introduction to Medical Statistics*, 4th ed. Oxford, UK: Oxford University Press, 2015.

Blum, A., Hopcroft, J., and Kannan, R., *Foundations of Data Science* (Ithaca, NY: Department of Computer Science, Cornell University, 2016).

Booz Allen Hamilton, *The Field Guide to Data Science* (McLean, VA: Booz Allen Hamilton Inc, 2015).

Bughin, J., Hazan, E., Ramaswamy, S., Chui, M. et al., *Artificial Intelligence: The Next Digital Frontier?* (New York: McKinsey & Company, 2017).

Campbell, M., Machin, D. and Walters, S., *Medical Statistics: A Textbook for the Health Sciences*. Chichester, UK: John Wiley & Sons Ltd., 2007.

Castledine, J., *A Kill Switch for the Random Actor* (Derby, UK: University of Derby, 2018).

Cukierski, W. and Hamner, B., *Just the Basics: Core Data Science Skills* (San Francisco, CA: Kaggle Inc., 2013).

D'Eon, P., "Meeting the Challenge of Troubleshooting Complex Aircraft Systems," Farnborough International Airshow Official Show Catalogue 2016, Farnborough, UK: Farnborough International Ltd, 2016), 90-91.

Daigle, M. and Bregon, A. eds., *PHM 2015: Proceedings of the Annual Conference of the PHM Society 2015*, Coronado Island Marriott, Coronado, CA, October 18-24, 2015 (New York: Prognostics and Health Management Society).

Daigle, M. and Bregon, A. eds., *PHM 2016: Proceedings of the Annual Conference of the PHM Society 2016*, Denver Embassy Suites, Downtown Conference Center, Denver, CO, October 3-6, 2016 (New York: Prognostics and Health Management Society).

Dibsdale, C., "Introduction of Predictive Technology," Coventry, UK: Warwick Manufacturing Group, Warwick University, 2012.

Diggle, P., "Statistics: A Data Science for the 21st Century," *Journal of the Royal Statistical Society Series A: Statistics in Society* 178, Part 4 (2015): 793-813.

Eballard, I. and Bregon, A. eds., *PHME 2016: Proceedings of the 3rd European Conference of the PHM Society 2016*, Euskalduna Conference Center, Bilbao, Spain, July 5-8, 2016 (New York: Prognostics and Health Management Society).

Efron, B. and Hastie, T., *Computer Age Statistical Inference: Algorithms, Inference and Data Science* (Cambridge, UK: Cambridge University Press, 2016).

Foslien, W., Guralnik, V., and Zita Haigh, K., "Data Mining for Space Applications," *SpaceOPS 2004 Conference*, Centre Mont Royal, Montréal, Québec, Canada, May 17-21, 2004 (Reston, VA: American Institute of Aeronautics and Astronautics).

Ganas, S., "Data Mining and Predictive Modeling with Excel® 2007," *Casualty Actuarial Society Forum*, Spring 2010, 1-17.

Hammer, Ø. and Harper, D., *Paleontological Data Analysis* (New Delhi, India: Wiley India Pvt. Ltd, 2014).

Henke, N., Bughin, J., Chui, M., Manyika, J. et al., "The Age of Analytics: Competing in a Data-Driven World - Executive Summary," New York: McKinsey & Company, 2016a.

Henke, N., Bughin, J., Chui, M., Manyika, J. et al., "The Age of Analytics: Competing in a Data-Driven World – Full Report," New York: McKinsey & Company, 2016b.

ISPredict, "Predictive Intelligence: Reduce Costs by Self-Learning Predictive Analytics Solution – Flyer," Saarbrücken, Germany: ISPredict GmbH, 2015.

Kuhn, M. and Johnson, K., *Applied Predictive Modeling* (New York: Springer Science+Business Media LLC, 2013).

Langmore, I. and Krasner, D., "Applied Data Science," New York: Columbia University, 2013.

Linoff, G., *Data Analysis Using SQL and Excel®* (Indianapolis, IN: Wiley Publishing Inc, 2008).

MacFarlane, D. and Parlikad, A., "Big and Clever: A System-Wide Approach to Data Analytics," in: Stirling, W. ed., *UK Manufacturing Review 2015-2016* (Teddington, UK: Stirling Media Ltd, 2016), 90.

* Matthews, R., *Chancing It: The Laws of Chance and How They Can Work for You* (London, UK: Profile Books, 2016).

Morris, D., *Bayes Theorem: A Visual Introduction for Beginners*, Kindle edition (City of Legazpi, Philippines: Blue Windmill Media, 2016).

Morrison, D., Foslien, W., MacArthur, W., and Jofriet, P., "The Early Event Detection Toolkit: White Paper," Phoenix, AZ: Honeywell Process Solutions/Honeywell Inc, 2006.

Nisbett, R., *Mindware: Tools for Smart Thinking* (London, UK: Penguin Books/Allen Lane, 2015).

Peng, R., *R Programming for Data Science* (Raleigh, NC: Lulu Press Inc, 2015).

* Russell, S. and Norvig, P., *Artificial Intelligence: A Modern Approach* (Harlow, UK: Pearson Education Ltd, 2010).

* Savage, S., *The Flaw of Averages: Why We Underestimate Risk in the Face of Uncertainty* (Hoboken, NJ: John Wiley & Sons Inc, 2012).

Shmueli, G., Patel, N., and Bruce, P., "Data Mining in Excel®: Lecture Notes and Cases (Draft)," Arlington, VA: Resampling Stats Inc, 2005.

Stigler, S., *Seven Pillars of Statistical Wisdom* (Cambridge, MA: Harvard University Press, 2016).

Strickland, J., *Data Science and Analytics for Ordinary People* (Raleigh, NC: Lulu Press Inc, 2014a).

Strickland, J., *Predictive Analytics Using R* (Raleigh, NC: Lulu Press Inc, 2014b).

Strickland, J., *Data Analytics Using Open-Source Tools* (Raleigh, NC: Lulu Press Inc, 2016).

Tessella, *Five Rules to Deliver Value from Data Analytics: An Executives Guide* (Abingdon, UK: Tessella plc, 2016).

Tessella, *Maximizing Value from AI: The Digital Transformers Guide* (Abingdon, UK: Tessella plc, 2017).

* Tetlock, P. and Gardner, D., *Superforecasting* (London, UK: Random House Books, 2015).

* Vose, D., *Risk Analysis: A Quantitative Guide*, 3rd ed. (Chichester, UK: John Wiley & Sons Ltd, 2008).

Vose, D., "Probability and Uncertainty," Sint-Amandsberg, Belgium: Vose Software, 2018a.

Vose, D., "The Perplexing Math of Uncertainty," Sint-Amandsberg, Belgium: Vose Software, 2018c.

Winters, P., Adae, I., and Silipo, R., "Anomaly Detection in Predictive Maintenance: Anomaly Detection with Time Series Analysis," Zürich, Switzerland: KNIME.com AG, 2014.

Zita Haigh, K., Foslien, W., and Guralnik, V., "Visual Query Language - Finding Patterns and Relationships among Time Series Data," *Seventh Workshop on Mining Scientific and Engineering Datasets, SIAM International Conference on Data Mining*, Lake Buena Vista, FL, April 24, 2004 (Philadelphia, PA: Society for Industrial and Applied Mathematics (SIAM)).

Websites

Booz Allen Hamilton Inc.	www.boozallen.com
Flaw of Averages	www.flawofaverages.com
Humalytica Analytics	www.humalytica.com
IS Predict GmbH	www.ispredict.com
Kaggle	www.kaggle.com
MapR Technologies, Inc.	http://mapr.com
Nutonian, Inc. (Eureqa®)	www.nutonian.com
PAST Software	http://folk.uio.no/ohammer/past
Royal Statistical Society	www.rss.org.uk
Society for Industrial and Applied Mathematics	www.siam.org
Tiberius Data Mining	www.tiberius.biz
Trusted Analytics Platform	http://01.org/tap
Vose Software	www.vosesoftware.com

CHAPTER

9

Data Gathering and Software Architectures

CHAPTER 9

Introduction

This chapter gives a very high-level overview of data gathering and software architectures for asset management, starting with how asset management systems fit with the other systems that are required to run a business. A discussion of data gathering strategy is followed by reviews of some current ideas and developments in machine connectivity and knowledge management and some of the software that can be used for gathering and managing asset data and information. A short discussion on big data concludes the chapter.

Software Schematics

Figure 9.1 shows a typical "jigsaw" of the systems required to run an organization. All of these are required in some form by any business and can be implemented with anything from pencil, paper, and pocket calculators and Microsoft Excel® spreadsheets to fully integrated Enterprise Resource Planning (ERP) systems, such as those marketed by SAP (www.sap.com), IFS (www.ifsworld.com), Infor (www.infor.com), and others. Asset management and asset information systems are small parts of this whole, as shown in the bottom right of Figure 9.1.

Figure 9.2 shows the flow of typical asset management information, with the asset life cycle on the left flowing from top to bottom and the supporting asset management functions and IT infrastructure (including Computerized Maintenance Management Systems (CMMS) such as IBM's Maximo® (www.ibm.com/software/products/en/maximoassetmanagement), service management systems such as those marketed by the Zafire group (www.zafire.com), and ERP systems on the right. CMMS systems are discussed in more detail in Chapter 10. Figure 9.3 shows a typical asset data flow schematic, illustrating

FIGURE 9.1 System implementation components and dependencies (Decision Evaluation Ltd. (www.decisionevaluation.co.uk), under contract to Intelligent Energy Ltd.: reproduced with permission).

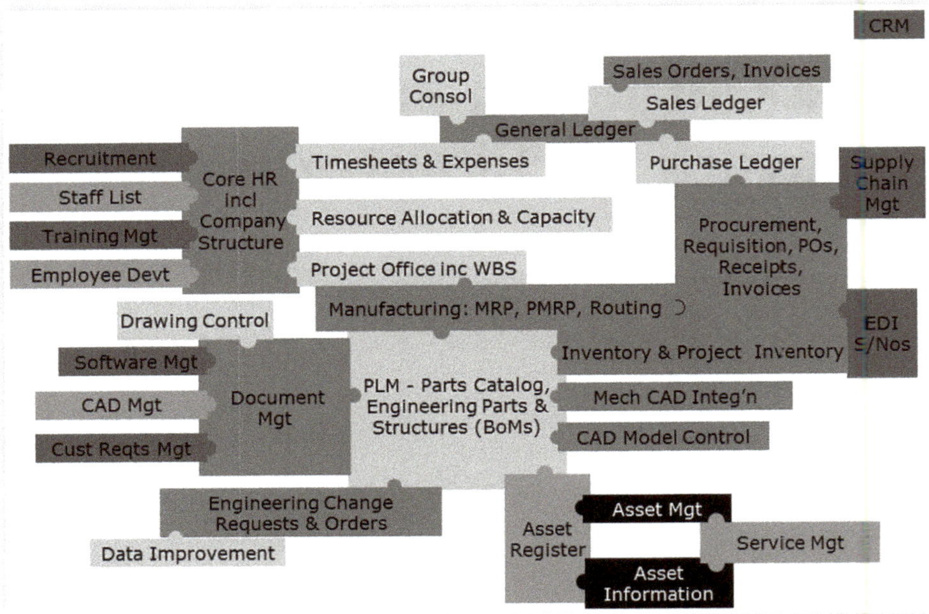

FIGURE 9.2 Asset management information flow schematic (Intelligent Energy Ltd.: reproduced with permission).

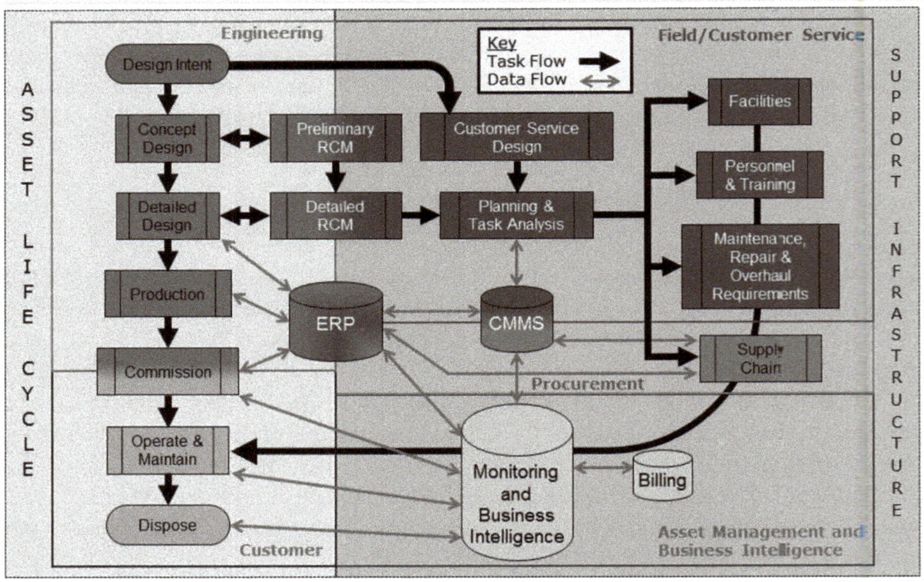

FIGURE 9.3 Asset data and information flow schematic (Intelligent Energy Ltd.: reproduced with permission).

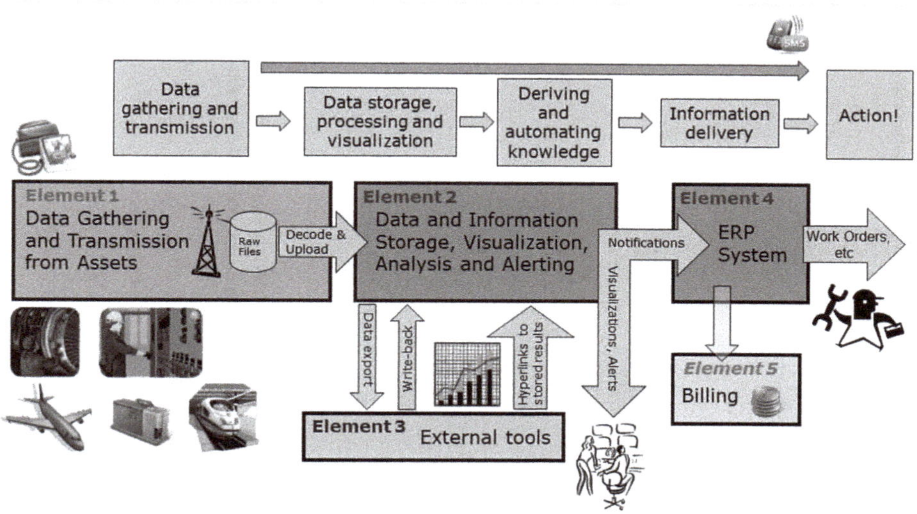

the systems needed to convert data transmitted from assets in the field into actionable information. Data processing and visualization have been discussed in Chapters 4, 6, and 7.

Data Gathering

Figure 9.4 shows a template that can be used to develop a rational data gathering and transmittal strategy from in-service assets. Table 1 in Figure 9.4 prompts for the business case and number of assets expected to be deployed to be summarized, which scopes the problem. Table 2 in Figure 9.4 prompts for summary specifications of the parameters, data gathering frequencies, data transfer mechanisms, and on- and off-board data processing that need to be produced for each of the five typical data usage scenarios:

- Billing the customer for asset usage and performance.
- Operations support.
- Maintenance support.
- Engineering data used to gain detailed understanding of asset in-service performance.
- Other data, as required to need business and customer needs.

Each of these data usage scenarios can run to a different "heartbeat," with different numbers and types of parameters, different internal and external data consumers, possibly different external system integration requirements, different requirements for data delivery (from real-time data transmittal via secure remote wireless communication links to retrieval of bulk data from a USB memory stick that can be removed during a site visit), and different asset data sources (e.g., the asset's control system, dedicated data gathering hardware, third-party certified meter, etc.). Provision may have to be made for the asset to gather other data from its environment, or any surrounding equipment it interacts with. Table 3 in Figure 9.4 prompts for the costs and benefits of asset data gathering and processing to be summarized.

FIGURE 9.4 Data gathering and transmittal requirements capture template (Intelligent Energy Ltd.: reproduced with permission).

Table 1

Summary of Business Case for collecting and using Asset Data (include views on how this may evolve over time)					
No of Assets	Year 1:	Year 2:	Year 3:	Year 5:	Year 10:

Table 2

Data Usage	Parameter List	Data Gathering Frequency	Data Transfer Mechanism	On-Board Processing Needs	Off-Board Processing Needs
Billing					
Operations					
Maintenance					
Engineering					
Other					

Table 3

Costs (include views on how these evolve over time)	Benefits (include views on how these evolve over time)

Books and Papers

The short but excellent book by Max Shron (*Shron 2014*) should be read by anyone concerned with turning data into information. The book discusses the four parts of a data project (context, need, vision, and outcome) and explains using simple case studies why blindly applying analytical routines to datasets rarely solves real analytical problems. The important point is that there is a need to understand *why* an analysis needs to be done before getting to grips with *how* to do it. The article by Robert DiStefano and Stephen Thomas (*DiStefano and Thomas 2011*) looks at the issues surrounding data integrity and claims that finding information that can be trusted is increasingly difficult despite the availability of more data. The book by Robert Plant and Stephen Murrell (*Plant and Murrell 2007*) is a compendium of technological terms written for the nontechnical executive, reviewing important aspects of IT from a business perspective, and pointing out advantages, disadvantages, and business value propositions in a succinct way.

Machine to Machine (M2M)

Machine to Machine (M2M) refers to technologies that allow both wireless and wired systems to communicate with other devices of the same type: it is a key element of the industrial internet. Axeda Corp. has produced a simple guide to M2M (*Axeda 2012*), examples of Internet Of Things (IoT) value propositions and a Return On Investment (ROI) model for building a business case and tracking results related to IoT initiatives (*Axeda 2014a*) and a vision of the machine of the future (*Axeda 2014b*), which will be serviceable, trackable, informative, self-healing, and integrated both with other machines and with business systems.

The book by David Bailey and Edwin Wright (*Bailey and Wright 2003*) and the books by Steve Mackay and others (*Park and Mackay 2003*; *Park et al. 2003*; *Reynders et al. 2005*) go into considerable detail on the practicalities of M2M, while the article by Hany Fouda (*Fouda 2010*) discusses how integrating wireless instrumentation with Supervisory Control And Data Acquisition (SCADA) systems can drive operational efficiency and reduce deployment costs. National Instruments (www.ni.com) produces a huge range of data gathering hardware and software, using a graphical system design approach that leverages productive software and reconfigurable hardware platforms along with a community of developers and applications to simplify system development and arrive at solutions faster.

One key set of standards is Open Productivity and Connectivity (OPC®), promoted by the OPC® Foundation (www.opcfoundation.org) which is dedicated to ensuring interoperability in automation by creating and maintaining open specifications that standardize the communication of acquired process data, alarm and event records, historical data, and batch data to multi-vendor enterprise systems and between production devices. Darek Kominek and Randy Kondor have written good overviews of OPC® (*Kominek 2009*; *Kondor 2010*: www.matrikonopc.com) while the book by Wolfgang Mahnke and others (*Mahnke et al. 2009*) goes into much more detail.

The Weightless™ Special Interest Group (www.weightless.org) is promoting a new M2M standard that makes use of the "white space" that is appearing in the radio frequency spectrum following the digitization of TV signals. William Webb's book and paper (*Webb 2012a, 2012b*) describe the background and technology.

The ZigBee Alliance (www.zigbee.org) is promoting a short-range M2M technology for use in buildings (including smart energy, lighting, remote control, and home automation), healthcare, telecoms, and retail. Their presentation (*ZigBee 2009*) and the presentation by Jay Hendrix and Jim Kohl (*Hendrix and Kohl 2009*) provide an overview, while the paper by Anshul Agarwal and others (*Agarwal et al. 2013*) goes into more detail. David Webster has edited a comprehensive ZigBee resource guide (*Webster 2013*) which is regularly updated.

Finally, MIMOSA (www.mimosa.org) is a not-for-profit trade association dedicated to developing and encouraging the adoption of open information standards for operations and maintenance in manufacturing, fleet, and facility environments. MIMOSA's open standards enable collaborative asset life cycle management in both commercial and military applications.

The Industrial Internet and the IoT

One of the key enablers of much of what has been discussed so far is the easy availability of copious quantities of raw data that can be transmitted from assets virtually anywhere. The papers by Peter Evans and Marco Annunziata (*Evans and Annunziata 2012; Annunziata and Evans 2013*: www.ge-ip.com) and the booklet by Jon Bruner (*Bruner 2013*) discuss the profound transformation to global industry that this will produce by connecting more intelligent machines, advanced analytics, and people at work. This deeper meshing of the digital world with the world of machines has the potential to bring enormous economic benefits, estimated to be as much as $10-15 trillion of global GDP over the next 20 years. Examples from the medical, transportation, and energy fields are used to illustrate the points made. The e-book by Tony Paine of Kepware (*Paine 2015*: www.kepware.com) explores the industrial internet of things (IIoT), describes the benefits of internet-enabling all hardware and software components that comprise an automation system, and delves into the challenges the industry must overcome for the IIoT to be successful, while the presentation by Andreas Schroeder (*Schroeder 2015*) reviews the technical, organizational, and strategic issues of the IoT and manufacturing services.

The article by Ben Sampson (*Sampson 2015*), the book by Daniel Kellmereit and Daniel Obodovski (*Kellmereit and Obodovski 2013*: http://thesilentintelligence.com), and the book by Samuel Greengard (*Greengard 2015*) are excellent layperson's guides to the IoT, while the book by Michael Marcovici (*Marcovici 2014*) is a collection of articles addressing many IoT issues in more detail. The paper by Tomás Sánchez López and others (*Sánchez López et al. 2012*) looks at some of the potential benefits from the IoT, including improved management of global supply chain logistics, product counterfeit detection, manufacturing automation, smart homes and appliances, e-government, improved integrated vehicle health management, and e-health: it also examines the technologies that will be fundamental for realizing the IoT concept and proposes an architecture that integrates them into a single platform. The presentation by Alicia Asín and David Gascón (*Asín and Gascón 2013*: www.libelium.com) lists some of the potential applications of smart sensor technologies.

The case for an industrial big data platform is made in a paper by GE (*GE Software 2013*) while the impact on businesses is discussed more fully in the paper by Clint Witchalls and James Chambers (*Witchalls and Chambers 2013*). The paper by James Watson and Jason Sumner (*Watson and Sumner 2012*), the articles by Michael Porter and James Heppelmann (*Porter and Heppelmann 2014, 2015*), and the article by Marco Iansiti and Karim Lakhani (*Iansiti and Lakhani 2014*) give good overviews of IoT technology and its potential business and competitive impacts, while the reports by James Manyika and others (*Manyika et al. 2015a, 2015b, 2015c*: www.mckinsey.com) go into more detail on the business opportunities that IoT technologies offer.

The paper by Oxford Economics (*Oxford Economics 2014*: www.oxfordeconomics.com) gives the results of a survey conducted with 300 manufacturing executives worldwide, which shows that the smart connected products revolution is well under way but remains in its early stages. Manufacturers are still rethinking their products, services, and processes for the new era and most of the gains anticipated remain "up for grabs." The white papers by Harbor Research (*Harbor Research 2015a, 2015b*: www.harborresearch.com) discuss smart systems and services growth opportunities and describe a comprehensive road map for technology and systems development for the IoT. The paper by ThingWorx (*ThingWorx 2011*: www.thingworx.com) discusses the development of their connected systems platform, while the article by Control Design (*Control Design 2015*: www.controldesign.com) looks at how machine builders of all sizes can improve fleet efficiency, accelerate product development, reduce costs, and improve customer satisfaction and profitability by plugging into the IoT.

A survey by Vitria Technology Inc. (*Vitria 2015*: www.vitria.com) concludes that maintenance and monitoring are top business needs and that real-time analytics and streaming analytics are becoming mainstream key strategic investment areas for IoT initiatives. A paper by Datawatch Corp. (*Datawatch 2014*: www.datawatch.com) reviews the use of streaming data analytics and visual data discovery to enable both operators and businesses to take appropriate action faster and make better decisions quicker.

The presentation by Peter Niblett and Gari Singh (*Niblett and Singh 2014*) gives another overview of the IoT and introduces a number of "starter kits" based around the IBM IoT Quickstart program and ARM mbed (http://mbed.org), Raspberry PI (www.raspberrypi.org), and other platforms. The Hub of All Things (HAT) project (*Ng 2013*: www.hubofallthings.org) uses the IoT to create a repository of personal data that will encourage personal data ownership and engineer a market for personal data exchange, allowing individuals to view and understand their behaviors and trade their personal data for future personalized products and services.

The paper by BloomReach (*BloomReach 2015*: http://bloomreach.com) traces the growing importance of machines in solving business problems and the need to resist being blinded by the power of computing: it provides examples of solutions that combine the best of humans and machines and suggests guidelines for deciding how to balance human skills and machine capabilities to best tackle tasks. Finally, the article by Tereza Pultarova (*Pultarova 2015*) points out that current IoT technologies have basic security flaws that need to be addressed before they can be adopted widely, citing several worrying examples of simple "hacks."

Knowledge Management

Knowledge management is a multidisciplined approach to the process of capturing, developing, sharing, and effectively using organizational knowledge and enabling achievement of organizational objectives by making best use of knowledge which, by its very nature, resides mainly in the minds of managers and employees and is therefore ephemeral: it walks out of the door at the end of every working day. The book by Thomas Davenport and Laurence Prusak (*Davenport and Prusak 2000*) provides an excellent overview of the subject and its importance, while the presentation by Charlie Dibsdale (*Dibsdale 2015*) describes the differences between explicit and tacit knowledge and the problems of encoding the latter. An excellent diagram by Andy Harrison (*Harrison 2015*) gives Rolls-Royce's view of knowledge management.

The books by Melissie Clemmons Rumizen (*Clemmons Rumizen 2002*) and Wendi Bukowitz and Ruth Williams (*Bukowitz and Williams 1999*) give good introductions to knowledge management strategies and implementation, including cultural, IT, and sustainability issues: the latter book contains a number of checklists and case studies. The books by Kenneth and Jane Laudon (*Laudon and Laudon 2010*) and Paul Bocij and others (*Bocij et al. 2008*) are comprehensive descriptions of all aspects of the technology, deployment, and management of business information systems, including numerous case studies. The book by George Siemens (*Siemens 2006*: www.learning-theories.com/connectivism-siemens-downes) and the series of articles by Stephen Downes (*Downes 2012*) describe types of learning and a "connectivist" view of knowledge management, in which knowledge is seen as a flow of information around networks of computers and individuals rather than merely a set of static repositories.

A major issue with the use of software systems is the faith that the users have (or do not have) in what is being presented to them and their ability to check underlying data and information reliability and trustworthiness. The STRAPP project (*Marshall et al. 2011*; *Townend et al. 2013*; *Venters 2014*) developed, prototyped, and evaluated innovative uses of provenance data to enhance the decision-making process: the overall goal was to deliver a generic framework for building provenance-based, personalized trusted digital spaces for timely and confident decision-making, together with a demonstrator system that supports decision-making in a secure, dependable, and personalized way.

The Rizolva toolkit being produced by Steve Magraw (*Magraw 2014, 2015*) will make high-quality, fully researched, referenced, and accredited knowledge available that meets the latest government competency standards by providing a business problem-solving portal that will aid organizational learning, development, and productivity where and when needed.

Data Historians

Data historians are designed for the acquisition and display of production and process data. They are built around the display of time series data and focus on answering questions and helping with decisions that companies typically need to address in real time. Typical users operate chemical and energy process plant, although their use is expanding beyond these fields. The paper by GE (*GE Intelligent Platforms 2012*) looks at the advantages of such systems over standard relational databases, while the presentations by John Daniels (*Daniels 2012*) and OSIsoft (*OSIsoft 2010*: www.osisoft.com) give an overview of the PI System™, widely regarded as one of the most comprehensive and capable data historians currently available. The PI System™ has been in existence for nearly four decades, has been installed in over 15,000 locations in over 110 countries, and is used by most major manufacturing and process companies. It has the ability to:

- Gather very large numbers of analogue and event data parameters as frequently as required in real time from many sources using almost all standard industry protocols.
- Store and archive the data securely using the latest virtualized and cloud-based IT technologies, controlling access to it and compressing it if needed to increase storage and retrieval efficiency.
- Organize the data in a logical asset hierarchy, making it easy to understand and retrieve.

- Distribute the data to other applications using standard protocols.
- Analyze the data using complex event processing and user-defined logic.
- Deliver customized alerts and notifications via a number of routes, with escalation if necessary.
- Display the data in a number of user-definable ways, customizable as required to meet the needs of operations, engineering, and business analysts.

Other producers of data historians include GE (with their Proficy™ historian: www.ge-ip.com/ products/proficy-historian/p2420), Matrikon Inc. (part of Honeywell Inc.: www.matrikon.com), Wonderware UK (part of the Invensys Group: www.wonderware.com), Rockwell Automation (www.rockwellautomation.com, whose FactoryTalk® services platform (*Rockwell Automation 2014*) is based on OSIsoft's PI System™), Canary Labs Inc. (www.canarylabs.com), and M.A.C. Solutions (UK) Ltd. (www.mac-solutions.co.uk).

The CoNVerge platform (www.nvable.com/introducting-converge) produced by NVable Ltd. (www.nvable.com) is designed to receive manual data through mobile applications in addition to stream data. Both sets of data can be analyzed together to provide business insight and act as a data integrator for other business systems. CoNVerge is being used in the aviation sector and is currently being trialled with a merchant navy operator. GE have also opened up their recently released Predix™ platform to other users (*GE Software 2015*, www.ge.com/digital/predix).

Big Data

Big data is currently the subject of much discussion and "hype," as both new and established companies fight for attention from a market still trying to get to grips with the concept. The article by Krish Satiah (*Satiah 2013*) discusses the issue that, while ever-increasing amounts of data are being generated and accumulated in businesses today, it is generally acknowledged that this does not necessarily result in an improvement in reliable information on which to base good decisions: in many cases the opposite is happening and finding information that can be trusted is becoming increasingly difficult. The book and presentation by Thomas Davenport (*Davenport 2014a, 2014b, 2014c*) look at the myths surrounding big data and the opportunities it brings, while the excellent book by Jules Berman (*Berman 2013*) explains how big data is characterized by volume, velocity, and variety and cuts through the mystique to give a very balanced view of what can and cannot be achieved with very large datasets and their associated analytical tools. The book also emphasizes the need for data quality and integrity, as well as discussing societal and legal issues. The presentation by Laura Colvine (*Colvine 2012*) summarizes many of the issues surrounding big data.

The book by Viktor Mayer-Schönberger and Kenneth Cukier (*Mayer-Schönberger and Cukier 2013*) is an excellent layperson's guide to the subject, while Nate Silver's popular book (*Silver 2012*) looks at the issues surrounding prediction and the problems of distinguishing signals from noise in large datasets. Jeff Zakrzewski's presentation (*Zakrzewski 2012*) provides a wide-ranging overview of the advent of big data and its technologies, vendors, and potential uses. The book by Judith Hurwitz and others (*Hurwitz et al. 2013*) goes into some detail on big data technologies, management, and implementation, while the series of books published by O'Reilly Media (*O'Reilly 2011, 2012*; *Webb and O'Brien 2014*; www.oreilly.com) discuss the latest trends in this subject. The presentation by Eugen Molnár and others (*Molnár et al. 2014*) discusses the application of big data to servitization, including social media data, text analytics, and speech analysis.

Stephen Few's very perceptive article (*Few 2013*: www.perceptualedge.com) takes a skeptic's view of big data, arguing that vendors focus on database sizes and technologies and ignore the fact that the value of data lies in the way it is used and understood. Indeed, the availability of too much data may actually be counterproductive. Kate Crawford's article (*Crawford 2013*) points out that big data may be subject to hidden biases, which are either not apparent or assumed to be absent purely because of the size of the database. Cathy O'Neil's booklet (*O'Neil 2013*) offers some observations as to why big data has been hyped up so much and argues that data skepticism, not cynicism, is good for both creativity and business.

IBM (www.ibm.com) is one of the leaders in big data analytical and database technologies. The paper by David Kiron and others (*Kiron et al. 2011*) discusses the growing divide between companies that see the value of business analytics and are transforming themselves to take advantage of these newfound opportunities and those that have yet to embrace them, while Steve LaValle and others (*LaValle et al. 2010*) surveyed a global sample of nearly, 3,000 managers and analysts and interviewed academic and subject matter experts to determine recommendations on how organizations can bolster their analytics capabilities to achieve long-term advantage. The paper by Michael Schroeck and others (*Schroeck et al. 2012*) looks at how organizations are beginning to use such technologies for business advantage, while an IBM presentation looks at how their Watson supercomputer can analyze big data to benefit healthcare, financial, and government services (*IBM 2011*). The book by Paul Zikopoulos and others (*Zikopoulos et al. 2013*) discusses the IBM big data platform in some detail. Microsoft Azure (www.microsoft.com) is a cloud computing platform for building, deploying, and managing analysis applications and services through a global network of hosted data centers: it supports many different programming languages, tools, and frameworks.

Wipro (www.wipro.com) have also produced a number of interesting business and technical articles on the subject (*Jain and Pallia 2011*; *Pallia and Prabhu 2011*; *Lawnin and Yurkanin 2011*; *Thakkar 2011*). Mike Loukides looks at the advent of data science in his booklet (*Loukides 2010*), which argues that companies and people need to turn data into products in order to prosper, while the booklet by Patil (*Patil 2011*) looks at the roles and characteristics of data scientists and the paper by Jim Giles and Gilda Stahl (*Giles and Stahl 2013*) explains how a company can create a data-driven culture. The paper by Philipp Hartmann and others (*Hartmann et al. 2014*) reports on a study which provides a series of pointers that may be particularly helpful to companies already leveraging "big data" for their businesses or planning to do so. It presents a data-driven business model framework that represents a basis for the analysis and clustering of business models.

The book by Bruce Ratner (*Ratner 2012*) contains essays offering detailed background, discussion, and illustration of specific methods for solving the most commonly experienced problems in predictive modelling and analysis of big data, while the book by Allan Izenman (*Izenman 2008*) discusses the current state of multivariate statistical analysis in an age of high-speed computation and large datasets, mixing new algorithmic techniques for analyzing large multivariate datasets with some of the more classical multivariate methods. The presentation by Stuart Gillen (*Gillen 2013*) looks at the use of big data in condition monitoring. The book by Jaap Bloem and others (*Bloem et al. 2012*) sheds light on the big data phenomenon, providing many suggestions for how a company can determine its specific big data potential to gain insight in what exactly makes its customers tick: the topics of privacy, technology, and the law are also covered. The documentary by Liz Bonnin and others (*Bonnin et al. 2014*) gives a layperson's view of the uses of big data, from condition monitoring of Rolls-Royce aeroengines and pattern matching of brain scan data by Cybula Ltd. to the privacy issues raised by the use of data by retailers, search engines, and others.

The infographic by Microsoft (*Microsoft 2015*: www.microsoft.com) and "cheat sheet" by Steve Cassidy (*Cassidy 2014*) give high-level overviews of predictive analytics and the impact of big data on business, while the excellent book and case studies by Bernard Marr (*Marr 2015a, 2015b*: www.ap-institute.com), the publications by Raconteur (*Fox-Brewster et al. 2014*; *Davis et al. 2015*: www.raconteur.net), and the reports by James Manyika and others (*Manyika et al. 2011a, 2011b*: www.mckinsey.com) go into more detail. Two articles by Nicholas Clarke (*Clarke 2014a, 2014b*: www.tessella.com) discuss some of the issues that can arise if big data techniques are thought about or used inappropriately, while another article on analytics maturity (*Clarke 2015*) details the approaches that must be taken to use big data techniques effectively. Three high-level presentations by Datameer (*Datameer 2014a, 2014b, 2014c*: www.datameer.com) detail the benefits of using big data databases and techniques in business, while an excellent booklet by the same company (*Datameer 2015*) discusses what to look for when selecting a big data analytics solution. The paper by Eileen McNulty-Holmes and others of Dataconomy (*McNulty-Holmes et al. 2014*: http://dataconomy.com) discusses the development of big data and the associated database and analytical techniques that are available, as well as giving a few case studies, while the biography by Charles Morgan (*Morgan 2015*) is an interesting history of early business developments written by one of the pioneers. A paper by the Economist Intelligence Unit/HSBC Bank (*EIU/HSBC 2015*: www.eiu.com) looks at how companies are using cloud computing to reinvent business models.

The book by John Kelly and Steve Hamm (*Kelly and Hamm 2013*) introduces the concept of "cognitive systems" and describes the work being done by IBM and others to create machines that sense, learn, reason, and interact with people in new ways to provide insight and advice.

The problems of keeping data that is redundant, obsolete, or trivial (including legal, compliance, and security risks) and possible software solutions and mitigation strategies are discussed in presentations by Adele Carboni and Amir Jaibaji (*Carboni and Jaibaji 2014*), IBM Corp. (*IBM 2014*), and Viewpointe Archive Services LLC (*Viewpointe 2014*). The dangers of big data are eloquently highlighted in a novel by Michelle Miller (*Miller 2015*), a blog entry by Kieran Healy (*Healy 2013*: http://kieranhealy.org), and the book by Edward Lucas (*Lucas 2015*).

CHAPTER 9

References

* denotes sources that the author found particularly helpful and/or inspiring.

Agarwal, A., Agarwal, M., Vyas, M., and Sharma, R., "A Study of ZigBee® Technology," *IJRITCC Journal* 1, no. 4 (April 2013): 287-292. Pg. 168

* Annunziata, M. and Evans, P., *The Industrial Internet@Work* (Fairfield, CT: General Electric Company, 2013). Pg. 168

Asín, A. and Gascón, D., *50 Sensor Applications for a Smarter World* (Zaragoza, Spain: Libelium Comunicaciones Distribuidas SL, 2013). Pg. 168

Axeda, *M2M 101: The Basics of Machine-to-Machine Communication* (Foxboro, MA: Axeda Corp., 2012). Pg. 167

Axeda, *The Business Case for Internet of Things Initiatives* (Foxboro, MA: Axeda Corp., 2014a). Pg. 167

Axeda, *The Machine of the Future: An Executive's Guide* (Foxboro, MA: Axeda Corp., 2014b). Pg. 167

Bailey, D. and Wright, E., *Practical SCADA for Industry* (Oxford, UK: Newnes/Elsevier/IDC Technologies, 2003). Pg. 167

* Berman, J., *Principles of Big Data: Preparing, Sharing and Analyzing Complex Information* (San Francisco, CA: Morgan Kaufmann, 2013). Pg. 171

Bloem, J., van Doorn, M., Duivestein, S., van Manen, T. et al., *No More Secrets with Big Data Analytics* (Vianen, the Netherlands: Sogeti Trend Lab VINT, 2013). Pg. 172

* BloomReach, *Why Can't We Be Friends? The Case for Man + Machine* (Mountain View, CA: BloomReach Inc., 2015). Pg. 169

Bocij, P., Greasley, A., and Hickie, S., *Business Information Systems: Technology, Development and Management for the E-Business*, 4th ed. (Harlow, UK: Pearson Education Ltd, 2008). Pg. 170

Bonnin, L., Philbin, M., and Stansfield, J. (Presenters), "Bang Goes the Theory: Big Data," BBC1 Television, UK, March 24, 2014. Pg. 172

Bruner, J., *Industrial Internet: The Machines Are Talking* (Sebastopol, CA: O'Reilly Media, 2013). Pg. 168

Bukowitz, W. and Williams, R., *The Knowledge Management Fieldbook* (Harlow, UK: Pearson Education Ltd, 1999). Pg. 170

Carboni, A. and Jaibaji, A., "Legacy Data Clean up and Sharepoint Governance," *IBM Information on Demand 2013*, Mandalay Bay Resort, Las Vegas, NV, November 3-7, 2014 (Armonk, NY: IBM Corp.). Pg. 173

Cassidy, S., "Big Data Cheat Sheet," *PC Pro*, Issue 242, 108, December 2014. Pg. 173

Clarke, N., *Braving the Analytics Minefield* (Abingdon, UK: Tessella plc, 2014a). Pg. 173

Clarke, N., *Subtleties under the Surface of Analytics* (Abingdon, UK: Tessella plc, 2014b). Pg. 173

Clarke, N., *Target Analytics Maturity to Beat the Big Data Backlash* (Abingdon, UK: Tessella plc, 2015). Pg. 173

Clemmons Rumizen, M., *The Complete Idiot's Guide to Knowledge Management* (Indianapolis, IN: Alpha Books, 2002). Pg. 170

Colvine, L., *Cloud and Big Data—Unlocking the Potential* (Portsmouth, UK: IBM Corp., 2012). Pg. 171

Control Design, *The OEM's Guide to Integrating Remote Monitoring* (Schaumburg, IL: Control Design for Machine Builders/Putnam Media, 2015). Pg. 169

Crawford, K., *The Hidden Biases in Big Data* (Watertown, MA: Harvard Business Review, 2013). Pg. 172

* Daniels, J., "PI System™ Overview," Unpublished presentation, San Leandro, CA: OSIsoft LLC, 2012. Pg. 170

Datameer, *Three Reasons Your Data Analytics Strategy Must Include Hadoop* (San Francisco, CA: Datameer Inc., 2014a). Pg. 173

Datameer, *Top Five High-Impact Use Cases for Big Data Analytics* (San Francisco, CA: Datameer Inc., 2014b). Pg. 173

Datameer, *Why Big Data Analytics?* (San Francisco, CA: Datameer Inc., 2014c). Pg. 173

* Datameer, *Big Data Analytics Buyer's Guide* (San Francisco, CA: Datameer Inc., 2015). Pg. 173

Datawatch, "Industrial Analytics Powered by the Internet of Things: The Next Wave of Business Transformation," Issue 2, August 2014, Chelmsford, MA: Datawatch Corp. Pg. 170

Davenport, T., *Big Data @ Work: Dispelling the Myths, Uncovering the Opportunities* (Boston, MA: Harvard Business School Press, 2014a). Pg. 171

Davenport, T., "Big Data at Work: Dispelling the Myths, Uncovering the Opportunities," Harvard Business School webinar notes, March 3, 2014b. Pg. 171

Davenport, T., "Big Data at Work: Dispelling the Myths, Uncovering the Opportunities," Harvard Business School webinar, March 3, 2014c. Pg. 171

Davenport, T. and Prusak, L., *Working Knowledge: How Organizations Manage What They Know*, 2nd ed. (Boston, MA: Harvard Business School Press, 2000). Pg. 169

Davis, H., du Preez, D., King, L., Matthews, D. et al., *The Data Economy* (London, UK: Raconteur/CIMdata/Times Newspapers Ltd, June 23, 2015). Pg. 173

Dibsdale, C., "More Effective Knowledge Sharing for Experts Who Identify Insights from Data," Unpublished Ph.D. proposal presentation, Cranfield University, Cranfield, UK, 2015a. Pg. 169

DiStefano, R. and Thomas, S., "Business Case for Data Integrity," *Uptime* 6, no. 46 (April/May 2011): 13-19. Pg. 167

Downes, S., *Connectivism and Connected Knowledge* (Ottawa, Ontario: National Research Council Canada, 2012). Pg. 170

EIU/HSBC, *Harnessing Cloud Technology: How Multinationals Are Using the Cloud to Reinvent Business Models* (London, UK: Economist Intelligence Unit Ltd, 2015). Pg. 173

* Evans, P. and Annunziata, M., *Industrial Internet: Pushing the Boundaries of Minds and Machines* (Fairfield, CT: General Electric Company, 2012). Pg. 168

* Few, S., *Big Data, Big Ruse* (Berkeley CA: Perceptual Edge, 2013. Pg. 172

Fouda, H., "Improving SCADA Operations Using Wireless Instrumentation," *Maintenance & Engineering* 10, no. 6 (November/December 2010): 45-48. Pg. 167

Fox-Brewster, T., du Preez, D., Fildes, N., Lamb, J. et al., *The Data Economy* (London, UK: Raconteur/CIMdata/Times Newspapers Ltd, September 9, 2014). Pg. 173

GE Intelligent Platforms, "The Advantages of Enterprise Historians vs. Relational Databases: Comparing Two Approaches for Data Collection and Optimized Process Operations," Charlottesville, VA: GE Intelligent Platforms Inc., 2012. Pg. 170

GE Software, *The Case for an Industrial Big Data Platform: Laying the Groundwork for the New Industrial Age* (San Ramon, CA: GE Software, 2013). Pg. 168

GE Software, *Your Cloud Platform for the Industrial Internet* (San Ramon, CA: GE Software, 2015). Pg. 171

Giles, J. and Stahl, G., *Fostering a Data-Driven Culture* (London, UK: Economist Intelligence Unit Ltd, 2013). Pg. 172

Gillen, S., *Managing Big Data* (Austin, TX: National Instruments Corp., 2013). Pg. 172

* Greengard, S., *The Internet of Things* (Cambridge, MA: MIT Press, 2015). Pg. 168

Harbor Research, *Smart Systems and Services Growth Opportunities* (Boulder, CO: Harbor Research, 2015a). Pg. 169

Harbor Research, *Smart Systems Manifesto: Road Map for the Internet of Things* (Boulder, CO: Harbor Research, 2015b). Pg. 169

Harrison, A., "Service Data and Knowledge," *Service Community Birmingham Special Event*, Aston Business School, Birmingham, UK, September 30, 2015 (Derby, UK: Rolls-Royce plc). Pg. 170

Hartmann, P., Zaki, M., Feldmann, N., and Neely, A. *Big Data for Big Business—A Taxonomy of Data-Driven Business Models Used by Start-Up Firms* (Cambridge, UK: Cambridge Service Alliance, Cambridge University, 2014). Pg. 172

Healy, K., "Using Metadata to Find Paul Revere," http://kieranhealy.org, June 2013. Pg. 173

Hendrix, J. and Kohl, J., *ZigBee® Overview* (San Ramon, CA: ZigBee® Alliance, 2009). Pg. 168

Hurwitz, J., Nugent, A., Halper, F., and Kaufman, M., *Big Data for Dummies* (Hoboken, NJ: John Wiley & Sons Inc., 2013). Pg. 171

Iansiti, M. and Lakhani, K., "Digital Ubiquity: How Connections, Sensors and Data Are Revolutionizing Business," *Harvard Business Review* 92, no. 11 (2014): 90-99. Pg. 168

IBM, IBM Watson Presentation, Armonk, NY: IBM Corp., 2011. Pg. 172

IBM, *Information Economics: Improve Information Economics, Cut Costs and Reduce Risks*, (Armonk, NY: IBM Corp., 2014). Pg. 173

Izenman, A., *Modern Multivariate Statistical Techniques: Regression, Classification and Manifold Learning*, (New York, NY: Springer-Verlag, 2008). Pg. 172

Jain, N. and Pallia, S., "Predictive Analytics: Future-Proofing Business," Winsights, Bangalore, India: Wipro Council for Industry Research, October-December 2011, 27-29. Bangalore, India: Wipro Ltd. Pg. 172

* Kellmereit, D. and Obodovski, D., *The Silent Intelligence: the Internet of Things* (San Francisco, CA: DND Ventures LLC, 2013). Pg. 168

Kelly, J. and Hamm, S., *Smart Machines: IBM's Watson and the Era of Cognitive Computing*, (New York, NY: Columbia University Press, 2013). Pg. 173

Kiron, D., Shockley, R., Kruschwitz, N., Finch, G. et al., *Analytics: The Widening Divide—How Companies Are Achieving Competitive Advantage through Analytics* (Somers, NY: IBM Global Business Services/MIT Sloan Management Review, 2011). Pg. 172

Kominek, D., *OPC®: The Ins and Outs to What It's About* (Edmonton, Alberta: MatrikonOPC, 2009). Pg. 167

Kondor, R., *OPC® Tutorial* (Edmonton, Alberta: MatrikonOPC, 2010). Pg. 167

Laudon, K. and Laudon, J., *Management Information Systems*, 11th ed. (Upper Saddle River, NJ: Pearson, 2010). Pg. 170

LaValle, S., Hopkins, M., Lesser, E., Shockley, R. et al., *Analytics: The New Path to Value—How the Smartest Organizations Are Embedding Analytics to Transform Insights into Action* (Somers, NY: IBM Global Business Services/MIT Sloan Management Review, 2010). Pg. 172

Lawnin, J. and Yurkanin, J., *Eating the Data Elephant*, (Bangalore, India: Wipro Technologies, 2011). Pg. 172

Loukides, M., *What is Data Science?*, (Sebastopol, CA: O'Reilly Media, 2010). Pg. 172

Lucas, E., *Cyberphobia: Identity, Trust, Security and the Internet* (London, UK: Bloomsbury Publishing, 2015). Pg. 173

Magraw, S., *Rizolva Business Benefits: Executive Summary* (Stalybridge, UK: Rizolva Ltd, 2014). Pg. 170

Magraw, S., *Rizolva Framework—Executive Summary* (Stalybridge, UK: Rizolva Ltd, 2015). Pg. 170

Mahnke, W., Leitner, S., and Damm, M., *OPC® Unified Architecture* (Berlin, Germany: Springer-Verlag, 2009). Pg. 167

Manyika, J., Chui, M., Brown, B., Bughin, J. et al., *Big Data: the Next Frontier for Innovation, Competition and Productivity*, Executive Summary (New York, NY: McKinsey & Company, 2011a). Pg. 173

Manyika, J., Chui, M., Brown, B., Bughin, J. et al., *Big Data: The Next Frontier for Innovation, Competition and Productivity*, Full Report (New York, NY: McKinsey & Company, 2011b). Pg. 173

Manyika, J., Chui. M., Bisson, P., Woetzel, J. et al., *The Internet of Things: Mapping the Value beyond the Hype*, Briefing (New York, NY: McKinsey & Company, 2015a). Pg. 168

Manyika, J., Chui. M., Bisson, P., Woetzel, J. et al., *The Internet of Things: Mapping the Value Beyond the Hype*, Executive Summary (New York, NY: McKinsey & Company, 2015b). Pg. 168

Manyika, J., Chui. M., Bisson, P., Woetzel, J. et al., *The Internet of Things: Mapping the Value Beyond the Hype*, Full Report (New York, NY: McKinsey & Company, 2015c). Pg. 168

Marcovici, M., *The Internet of Things* (Norderstedt, Germany: Books on Demand GmbH, 2014). Pg. 168

* Marr, B., *Big Data Case Study Collection: 7 Amazing Companies that Really Get Big Data* (Chichester, UK: John Wiley & Sons Ltd, 2015b). Pg. 173

* Marr, B., *Big Data: Using Smart Big Data, Analytics and Metrics to Make Better Decisions and Improve Performance* (Chichester, UK: John Wiley & Sons Ltd, 2015a). Pg. 173

Marshall, A., Townend, P., Venters, C., Lau, L. et al., *STRAPP: Trusted Digital Spaces through Timely Reliable and Personalised Provenance* (Leeds, UK: University of Leeds, 2011). Pg. 170

* Mayer-Schönberger, V. and Cukier, K., *Big Data: A Revolution that Will Transform How We Live, Work and Think* (London, UK: John Murray, 2013). Pg. 171

McNulty-Holmes, E., Patil, R., Shah, F. and Gray, D., *Understanding Big Data—A Beginners Guide to Data Science and the Business Applications* (Berlin, Germany: Dataconomy Media GmbH, 2014). Pg. 173

Microsoft, *Predictive Analytics Primer* (Redmond, WA: Microsoft Corp., 2015). Pg. 173

Miller, M., *The Underwriting* (Melbourne, Australia: Text Publishing Co., 2015). Pg. 173

Molnár, E., Kryvinska, N., and Greguš, M., "Customer-Driven Big Data Analytics for the Companies' Servitization," *Spring Servitization Conference 2014*, Aston Business School, Birmingham, UK, May 12-14, 2014 (Birmingham, UK: Aston Business School, Aston University). Pg. 171

Morgan, C., *Matters of Life and Data: A Memoir* (New York, NY: Morgan James Publishing, 2015). Pg. 173

Ng, I., "The H.A.T: Seeing Our Own Behaviours from the Way We Consume Products," Coventry, UK: Warwick Manufacturing Group, Warwick University, 2013. Pg. 169

Niblett, P. and Singh, G., "Connecting Devices to the Internet of Things," *IBM Impact 2014*, The Venetian, Las Vegas, NV, April 27-May 1, 2014 (Armonk, NY: IBM Corp.). Pg. 169

O'Neil, C., *On Being a Data Skeptic* (Sebastopol, CA: O'Reilly Media, 2013). Pg. 172

O'Reilly, *Big Data Now—Current Perspectives from O'Reilly Media* (Sebastopol, CA: O'Reilly Media, 2012). Pg. 171

O'Reilly, *Big Data Now—Current Perspectives from O'Reilly Radar* (Sebastopol, CA: O'Reilly Media, 2011). Pg. 171

* OSIsoft, *The Power of PI®: Real-Time Information Throughout the Enterprise* (San Leandro, CA: OSIsoft LLC, 2010). Pg. 170

Oxford Economics, *Smart Connected Products: Manufacturing's Next Transformation* (Oxford, UK: Oxford Economics, 2014). Pg. 169

Paine, T. *Industrial Internet of Things and Communications at the Edge* (Portland, ME: Kepware Inc., 2015). Pg. 168

Pallia, S. and Prabhu, J., "Information as an Enterprise Asset," Winsights, Bangalore, India: Wipro Council for Industry Research, October-December 2011, 14-17. Bangalore, India: Wipro Ltd. Pg. 172

Park, J. and Mackay, S., *Practical Data Acquisition for Instrumentation and Control Systems* (Oxford, UK: Newnes/Elsevier/IDC Technologies, 2003). Pg. 167

Park, J., Mackay, S., and Wright, E., *Practical Data Communications for Instrumentation and Control* (Oxford, UK: Newnes/Elsevier/IDC Technologies, 2003). Pg. 167

Patil, D., *Building Data Science Teams: The Skills, Tools and Perspectives behind Great Data Science Groups* (Sebastopol, CA: O'Reilly Media, 2011). Pg. 172

Plant, R. and Murrell, S., *An Executive's Guide to Information Technology: Principles, Business Models and Terminology* (Cambridge, UK: Cambridge University Press, 2007). Pg. 167

Porter, M. and Heppelmann, J., "How Smart Connected Products Are Transforming Competition," *Harvard Business Review* 92, no. 11 (2014): 64-88. Pg. 168

Porter, M. and Heppelmann, J., "How Smart Connected Products Are Transforming Companies," *Harvard Business Review* 93, no. 10 (2015): 96-114. Pg. 168

Pultarova, T., "'Immature' Internet of Things Hackable with Primitive Methods," *Engineering and Technology* 10, no. 3 (April 2015): 2. Pg. 169

Ratner, B., *Statistical and Machine-Learning Data Mining: Techniques for Better Predictive Modeling and Analysis of Big Data* (Boca Raton, FL: CRC Press/Taylor & Francis, 2012). Pg. 172

Reynders, D., Mackay, S., and Wright, E., *Practical Industrial Data Communications: Best Practice Techniques* (Oxford, UK: Newnes/Elsevier/IDC Technologies, 2005). Pg. 167

Rockwell Automation, *What's New—FactoryTalk® Historian Site Edition Version 4.0* (Milwaukee, WI: Rockwell Automation Inc., 2014). Pg. 171

Sampson, B., "Casting the Net Wider," *Professional Engineering* (January 2015): 52-64. (London, UK: Professional Engineering Publishing Ltd). Pg. 168

Sánchez López, T., Ranasinghe, D., Harrison, M., and McFarlane, D., *Using Smart Objects to Build the Internet of Things* (Cambridge, UK: Cambridge Service Alliance, Cambridge University/IEEE Internet Computing, 2012). Pg. 168

Satiah, K., "From Data to Intelligence: Unlocking the Value in Your Organisation's Information Assets," *Asset Management & Maintenance Journal* 26, no. 4 (July 2013): 20-24. Pg. 171

Schroeck, M., Shockley, R., Smart, J., Romero-Morales, D. et al., *Analytics: the Real-World Use of Big Data—How Innovative Enterprises Extract Value from Uncertain Data* (Somers, NY: IBM Global Business Services/Saïd Business School, 2012). Pg. 172

Schroeder, A., "IOT and Its Implications for Manufacturing Services," *Service Community Birmingham Special Event*, Aston Business School, Birmingham, UK, September 30, 2015. (Birmingham, UK: Aston Business School, Aston University). Pg. 168

* Shron, M., *Thinking with Data* (Sebastopol, CA: O'Reilly Media, 2014). Pg. 167

Siemens, G., *Knowing Knowledge* (Raleigh, NC: Lulu Press Inc., 2006). Pg. 170

Silver, N., *The Signal and the Noise* (London, UK: Penguin Group, 2012). Pg. 171

Thakkar, J., *Smart Data Diagnostics—Key to Information Accuracy for Utilities* (Bangalore, India: Wipro Technologies, 2011). Pg. 172

ThingWorx, *The Birth of the Connected Platform* (Exton, PA: ThingWorx, 2011). Pg. 169

Townend, P., Webster, D., Venters, C., Dimitrova, V. et al., "Personalised Provenance Reasoning Models and Risk Assessment in Business Systems: A Case Study," *2013 IEEE 7th International Symposium on Service Oriented System Engineering (SOSE)*, Hotel Sofitel, San Francisco, CA, March 25-28, 2013 (Red Hook, NY: Curran Associates Inc.). Pg. 170

Venters, C., Austin, J., Dibsdale, C., Dimitrova, V. et al., *To Trust or Not to Trust: Developing Trusted Digital Spaces through Timely Reliable and Personalized Provenance* (London, UK: University of Middlesex, 2014). Pg. 170

Viewpointe, *Dark Data, Big Data, Your Data: Creating an Action Plan for Information Governance* (New York, NY: Viewpointe Archive Services, LLC, 2014) Pg. 173

Vitria, "Real-Time Big Data Analytics + Internet of Things = Value Creation," Sunnyvale, CA: Vitria Technology Inc., 2015. Pg. 169

Watson, J. and Sumner, J., *Rise of the Machines: Moving from Hype to Reality in the Burgeoning Market for Machine-to-Machine Communication* (London, UK: Economist Intelligence Unit Ltd, 2012). Pg. 168

Webb, W., *Understanding Weightless™: Technology, Equipment and Network Deployment for M2M Communications in White Space* (Cambridge, UK: Cambridge University Press, 2012a). Pg. 167

Webb, W., *Weightless™: The Technology to Finally Realise the M2M Vision* (Histon, UK: Neul Ltd, 2012b). Pg. 167

Webb, J. and O'Brien, T. eds., *Big Data Now: Current Perspectives from O'Reilly Media* (Sebastopol, CA: O'Reilly Media, 2014). Pg. 171

Webster, D. eds., *ZigBee® Resource Guide*, (Greenwood Village, CO: Webcom Communications Corp., 2013). Pg. 167

Witchalls, C. and Chambers, J., *The Internet of Things Business Index: A Quiet Revolution Gathers Pace* (London, UK: Economist Intelligence Unit Ltd, 2013). Pg. 168

Zakrzewski, J., *Taming the Beast of Big Data* (Dayton, OH: Sogeti USA, 2012). Pg. 171

ZigBee, *ZigBee® Overview* (San Ramon, CA: ZigBee® Alliance, 2009). Pg. 167

Zikopoulos, P., deRoos, D., Parasuraman, K., Deutsch, T. et al., *Harness the Power of Big Data* (New York, NY: McGraw-Hill, 2013). Pg. 172

Websites

Advanced Performance Institute Pg. 173	www.ap-institute.com
ARM mbed Pg. 169	http://mbed.org
BloomReach Inc. Pg. 169	http://bloomreach.com
Canary Labs Inc. Pg. 171	www.canarylabs.com
Connectivism Pg. 170	www.learning-theories.com/connectivism-siemens-downes
Control Design Pg. 169	www.controldesign.com
CoNVerge Pg. 171	www.nvable.com/introducting-converge
Dataconomy Pg. 173	http://dataconomy.com
Datameer Inc. Pg. 173	www.datameer.com
Datawatch Corp. Pg. 169	www.datawatch.com
Decision Evaluation Ltd. Pg. 164	www.decisionevaluation.co.uk
Economist Intelligence Unit Pg. 173	www.eiu.com
GE Intelligent Platforms Pg. 168	www.ge-ip.com
GE Predix™ Pg. 171	www.ge.com/digital/predix
GE Proficy™ Pg. 171	www.ge-ip.com/products/proficy-historian/p2420
Harbor Research Pg. 169	www.harborresearch.com
Hub of All Things Pg. 169	www.hubofallthings.org
IBM Pg. 172	www.ibm.com
IBM (Maximo®) Pg. 163	www.ibm.com/software/products/en/maximoassetmanagement
IFS A.B. Pg. 163	www.ifsworld.com
Infor Pg. 163	www.infor.com
Kepware Inc. Pg. 168	www.kepware.com
Kieran Healy Pg. 173	http://kieranhealy.org
Libelium Comunicaciones Distribuidas SL Pg. 168	www.libelium.com
M.A.C. Solutions (UK) Ltd. Pg. 171	www.mac-solutions.co.uk
Matrikon Pg. 171	www.matrikon.com
MatrikonOPC Pg. 167	www.matrikonopc.com
McKinsey & Co. Pg. 169	www.mckinsey.com
Microsoft Corp. Pg. 172	www.microsoft.com
MIMOSA Open Systems Alliance Pg. 168	www.mimosa.org
National Instruments Pg. 167	www.ni.com
NVable Ltd. Pg. 171	www.nvable.com
O'Reilly Media Inc. Pg. 171	www.oreilly.com
OPC® Foundation Pg. 167	www.opcfoundation.org
OSIsoft LLC Pg. 170	www.osisoft.com

Oxford Economics Pg. 169	www.oxfordeconomics.com
Perceptual Edge Pg. 172	www.perceptualedge.com
Raconteur Pg. 173	www.raconteur.net
Raspberry PI Pg. 169	www.raspberrypi.org
Rockwell Automation Inc. Pg. 171	www.rockwellautomation.com
SAP AG Pg. 163	www.sap.com
Tessella plc. Pg. 173	www.tessella.com
The Silent Intelligence Pg. 168	http://thesilentintelligence.com
ThingWorx Pg. 169	www.thingworx.com
Vitria Technology Inc. Pg. 169	www.vitria.com
Weightless™ Special Interest Group Pg. 167	www.weightless.org
Wipro Ltd. Pg. 172	www.wipro.com
Wonderware UK Pg. 171	www.wonderware.com
Zafire Group Pg. 163	www.zafire.com
ZigBee® Alliance Pg. 167	www.zigbee.org

Additional Material

References

* denotes sources that the author found particularly helpful and/or inspiring.

Anderson, A. and Semmelroth, D., *Statistics for Big Data for Dummies* (Hoboken, NJ: John Wiley & Sons Inc., 2015).

Andrejevic, M. *Infoglut: How Too Much Information is Changing the Way We Think and Know* (Abingdon, UK: Routledge/Taylor & Francis Group, 2013).

Axelsson, L.-U., Singh, V., Visser, W., and Braun, R., "An Innovative Fleet Condition Monitoring Concept for a 2MW Gas Turbine, in *9th International Gas Turbine Conference: The Future of Gas Turbine Technology*, Le Plaza Hotel, Brussels, Belgium, October 10-11, 2018. Brussels, Belgium: European Turbine Network (ETN) Global, 2018.

Barnes, J., *Microsoft Azure Essentials: Azure Machine Learning*, Kindle Edition (Redmond, WA: Microsoft Press, 2015).

Brown, M., *Data Discovery for Dummies*, Teradata Special Edition (Hoboken, NJ: John Wiley & Sons Inc., 2014).

Calhoun, K., *Health Management at Rolls-Royce* (Indianapolis, IN: Rolls-Royce Corporation, 2009).

Collier, M. and Shahan, R., *Microsoft Azure Essentials: Fundamentals of Azure*, Kindle Edition (Redmond, WA: Microsoft Press, 2015).

* Demarest, G. and Scott, J., *Architects Guide to Implementing a Digital Transformation* (Santa Clara, CA: MapR Technologies, Inc., 2016).

Denis, M., "The Internet of Flying Things: Part 1," *AircraftIT MRO eJournal* 4, mo. 5 (August/September 2015a): 34-37.

Denis, M. "The Internet of Flying Things: Part 2," *AircraftIT MRO eJournal* 4, no. 6 (October/November 2015b): 28-33.

Dibsdale, C., "IoT in the Mining Industry," *IoT Day*, Cranfield University, Cranfield, UK, November 25, 2015 (Fareham, UK: Ox Mountain Ltd).

Dibsdale, C., "Standards and Systems For Information Management (through Life Engineering)," Pontypool, UK: Cedibs Predictive Technologies Consulting Ltd. 2015c.

Economist, The Industrial Internet of Things: Machine Learning," *The Economist*, November 21, 2015. London, UK: Economist Newspaper Ltd.

Elliott, S., Plant, A., Smith, M., Young, M. et al. *The Connected Train* (Bezons, France: Atos SE, 2014).

* EMC Education Services, *Data Science and Big Data Analytics: Discovering, Analysing, Visualizing and Presenting Data*(Indianapolis, IN: John Wiley & Sons Inc., 2015).

* Few, S., *Signal: Understanding What Matters in a World of Noise* (Burlingame, CA: Analytics Press, 2015).

* Few, S., *Big Data, Big Dupe: A Little Book about a Big Bunch of Nonsense* (El Dorado Hills, CA: Analytics Press, 2018).

GE Fanuc, *Proficy™ Software: A Powerful Production System that Delivers Results* (Charlottesville, VA: GE Intelligent Platforms Inc., 2009).

GE, *Proficy™ Automation Software: Enabling Real-Time Operational Intelligence* (Charlottesville, VA: GE Intelligent Platforms Inc., 2014).

* Gent, E., "Big Data's Dark Side," *Engineering and Technology* 11, no. 9 (October 2016): 32-35 (Stevenage, UK: Institution of Engineering and Technology).

HBM, *Measurement Data Acquisition* (Munich, Germany: Süddeutscher Verlag onpact GmbH, 2012).

Hewlett-Packard, *HP Communications and Media Solutions IoT Solution Set* (Houston, TX: Hewlett-Packard Development Co. LP, 2012).

Hicklin, J., Shurvinton, B., and Beard, G., *The Internet of Things for Dummies*, CGI Special Edition (Chichester, UK: John Wiley & Sons Ltd, 2015).

Holland, A., "BeanIoT™. Wearable. Deployable. Enjoyable," *IoT Day*, Cranfield University, Cranfield, UK, November 25, 2015 (Swaffham Bulbeck, UK: RF Module & Optical Design Ltd).

Intel, *TAP Technical Brief* (Santa Clara, CA: Intel Corp., 2015).

Kayser, H. and Gaus, N., *Siemens Digitalization Strategy & Sinanalytics Platform* (Munich, Germany: Siemens AG, 2016).

Kerravala, Z. and Miller, L., *Digital Transformation for Dummies*, Mitel Special Edition (Hoboken, NJ: John Wiley & Sons Inc., 2017).

King, S., *Big Data Analytics and Its Applications* (Derby, UK: Rolls-Royce plc, 2016).

* MapR, *6 Elements of Big Data Security* (Santa Clara, CA: MapR Technologies, Inc., 2017).

MapR and AtScale, *2016 Big Data Maturity Survey* Santa Clara, CA/San Mateo, CA: MapR Technologies, Inc./AtScale Inc., 2017).

Marr, B., *Big Data for Small Business for Dummies* (Chichester, UK: John Wiley & Sons Ltd, 2016a).

* Marr, B., *Big Data in Practice: How 45 Successful Companies Used Big Data Analytics to Deliver Extraordinary Results* (New Delhi, India: Wiley India Pvt. Ltd, 2016b).

McKeown, M., *Microsoft Azure Essentials: Azure Automation*, Kindle Edition (Redmond, WA: Microsoft Press, 2015).

Morris, J., "GE Turns Big Data into Personal Healthcare for Engines," Aviation Week and Space Technology/Air Transport World/Speednews Farnborough International Airshow News, July 14, 2016, 10.

Mullis, P., "Internet of Things," *IoT Day*, Cranfield University, Cranfield, UK, November 25, 2015. (Bracknell, UK: HP Enterprise Services UK Ltd).

Nexiona, *Miimetiq IoT Composer: Brochure*, (Barcelona, Spain: Nexiona Connectocrats, SL, 2016).

Norbury, A., "Industry 4.0—Vision to Reality," *IoT Day*, Cranfield University, Cranfield, UK, November 25, 2015 (Munich, Germany: Siemens AG).

Noventum, *Manufacturer's Advanced Services: IoT as the Key to Profitability and Growth* (Nicosia, Cyprus: Noventum Service Management Ltd, 2016).

OPC Foundation, *OPC Unified Architecture: Interoperability for Industrie 4.0 and the Internet of Things* (Scottsdale, AZ: OPC Foundation, 2015).

Perry, M., *Evaluating and Choosing an IoT Platform* (Sebastopol, CA: O'Reilly Media, 2016).

Pool, R., "The First Sign of Trouble," *E&T Test & Measurement Supplement*, September 2016.

Rainey, R., *Microsoft Azure Essentials: Azure Web Apps for Developers*, Kindle Edition (Redmond, WA: Microsoft Press, 2015).

Roshchin, R., *Sinanalytics Enables Digitalization: Industrial Data Analytics* (Munich, Germany: Siemens AG, 2016).

Saunders, P. and Knolla, J., "Drinking from the Fire Hose—Modern Aircraft and Big Data," *AircraftIT MRO eJournal* 4, no. 5 (August/September 2015): 26-29.

Schwarzenbach, J., *The Data Zoo: How User Behaviours Affect Information Quality*. Litchfield, UK: Data and Process Advantage Ltd., 2013.

Simudyne, *Fact Sheet* (London, UK: Simudyne Ltd, 2015).

Somers, D., "Incredibly Overhyped Technology? IoT Data Analytics—The Myths Busted," *IoT Day*, Cranfield University, Cranfield, UK, November 25, 2015 (London, UK: Warwick Analytical Software Ltd).

Strauss, P., "IoT—The Internet of 'Truths, Trials and Technical Complexities,'" *IoT Day*, Cranfield University, Cranfield, UK, November 25, 2015 (Milton Keynes, UK: Illumitas).

Vanson Bourne and ServiceMax, *Asset and Service Data Gravity*. Newbury, UK and Pleasanton, CA: Vanson Bourne Ltd. and ServiceMax Inc., 2018.

Vermesan, O. and Friess, P. eds., *Internet of Things: From Research and Innovation to Market Deployment* (Aalborg, Denmark: River Publishers, 2014).

Vermesan, O. and Friess, P. eds., *Digitising the Industry: Internet of Things Connecting the Physical, Digital and Virtual Worlds* (Gistrup, Denmark: River Publishers, 2016).

Visser, W. and Singh, V., "An Innovative Fleet Condition Monitoring Concept for a 2MW Gas Turbine - Presentation," in *9th International Gas Turbine Conference: The Future of Gas Turbine Technology,* Le Plaza Hotel, Brussels, Belgium, October 10-11, 2018. Brussels, Belgium: European Turbine Network (ETN) Global.

* Vodafone, "Creating a Smarter World: How M2M and the Internet of Things Are Driving the Next Industrial Revolution," Newbury, UK: Vodafone Group, 2014.

Vodafone, "Vodafone IoT Barometer 2016,". Newbury, UK: Vodafone Group, 2016.

Vodafone, "Vodafone IoT Barometer 2017-18," Newbury, UK: Vodafone Group, 2017.

West, S., "Big Data: Small Firms—Big Problems!," *Big Data Innovation Conference 2016*, Park Inn, Frankfurt, Germany, November 16-17, 2016 (London, UK: Platinum Global Solutions Ltd).

* Wheatley, M., Visibility and Control, Stirling, W. eds., *UK Manufacturing Review 2015-2016* (Teddington, UK: Stirling Media Ltd, 2016), 80-83.

Websites

AircraftIT MRO	www.aircraftit.com/MRO/Index.aspx
Anodot	www.anodot.com
Data and Process Advantage Ltd.	www.dpadvantage.co.uk
Distence Oy.	www.distence.fi
European Turbine Network (ETN) Global	http://etn.global
GE Automation	www.geautomation.com
Hewlett Packard Enterprise	www.hpe.com
Intel Corp.	www.intel.com
Lawrence Griffiths	http://industrialinternet.co.uk
Nexiona Connectocrats S.L.	http://nexiona.com
Noventum Service Management Ltd.	www.noventum.eu
Purrmetrix	www.purrmetrix.com
RF Module & Optical Design Ltd.	www.beaniot.com, www.rfmod.com
Siemens AG	www.siemens.com
Simudyne	www.simudyne.com
Tibbo Technology Inc.	www.tibbo.com, http://aggregate.tibbo.com
Trak365	www.trak365.com
Vodafone Group	www.vodafone.co.uk
Warwick Analytical Software Ltd.	www.warwickanalytics.com

CHAPTER 10

The Practicalities

Introduction

The books by Ron Moore (*Moore 2004, 2006*) take a common-sense look at plant design, procurement, parts management, installation and maintenance, training, and implementing a Computerized Maintenance Management System (CMMS). They also look at which improvement tools to apply to an organization and when to do so. Howard Penrose's book (*Penrose 2008*) gives an executive overview of maintenance and its potential economic impact.

The article by Mike Knapp (*Knapp 2014*) shows how integrated control, data, and asset management systems can make more intelligent, predictive maintenance possible. Bob DiStefano's paper (*DiStefano 2005*) puts the technical and engineering aspects of maintenance and reliability into business terms to help communicate to top executives, and people not directly involved in maintenance, the tremendous business value associated with maintenance and reliability in a company. The book by Pat Kennedy and others (*Kennedy et al. 2008*) is a narrative case study on improving the performance of process plant. The article by Gary Fallaize (*Fallaize 2010*) discusses why risk assessment is still such a challenge for maintenance and engineering organizations.

There are several excellent magazines devoted to the subjects of maintenance and asset management that are worth subscribing to. *Maintenance & Engineering* (available from Conference Communications, Farnham, UK: www.maintenanceandengineering.com), edited by David Wilson, is a free bimonthly UK magazine that carries many in-depth technical articles that are relevant to today's professional engineers charged with managing and maintaining their company's physical assets, including topics such as asset management, computerized maintenance management, Condition Monitoring (CM), health and safety, outsourcing, premises management, predictive maintenance, training and apprenticeships, and more. A print version is available free to UK subscribers and the current edition and back numbers are available on the website in pdf format. *Asset Management & Maintenance Journal*, edited by Len Bradshaw, was a bimonthly electronic journal that covered a similar range of topics. Some back numbers are available in pdf format and an index is available (*Bradshaw 2014*). *Uptime* (available from Reliabilityweb.com, Fort Myers, FL: www.uptimemagazine.com), edited by Jenny Brunson, is a free bimonthly electronic magazine now available as an iPad "app" as well as in pdf format and through an internet browser for maintenance reliability professionals that provides case studies, tutorials, practical tips, news, book reviews, and interactive content: a print version is available free to U.S. subscribers and back numbers are downloadable. Finally, *Reliability EDGE* (available from ReliaSoft Publishing, Tucson, AZ: www.reliasoft.com), edited by Lisa Hacker, is a quarterly pdf publication aimed at the reliability professional.

Reliability

The presentation by Bruce Hawkins (*Hawkins 2011*) explains how reliability affects shareholder value: it includes a diagram showing the drivers of reliability in a business. The article by Andrew Fraser (*Fraser 2014a*) and presentations by him (*Fraser 2014b*) and Ron Moore (*Moore 2007, 2013*) look at the positive impacts reliability has on safety, cost-effectiveness, and operational excellence, while another article by Ron Moore (*Moore 2015*) reinforces the connections between safety and reliability in an organization.

The presentation by Mike Sondalini (*Sondalini 2013d*) explains the importance of reliability, the influence of single points of failure, and how reliability is driven by variation that is in turn driven by policies and work quality. He has also produced a number of other articles discussing reliability, available on the Lifetime Reliability Solutions website (www.lifetime-reliability.com), which make the points that many industrial companies blindly commit industrial suicide daily by leaping off "reliability cliffs" (*Sondalini 2012*) and machines only fail when their parts fail (*Sondalini 2013f*). He also lists the top four machine faults (*Sondalini 2013i*) and makes the point that, although we know exactly what needs to be done to get very reliable machines, companies still do not achieve this: the limitations are not technical, but organizational, cultural, and human factors related (*Sondalini 2013j*). Another article by Mike Sondalini (*Sondalini 2011*) shows that organizations can create outstanding equipment reliability, deliver high production uptime, and guarantee lower operational costs by removing the risk of failure from machines and equipment: the more risks production plants and equipment are exposed to, the more certain it is that low plant reliability and high maintenance costs will result. The paper by Mike Sondalini and Howard Witt (*Sondalini and Witt 2013*) concludes that the secret to remarkably long and trouble-free equipment lives is to keep parts and components at low stress, within good local environmental conditions, so there is little risk that they are unable to handle their design duty.

The article by Robert DiStefano and Larry Covino (*DiStefano and Covino 2007*) presents two case studies showing the importance of reliability to business performance. The article by Paul Lanthier (*Lanthier 2011*) makes the point that asset performance improvement initiatives, based on an increase in asset reliability, are an excellent way to maximize financial return from assets, since they provide significant and sustainable benefits for relatively low financial investment compared to capital expenditure alternatives. The article describes how to quantify these financial benefits, possible metrics to use to manage the initiative and includes a number of examples where such benefits have been achieved and provides a normalized compilation of results from work performed over the past ten years. The article by Doug Plucknette and Chris Colson (*Plucknette and Colson 2011*) shows how equipment reliability delivers low-cost, energy-efficient assets at plants around the world, while the article by Ricky Smith (*Smith 2012*) lays out an approach that can be used in a facility for improving equipment reliability in seven days. The article by Klaus Blache (*Blache 2010*) presents a benchmarking study of 217 North American companies, looking at reliability and maintenance practices and improvements.

The article by Fernando Vicente (*Vicente 2010*) demonstrates an approach that focuses on reliability, availability, and maintainability prediction, which helps detect components, equipment, and systems that require improvements and helps maintenance managers make the right decision when analyzing a centrifugal pump system in a gas plant: this process can be applied to many other components and systems across all industries. The article by Johnny Bofillos (*Bofillos 2012*) makes the point that, without a solid risk-based asset management strategy in place, implementing a new software solution may not be a solution at all and that there is no short-cut for learning and adopting reliability best practices.

The excellent book by Jean-Marie Flaus (*Flaus 2013*) gives an overview of the methods used for risk analysis in a variety of industrial sectors with a particular focus on the consideration of human aspects and provides a definition of all the fundamental notions associated with risks and risk management, as well as clearly placing the discipline of risk analysis within the broader context of risk management processes. The article by Brian Webster (*Webster 2012*) explains the conflict between traditional risk-calculation methods and distance methods, as well as the potential for poor business decisions that could result from using such distance methods. The articles by Terry Nelson (*Nelson 2011, 2012*) present risk and criticality and show that they are of great value in managing systems and processes, allowing preparation, proactivity and prevention of disruptive events. The article by Keith Mobley (*Mobley 2011*) gives an overview of risk, defining risk management as simply the identification, assessment, and prioritization of risks, followed by a coordinated and economical application of resources to minimize or control the probability of occurrence and the impact of adverse events as well as to maximize the realization of opportunities. The handbooks by Homayoon Dezfuli and others (*Dezfuli et al. 2010; Dezfuli et al. 2011:* www.nasa.gov) cover the National Aeronautics and Space Administration (NASA) approach to risk and risk management in some detail.

An excellent introductory text to reliability theory has been written by Patrick O'Connor and others (*O'Connor et al. 1995*); the book by Sue Cox and Robin Tait (*Cox and Tait 1998*) is also very accessible. An introduction to Reliability Block Diagrams (RBDs) is provided in a paper from ITEM Software Inc. (*ITEM Software 2007*). Readers who are comfortable with a more academic approach are referred to the books by Hongzhou Wang and Hoang Pham (*Wang and Pham 2006*) and John Andrews and Thomas "Bob" Moss (*Andrews and Moss 2002*).

Optimal Maintenance Decisions Inc. (OMDEC) is a leader in the field of reliability analysis. The book by Andrew Jardine and Albert Tsang (*Jardine and Tsang 2006*) presents the theory of maintenance, replacement, and reliability in some detail and introduces the OMDEC tools. The presentation by Klaus Krüppel and Tony Lawton (*Krüppel and Lawton 2008*) provides an overview, as does the paper by Ben Stevens (*Stevens 2008*). There is also a great deal of information available on the OMDEC website (www.omdec.com), including an interesting discussion about how difficult it is to determine the onset of failure (*OMDEC 2010*). The article by Ricky Smith (*Smith 2013b*) presents the basics of the P-F curve.

Some of the pitfalls of analyzing reliability, including poor use of language and the "blame game," are discussed in a paper by ARMS Reliability (*ARMS Reliability 2012*: www.armsreliability.com), while the articles by Fred Schenkelberg (*Schenkelberg 2013a, 2013b*) discuss the actual meaning and proper use of Mean Time Between Failure (MTBF) and when to use other measures. The article by Steve Turner (*Turner 2010*) argues that, while computerization has put massive power and capability into the hands of people from all walks of life, the downside of this is that statistical methods that are available through cheap and superficially easy-to-use software packages can easily produce the wrong answers. When these tools are used by people with low statistical literacy, then major problems can occur, particularly if these methods are not competently applied to major facilities.

The article by George Karalexis and Ricky Smith (*Karalexis and Smith 2010*) looks at known best practices for managing asset reliability and reducing cost and risk, while the articles by Dan Miller (*Miller 2015a, 2015b*) look at the effects of setup and changeovers on asset reliability and the importance of carefully controlled changeover processes. The articles by Greg Williams (*Williams 2013, 2014*) challenge the distinctions between risk management as an empirically formal procedure-based discipline and risk management as an intuitive practice with respect to asset management functions. The article by Paul Dufresne (*Dufresne 2015*) reinforces the point that a disciplined approach to addressing the fundamentals of reliability will improve a company's competitive position. The article by Tacoma Zach (*Zach 2015*) looks at some commonly held myths and misperceptions about criticality that often prevent organizations from performing a comprehensive criticality analysis.

The articles by Tom Dabbs and Dan Pereira (*Dabbs and Pereira 2012, 2013*) look at the steps that should be taken to achieve sustainable pump reliability, while the articles by Nwaoha Chikezie (*Chikezie 2010, 2011*) look at the causes, control, and prevention of pump cavitation, a common cause of pump failure.

Fault Trees, FMEA, FMECA, and FRACAS

Fault trees are an important method for determining possible equipment faults, either at the equipment design stage or when it is in service. They are well described in a NASA handbook written by Michael Stamatelatos and others (*Stamatelatos et al. 2002*) and there are very good examples in the papers by Pratap Rama and others (*Rama et al. 2008a, 2008b*).

The article by Ray Garvey (*Garvey 2013*) gives common failure mechanisms and associated root causes of machine component damage. The articles by Larry Tyson (*Tyson 2011, 2012*) show how he and his team were able to take a complex situation, perform the failure mode analysis, filter or reduce those failures down to the most critical, and do a detailed Root Cause Analysis (RCA). The article by Jorge Kalocai (*Kalocai 2012*) discusses a Weibull-based method for failure mode characterization and remaining life expectancy estimation. The presentation by ARMS Reliability (*ARMS Reliability 2014a*) discusses how to improve RCA, while the article by David Gluzman (*Gluzman 2013*) offers a view on certain details of RCA and suggests modifications that are crucial for clarification of the process. The article by Carlos Pernett (*Pernett 2008*) lists the ten most common deficiencies in the implementation of RCA. The Paul Barringer and Associates Inc. website (www.barringer1.com) contains numerous resources for Weibull analysis; the website www.weibull.com is also a good source of information on all aspects of reliability engineering.

Another important method is Failure Modes and Effects Analysis (FMEA), in which possible equipment faults are recorded and a combination of design documents, service experience, and engineering judgement is used to list the resultant effects on the equipment and its surroundings. The book by Robin McDermott and others (*McDermott et al. 2009*) provides expert advice to help shorten the learning curve for teams to conduct effective and efficient FMEAs. Simon Mills' presentation (*Mills 2012*) describes a successful approach to FMEA. A related technique, Failure Modes, Effects and Criticality Analysis (FMECA), adds the criticalities of the effects of faults to the analysis; the NASA presentation on FMEAs (*NASA 2012*: www.nasa.gov) discusses this. The article by Rohit Banerji and Debajyoti Chakraborty (*Banerji and Chakraborty 2011*) also looks at life cycle FMECA, which focuses on forward-looking risk-based analysis in order to develop robust operational and financial profiles of assets as they live through their operate-and-maintain phase. The article by Doug Plucknette and others (*Plucknette et al. 2013*) examines the differences and similarities between Reliability-Centered Maintenance (RCM) and FMEA, pointing out that both methods can be effective and are not mutually exclusive.

When equipment is in service it is important to systematically document faults that occur and the corrective actions that were taken, in order to speed up the learning process and provide objective data for any required design changes: Failure Reporting, Analysis, and Corrective Action Systems (FRACAS) are widely used to do this. Details of the process and the system marketed by ReliaSoft (www.reliasoft.com) are given in *ReliaSoft 2008*, while the article by Ricky Smith and Bill Keeter (*Smith and Keeter 2011*) discusses the use of reports from maintenance systems or other specialized software to eliminate, mitigate, or control failures via FRACAS.

The article by Mahfoud Chafai and Larbi Refou (*Chafai and Refou 2008*) presents a FMECA of a rotary kiln drive of a cement plant, identifies the critical points, and offers an RCM strategy.

Root Cause Analysis (RCA) is a powerful tool to investigate issues that have affected the performance of people or systems within an organization. The paper by ARMS Reliability (*ARMS Reliability 2014b*) describes four simple steps for the creation of cause and effect charts that will support effective problem solving, while the article by Robert Latino (*Latino 2015*) refutes some of the arguments that may be made against RCA.

The article by Sander Hendriks and others (*Hendriks et al. 2015*) describes the development and implementation of Maintenance Criticality Reassessment (MCR), based on FMECA, which improves and consolidates the effectiveness and cost efficiency of the Preventative Maintenance (PM) of low-risk systems.

The article by ReliaSoft (*ReliaSoft 2015*) shows how to use modern database and internet technologies to manage the FRACAS testing, data collection, and problem-solving processes across multiple departments.

Alarms

The alarm systems fitted to many process plants, utility distribution systems, etc., are extremely comprehensive, which can lead to operator overload because of incorrect alarm limits, the absence of alarm prioritization, confusion between alarms provided for operational or maintenance reasons, confusion between alarms provided for different operational phases (start-up, steady-state, shutdown), and "cascades" of alarms that happen when the alarm indicating the root cause of a problem is swamped by dozens or hundreds of alarms from the other systems that are affected. GoalArt® (www.goalart.com) has produced a set of analysis and alarm management tools that vastly simplify the alarm management problem; these are described in a presentation by Jan Eric Larsson (*Larsson 2014*) and a booklet by GoalArt® (*GoalArt 2008*). The paper by Tim Butters and others (*Butters et al. 2014*: www.sabisu.co) presents a novel method for the identification of redundant or "bad actors" in alarm systems through the application of statistical cluster analysis, which allows alarm systems to be optimized to reduce the load on the operators by applying existing systems change management processes. Finally, the article by Matt Spurlock and Jeff Keen (*Spurlock and Keen 2012*) discusses the concepts behind setting alarm levels, pointing out that while the goal of predictive maintenance is to identify a potential failure high up on the failure curve, it is likely that one could create a false identification of failure without a full understanding of alarms.

Maintenance

The excellent book by Stuart Emmett and Paul Wheelhouse (*Emmett and Wheelhouse 2011*) discusses the essential tools for efficient and effective maintenance management, including purchasing, inventory, and warehousing; the evaluation of current procedures, cost, and inventory reduction; and continuous improvement. Another excellent book edited by John Campbell and James Reyes-Picknell (*Campbell and Reyes-Picknell 2015*) looks at trends in technology, reliability-maintenance improvements, and the challenges of finding qualified maintenance personnel due to an aging labor force, while another excellent book by Vee Narayan and others (*Narayan et al. 2012*) describes 42 practical case studies from their work experience from which they gained valuable insights into a wealth of maintenance and reliability best practices.

An excellent overview of all aspects of maintenance and reliability, including cultural and management issues, planning and scheduling, inventory management, and maintenance optimization, has been written by Ramesh Gulati (*Gulati 2013*); an accompanying workbook by him and Christopher Mears (*Gulati and Mears 2014*) emphasizes the important points. The article by Gurumurthy Anand and others (*Anand et al. 2008*) gives a history of maintenance, including a very comprehensive table comparing the different maintenance approaches. The article by John Atkinson (*Atkinson 2012*) discusses the various strategies available to today's maintenance managers, comparing reactive/breakdown maintenance, Planned Maintenance (PM), Condition-Based Maintenance (CBM), and proactive maintenance. The articles by Alan Friedman (*Friedman 2009a, 2009b, 2010*) discuss the reasons why many Predictive Maintenance (PdM) programs fail, while the article by Dave Koelzer (*Koelzer 2009*) explores how a small reduction in unplanned and emergency work can lead to significant cost reductions and improvement in resource availability and productivity.

The article by Bill Berneski (*Berneski 2011*) is an attempt to consolidate some existing but distinct concepts relating to maintenance periodicity selection and to provide some guidance on the best way to apply them. The article by Brad Peterson (*Peterson 2012*) makes the point that plant uptime and safety is a result of doing all the right things, while the article by John Ross (*Ross 2009*) looks at the issues surrounding the move from reactive to proactive maintenance. The article by James Davis (*Davis 2010*) looks at the benefits of applying a proactive performance culture to maintenance and inspection. The article by Kris Goly (*Goly 2010*) presents a business-based approach that has been utilized to implement PdM programs successfully throughout the world and across different industries. The article by Malcolm Hide (*Hide 2010*) discusses a three-step process that delivers a robust maintenance plan based on a clearly defined strategy which is easy to review and enables the implementation of changes when necessary, regardless of the size of the system.

The article by Dennis Belanger (*Belanger 2013*) argues that many improvement initiatives focus on the benefits they can provide, such as cost savings and improved productivity, but overlook the importance of discussing and identifying the enablers that must be in place to achieve the projected benefits. Taking the time to ensure that proper enablers are in place greatly improves an initiative's chance for sustained success: when enablers are left to chance, failure often results and that taking the time to understand, establish, and surround initiatives with a foundation of enablers will lead to success. The article by Rod Bennett (*Bennett 2008*) makes the point that the vast majority of equipment faults are self-inflicted and, therefore, avoidable by improvements in basic trade skills and practices.

Joel Levitt's books (*Levitt 2003, 2009*) are a mine of practical information, from simple task lists to choosing and setting up a CMMS and justifying maintenance activities in economic terms. The articles by Bruce Hawkins (*Hawkins 2007, 2008*) present some simple lessons about how to manage a maintenance program effectively.

Mike Sondalini's book (*Sondalini 2009b*) and the Lifetime Reliability Solutions website (www.lifetime-reliability.com) are very comprehensive resources on everything to do with maintenance and reliability. Readers who wish to get a full understanding of maintenance and reliability issues are urged to access and become familiar with this material, much of which is free to download. Other articles on the website include timeless advice from our forebears on successful maintenance and reliability (*Sondalini 2009a*); a list of ways to lower maintenance costs and increase reliability (*Sondalini 2013a*); a view that the focus of maintenance should be upon maintaining the "well-being" of the plant, so if the task is to "fix the machine," then maintenance has failed in its fundamental mission (*Sondalini 2013e*); a course on maintenance best practices (*Sondalini 2013g*); and a view that maintenance needs to be praised and glorified for the money it makes a company, not singled out for the money it spends (*Sondalini 2013h*). The article by Mike Sondalini (*Sondalini 2014*) discusses the use of time series charts and distributions to predict future plant operating and business performance.

The article by Ross Francis (*Francis 2008*) presents the case for ensuring that organizations are educated about the need for shutdowns to ensure uptime and the place of shutdowns in asset management strategies for ensuring asset integrity and operational capability into the future. The article explores time-saving measures in shutting down, planning and executing shutdowns, and starting up the plant after the work is complete and shows how basic processes, such as critical path analysis, progress reporting, schedule monitoring, meetings and forums, logistics, and managing change, have significant effects on successful shutdowns.

A powerful case for PdM is made in an article by Allied Reliability Inc. (*Allied Reliability 2006*: www.alliedreliability.com). The same company provides a high-level overview on how to set up a PdM program in *Allied Reliability 2010*. The paper by Aladon Ltd. (*Aladon 1999*) compares and contrasts old and new maintenance paradigms. Mike Neale's paper (*Neale 1996*) provides a brief review of some available techniques for plant maintenance and some guidance on typical maintenance expenditure, while Simon Mills' paper (*Mills 2007*) gives a high-level overview of setting up and managing a maintenance program. Richard Wurzbach's paper (*Wurzbach 2000*) describes a web-based tool for maintenance cost-benefit analysis.

Reliability Centered Maintenance (RCM) is a rigorous approach based on studies done by Stanley Nowlan and Howard Heap (*Nowlan and Heap 1978*) on civil airliner maintenance in the early 1970s, when it was determined that many failures were caused by, rather than prevented by, time-based maintenance. The core principles of this maintenance approach, which focuses on maintaining the functionality of the asset desired by the user, are described in the article by Richard Overman (*Overman 2009*), while it is described in detail in John Moubray's classic book (*Moubray 2001*). NASA has produced an RCM Guide (*NASA 2008*: www.nasa.gov) and there are courses available (e.g., *Schlumberger Sema 2004*). There have been attempts to short-cut what can seem to be an overly bureaucratic and detailed process, but these "streamlined RCM" methods have their drawbacks (*Moubray 2000*). Mike Sondalini cautions that RCM only works if the culture is correct, citing the airline industry as an example (*Sondalini 2013b*). The article by Umeet Bhachu (*Bhachu 2009*) argues that the correct approach to RCM is understanding it as a process that provides guidance in determining the predictive, preventative, and corrective actions that must be taken in order to ensure that a physical asset performs to its required expectations, rather than merely a set of specific rules: the basis of such actions and strategies takes into account the economics, environmental, safety, and operational criteria for the asset in the given operating circumstances and optimizes operational expenditures by rationalizing the maintenance decision-making process, shifting from a reactive model to a proactive maintenance model.

The article by Paul Castro (*Castro 2010*) shows that in order to optimize both maintenance risk and cost, the interrelationships between reliability, maintenance, and operations must be considered and leveraged to capitalize on the strengths of each. Reliability-Centered Operations (RCO) is an approach that optimizes these relationships through the application of a maintenance strategy built from failure analysis that will yield more expansive and cost-effective risk reduction tasks; this approach links the operators into the development and execution of this strategy. The article discusses the technical solution for a systematic, technology-based approach to develop a strategy and shows results for typical projects using the RCO approach. The article by Doug Plucknette (*Plucknette 2010*) answers the ten most commonly asked questions about RCM, while the article by Doug Plucknette and Paul Castro (*Plucknette and Castro 2010*) makes the point that one of the keys to getting the most from an RCM effort is having a cross-functional team that includes operations, maintenance, and reliability: it is imperative that these three groups work together closely and develop a joint vision and strategy to address reliability issues. This strategy will include a holistic approach to failure mitigation, resulting in improving work processes and communication channels and a more robust solution. The article by Michael Rezendes (*Rezendes 2009*) sets out to familiarize the reader with the concept of RCM, looking at the necessary abilities and mind-set needed by the actual person(s) performing the analysis, while the article by Anthony Smith and Tim Allen (*Smith and Allen 2011*) discusses the risks of using standard maintenance "templates" instead of carrying out a full analysis tailored to particular situations.

The papers by Mark Haarman (*Haarman 2002, 2011*: www.mainnovation.com) and the book and articles by him and others (*Haarman and Delahay 2004, 2013*; *Jonker and Haarman 2006*) from the Mainnovation consultancy give details of their Value-Driven Maintenance (VDM) theory and tools, which use the net present value of cash flows to understand the business drivers and quantify the value that maintenance activities provide to an organization. This methodology builds a bridge between traditional maintenance philosophies and managing by economic added value.

The mathematics of maintenance are introduced well in V. Narayan (*Narayan 2004*) and expanded on in J. Knezevic (*Knezevic 1997*), B. Dhillon (*Dhillon 2002*), and W. Blischke and D. Murthy (*Blischke and Murthy 2003*). There are several excellent PhD theses on aspects of maintenance, including the ones by Sulene Burnett (*Burnett 2013*), Tuomo Honkanen (*Honkanen 2004*), and Amir Al Shaalane (*Al Shaalane 2012*). The paper by Engineered Software Inc. (*Engineered Software 1999*: www.engineeredsoftware.com) looks at the use of statistics to schedule maintenance.

The article by Ben Stevens (*Stevens 2010*) looks at the combination of logic, statistics, and the application of well-accepted methods to improve maintenance decision-making, while the article by Paul Dean (*Dean 2010*) discusses the two types of indicators (lagging and leading) and the need to understand the fundamental nature and use of each type. The article by Jim Harper (*Harper 2011*) discusses how strategic maintenance reporting can facilitate sustained improvement, leading to smarter and more focused maintenance and ultimately cost reduction, while the article by Peter Todd (*Todd 2012*) discusses the contribution that maintenance and CM can make to business improvement initiatives. The article by Shane Daniel (*Daniel 2012*) explains how to make maintenance planning and scheduling as effective as it should be. The article by Ricky Smith (*Smith 2013a*) looks at the reasons why maintenance planning is not always as effective as it should be, while the article by Paul Castro (*Castro 2013*) lists the general leadership principles that result in successful maintenance and reliability programs.

The article by Mike Killick and Gary Thomas (*Killick and Thomas 2008*) introduces the principles of the maintenance planning framework, the shut calendar, and the planner's plan for best practice maintenance planning, while the article by Andy Page and George Karalexis (*Page and Karalexis 2009*) looks at how to carry out PM in a way that minimizes costs while still meeting equipment reliability expectations. The article by Peter Wilmott (*Wilmott 2010*) looks at how Total Productive Maintenance (TPM) can contribute to lean manufacturing.

The articles by Sauro Riccetti (*Riccetti 2011, 2014*) look at the processes used to design maintenance procedures for the food industry that bring under control all the critical factors that might contribute to end-product contamination and low equipment reliability and how to monitor food equipment critical parts for maintenance.

The articles by Paul Wheelhouse (*Wheelhouse 2013, 2015*) provide a summary of important considerations when designing Key Performance Indicators (KPIs) as part of any performance management system: technical, behavioral, and psychological aspects are all touched upon, as are important results from physics, statistics, information theory, and control theory to provide practical guidance. The article by Ron Moore (*Moore 2012*) discusses how the various improvement tools being offered on the market might relate to each other and the enabling practices or readiness that any given organization might need in order to effectively apply those tools. The book by Anthony Kelly (*Kelly 2002*) discusses a procedure for auditing and benchmarking maintenance functions and presents case studies from several industries.

A series of four articles by Sandy Dunn of Assetivity (*Assetivity 2014*; *Dunn 2014a, 2014b, 2015a*: www.assetivity.com.au) discusses methods to improve maintenance productivity through lean maintenance techniques. Another article by the same author (*Dunn 2015b*) discusses four core concepts that apply to the development of effective PM programs, no matter which approach is taken. The article by Robert Crotty (*Crotty 2015*) also shows how large Maintenance, Repair, and Overhaul (MRO) jobs can be significantly improved in terms of safety, cost, downtime, and quality by applying lean thinking. The article by Andy Page and Carey Repasz (*Page and Repasz 2015*) gives an in-depth review of PM, while the articles by Carlos Mario Perez Jaramillo (*Jaramillo 2014a, Jaramillo 2014b*) discuss the practical implementation of RCM. The article by Stephen Renshaw (*Renshaw 2015*) reviews the problems and benefits of using standard "templates" of tasks to develop maintenance plans, while the article by Paul Tomlingson (*Tomlingson 2015*) shows how the success of all maintenance functions is enhanced when a maintenance program is commonly understood across the entire operation; it is important to ensure that other departments that must support maintenance or utilize its services know how to do so.

The article by Marita del Carmen Garcia Lizarraga and Jyoti Sinha (*del Carmen Garcia Lizarraga and Sinha 2015*) describes the Reliability Maintenance Index for Equipment Profiling (RMI-EP) model which integrates information capable of giving insights into how a piece of equipment is behaving and enables the estimation of an index that indicates the need for maintenance and prioritizes any required maintenance tasks.

The white paper by Cyient (*Cyient 2015*) provides a good overview of how to successfully apply predictive maintenance techniques, using the rail industry as an example. The MSc thesis by Stephen Gauthier (*Gauthier 2006*) provides a stochastic modelling tool to assist in the component selection process for U.S. Army Aviation's Condition-Based Maintenance Plus (CBM+) program, using AH-64/UH-60 T701C helicopter engine data and information.

The excellent guidebook by Keith Pierce and others (*Pierce et al. 2015*) provides recommendations for rapidly implementing CBM solutions with popular CMMS and how to support them with high fidelity asset data. It defines and builds on the principles of CBM to show additional business value propositions related to CBM solutions and discusses the advantages of optimizing CMMS with both real-time and archived asset data using the OSIsoft PI System™. References to industry case studies are included.

The report published by Life Cycle Engineering (*LCE 2006*: www.lce.com) describes Preventative and Predictive Maintenance (PPM) as the regular and systematic application of engineering knowledge and maintenance attention to equipment and facilities to ensure their proper functionality and to reduce their rate of deterioration. In addition to dedicated engineering, PPM encompasses regular examination, inspection, lubrication, testing, and adjustments of equipment. PPM also provides the framework for all planned maintenance activity, including the generation of planned work orders to correct potential problems identified by inspection. The result is a proactive, rather than reactive, environment that optimizes equipment performance and life.

The article by Edwin Gutiérrez (*Gutiérrez 2015*) discusses the concept of Integral Asset Care (IAC), which brings together and integrates the optimum combination of methodologies, such as criticality analysis, risk-based inspection, RCM, FMEA, and cost risk optimization for the design of maintenance activities for dynamic, fixed, and electrical equipment and instruments.

The article by Geoff Walker (*Walker 2010*) reports on the difference that adopting the right type of maintenance strategy can make to business efficiency, productivity, and profit. The article by Tony Kelly (*Kelly 2011*) explains the dynamics of maintenance budgeting, while the article by John Gallimore (*Gallimore 2009*) shows that abandonment of PM or across-the-board task interval extensions is a possible but risky response to squeezed maintenance budgets: for a comparatively modest effort, up to 50% savings can be achieved without compromising plant performance or employee safety. The article by Jay Lee and Mohammed AbuAli (*Lee and AbuAli 2010*) presents recent advances in intelligent maintenance systems, as well as systematic methodologies and tools that have been effectively utilized to transform maintenance into innovative and productive service systems in a diverse set of industries: two brief case studies are provided to illustrate the lessons learned, with discussions for future service innovation impacts.

Two papers by Shannon Ackert (*Ackert 2010, 2011*) provide very good basic introductions to civil aircraft and aeroengine maintenance, while Simon Smith has written a paper (*Smith 2000*) on selecting the right equipment to inspect and overhaul in a chemical plant. The book by Richard "Doc" Palmer (*Palmer 2012*) provides proven planning and scheduling strategies that improve the performance of maintenance. The article by Terry Wireman (*Wireman 2011*) presents the common pitfalls that will be encountered when implementing the planner/scheduler function in a maintenance organization, while the article by Tarek Atout (*Atout 2011*) discusses the important role that the planner has in maintenance and describes the ideal person for the job. Finally, the article by Paul Swatkowski (*Swatkowski 2009*) is based directly on Steven Covey's famous book *Seven Habits of Highly Successful People*, with the examples modified to fit the maintenance world.

Spare Parts Management

The book by Phillip Slater (*Slater 2010a*: www.phillipslater.com, www.sparepartsknowhow.com) provides specific coverage of the issues faced in and requirements for managing engineering materials and spare parts and what to do to improve results, as well as including examples and real-life case studies to demonstrate the application of the concepts and ideas. The article by Phillip Slater (*Slater 2012*) argues that a properly managed supply of spare parts and materials can make a significant contribution to profitability, not only through its role in the minimization of downtime losses but also through the minimization of the costs of acquiring and holding spares inventory: advice is given on basic steps regarding categorization, stock levels, stock-outs, critical parts, excess holdings, and stores security to improve the effectiveness of this management task. The article by Phillip Slater and Joel Levitt (*Slater and Levitt 2014*) explains that the economic management of the inventory of spare parts is a form of insurance policy, mitigating the risks (to production, business reputation, safety, and the environment) of spares stock-out (the lack of spares when they are needed): the reasons for holding spares, the consequences of their unavailability in the event of plant breakdown, and the determination of spares holding and acquisition policy (in particular via a stock-out probability-consequence decision matrix) are discussed in detail. The article by Phillip Slater (*Slater 2009*) shows that when parts are held outside the official storeroom or inventory management system, the rest of the inventory holding for that part is affected, not only by reduced availability and access but also in less obvious ways relating to inventory levels, operational expenditure, and reliability programs. Another article by Phillip Slater (*Slater 2010b*) shows that the theoretical inventory control model and the actual situation can be sufficiently different to make the use of simplistic solutions not only pointless to operational goals and company finances but also even dangerous. A smart inventory solution ensures that the influence, impact, and complicating factors of all the elements of materials and inventory management are properly considered. The article by Supanee Arthasartsri and He Ren (*Arthasartsri and Ren 2008*) explains the inventory control process and gives some examples of airline inventory control strategies. Finally, the article by Phillip Slater (*Slater 2013*) shows how different views of materials and spare parts management can cause confusion and misunderstanding, to the detriment of the whole enterprise.

The article by Daniel DeWald (*DeWald 2011*) describes what a KPI is, gives the formulae and the priority for each KPI for MRO, and stores and discusses a criticality matrix, benchmarking activities, and expected outcomes when implementing performance measurements, while the article by Kris Goly (*Goly 2012*) lays out a simple process that will allow any organization to minimize the risk to the business of not having the right spares, while simultaneously minimizing the capital invested in spares. The article by Art Posey and Phillip Slater (*Posey and Slater 2010*) makes the point that many self-induced and premature failures are likely to be the result of poor materials handling and storage methods for spare parts and that materials management is the missing link in achieving reliability. The article by Ralph Rio (*Rio 2011*) makes the point that maintenance organizations tend to focus on work orders and equipment while giving lower priority to managing MRO inventory: the lack of inventory optimization results in higher inventory levels, carrying costs, and operational disruption. The article by Phillip Slater (*Slater 2008a*) looks at how to develop a high-performing spares stocking policy, while another article (*Slater 2008b*) makes the point that the definition of critical spares is usually more emotional than scientific and discusses the true meaning of the term. The article by Jeff Zieler (*Zieler 2012*) makes the point that the parts storeroom also provides functions that are absolutely critical to the maintenance operation which are so important that when the storeroom is operating in a best practices mode, the rest of the maintenance operation can excel.

Computerized Maintenance Management Systems (CMMS)

The article by Phil Taylor (*Taylor 2011*) suggests that it is essential to develop a strategy for maintenance software, considering what elements should be included and why this should be aligned to business and IT strategies. The article by Scott Welland (*Welland 2008*) makes the case for maintenance, operations, and IT to align together and develop a unified information and asset utilization strategy and that the technology to do this exists now. The article by Gottfried Roider (*Roider 2010*) discusses the primary requirement for efficient plant maintenance for prompt, detailed recording of maintenance activities: detailed documentation of maintenance activities in a Maintenance Planning System (MPS) is indispensable. All planned and unplanned maintenance activities are recorded in the MPS, providing more efficient monitoring for technical plants.

The book by Kishan Bagadia (*Bagadia 2006*) presents a clear, step-by-step approach for evaluating a company's maintenance operation needs, selecting the right CMMS, and implementing the system for optimal efficiency and cost-effectiveness. The articles by Joel Levitt (*Levitt 2007*) and Kamalapurka and others (*Kamalapurka et al. 2007*) present a detailed list of questions and requirements that a CMMS should satisfy, while the article by Jim Harper (*Harper 2010*) presents a number of important learning points from a successful CMMS implementation, reinforced by particular reference to real scenarios and behaviors encountered during implementation. The article by John Reeve (*Reeve 2009*) discusses the many reasons why the implementation of CMMS can fail, while the article by Real Asset Management (*Real Asset Management 2013*) is designed to assist in selecting a CMMS solution, highlighting valid reasons for considering a CMMS and offering advice for each stage of the process, equipping readers with the knowledge and tools needed to make an informed decision. The article by Terrence O'Hanlon (*O'Hanlon 2008*) presents the results of a large CMMS/EAM benchmarking survey which gives pointers to successful implementation, while the article by Tracy Smith and Clay Bush (*Smith and Bush 2010*) discusses the problem of data quality in CMMS and how poor data quality can make such systems ineffective. The article by Roger Evans (*Evans 2011*) explains why small companies need a CMMS system and the basic steps that need to be taken by a small company for the beneficial implementation of a CMMS.

Finally, the book by Martin Tate (*Tate 2015*: www.decisionevaluation.co.uk) describes a proven, rigorous, and robust method for software selection, while the presentation by Jeremy Dick of Integrate Systems Engineering (*Dick 2015*: www.integrate.biz) discusses the necessity for proper requirements capture and describes the IBM DOORS requirements capture tool.

Availability

One critical KPI that is often quoted for equipment operation is availability. The calculations of this measure need to be precisely specified in order to be meaningful. The reports the IEEE Power Engineering Society (*IEEE 2006*) and Francisco Infante and Niels Raben (*Infante and Raben 2010*: www.iec.ch) go into the necessary detail.

Condition Monitoring Techniques

Figure 10.1 summarizes some factors that influence CM and the parameters that can be measured on various equipment types.

Mark Sexton's excellent paper (*Sexton 2014*) presents the business and technical benefits of embedding CM at the start of design, rather than the usual approach of adding CM toward the end of the construction phase or during commissioning, using the London Crossrail project as an example. David Yardley's book (*Yardley 2002*) is a wide-ranging overview of CM, looking at the rationale, techniques, and applications across a range of industries. The article by Petri Nohynek (*Nohynek 2011*) states that CM is an essential component of any predictive maintenance regime, with more than 75 different types of nonintrusive CM techniques (oil particulate analysis, temperature monitoring, thermography, motor-current analysis, ultrasonics, etc.) available.

FIGURE 10.1 System factors influencing CM (*BS ISO 17359:2011*: reproduced with permission).

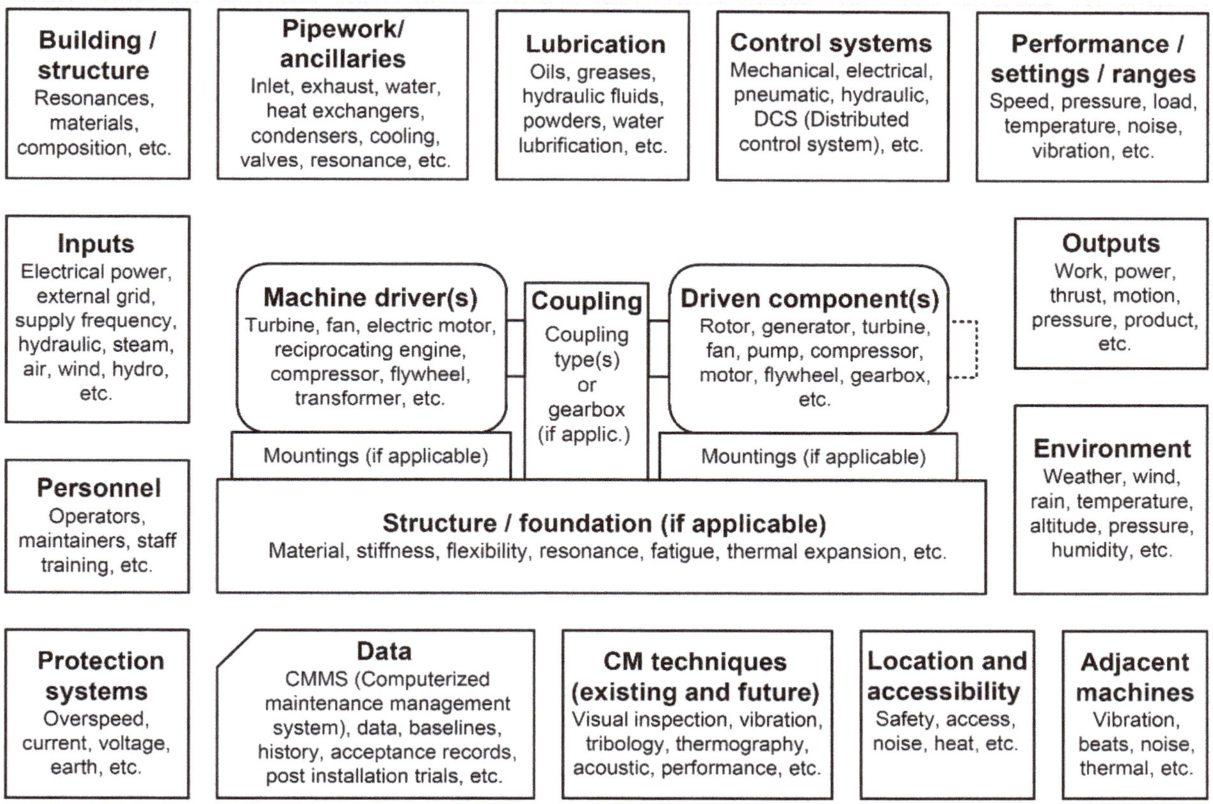

Figure 10.2 tabulates examples of CM parameters by machine type. The most widely used technique for rotating machinery, however, is still vibration monitoring, and this, allied to sophisticated data gathering and analysis systems, forms the basis for many of today's CM programs. The article by Jason Tranter and Dean Whittle (*Tranter and Whittle 2012*) looks at all the actions necessary to minimize machinery defects, while

FIGURE 10.2 Examples of CM parameters by machine type (source *BS ISO 17359:2011*: reproduced with permission).

Parameter	Machine type								
	Electric motor	Steam turbine	Aero gas turbine	Industrial gas turbine	Pump	Com-pressor	Electric generator	Reciprocating internal combustion engine	Fan
Temperature	•	•	•	•	•	•	•	•	•
Pressure		•	•	•	•	•		•	•
Pressure (head)					•				
Pressure ratio			•	•		•			
Pressure (vacuum)		•			•				
Air flow			•	•		•		•	•
Fuel flow			•	•				•	
Fluid flow		•			•	•			
Current	•						•		
Voltage	•						•		
Resistance	•						•		
Electrical phase	•						•		
Input power	•				•	•	•		•
Output power	•	•	•	•			•	•	
Noise	•	•	•	•	•	•	•	•	•
Vibration	•	•	•	•	•	•	•	•	•
Acoustic emission	•	•	•	•	•	•	•	•	•
Ultrasonics	•	•	•	•	•	•	•	•	•
Oil pressure	•	•	•	•	•	•	•	•	•
Oil consumption	•	•	•	•	•	•	•	•	•
Oil (tribology)	•	•	•	•	•	•	•	•	•
Thermography	•	•	•	•	•	•	•	•	•
Torque	•	•				•	•	•	
Speed	•	•	•	•	•	•	•	•	•
Length		•							
Angular position		•	•	•		•			
Efficiency (derived)		•	•	•	•	•		•	

• Indicates condition monitoring measurement parameter is applicable.

the article by Frank Vandijk (*Vandijk 2012*) discusses the role of online and off-line CM, and the article by John Bernet (*Bernet 2009*) looks at the development, use, and value of remote machine condition assessment. The article by Andrew Mellor (*Mellor 2013*) states that some dramatic claims have been made about the level of expected returns from CBM, with figures ranging from 5:1 to 20:1, whereas the experience is less dramatic, with many organizations making returns of less than 2:1. Nevertheless some organizations do achieve the significantly higher levels of savings promised. The article examines important strategies that distinguish the best practitioners from the others and serves as a checklist to extract maximum value out of a squeezed maintenance budget. The article by Dennis Shreve (*Shreve 2009*) focuses on the recent transition from a traditional portable, walk-around CM program to continuous surveillance systems, addressing areas for concern and presenting a specific case history and success story to show the advantages of increased awareness and improved reliability with online surveillance.

The article by Michael Tansley (*Tansley 2010*) explains that the real reason CM is carried out is to help manage the risk of operating an item of equipment by providing information that enables a rational decision to be made on the maintenance required. Maintenance is about managing the risk of operating equipment in a certain condition and under certain operating conditions, either from a "duty of care" perspective that demands the provision of safe equipment or from an operational perspective, ensuring that equipment is available and reliable. The article by Kate Hartigan (*Hartigan 2010*) discusses how, by using the latest CM and automatic lubrication systems for bearings, process manufacturers can reduce the risk and costs associated with unforeseen breakdowns to critical production plant and machinery.

A detailed review of the theoretical and practical aspects of online monitoring for performance assessment is presented in a series of three reports written by Wesley Hines and others (*Hines and Siebert 2006*; *Hines et al. 2008a, 2008b*: www.nrc.gov).

The Coxmoor series of CM books, now sadly discontinued, covers a wide range of techniques in both theoretical and practical detail. As well as a concise encyclopedia of the subject (*Hunt 2006*), there are books on acoustic emission and ultrasonics (*Holroyd 2000*), corrosion (*Rothwell and Tullmin 1999*), level, leakage and flow (*Hunt 2001*), load monitoring (*Scott 2003*), noise and acoustics (*Peters 2002*), oil analysis (*Evans and Hunt 2003*), thermography (*Thomas 1999*), vibration (*Reeves 1998*), and wear debris analysis (*Roylance and Hunt 1999*).

The article by Danny Vandeput (*Vandeput 2011*) describes the benefits of oil analysis, an often-overlooked technology that complements vibration analysis and provides valuable additional information, while the article by Martin Williamson (*Williamson 2013*) argues that recent technological advances in rapid on-site oil checking should encourage companies to make more use of oil condition information and analysis.

The article by Michael Herring (*Herring 2011*) argues that CM of electric motors should involve not only testing for bearing failure (via vibration analysis, oil analysis, etc.) but also a structured testing regime for electrical faults and the monitoring of motor efficiency: correct monitoring and adjustment of motor performance will improve reliability, extend the life of the motor, and reduce overall operating costs, since motors consume a large fraction of a facility's energy. The article by Michael Herring and Tony Ruane (*Herring and Ruane 2013*) looks at the concepts of electric motor monitoring, predictive maintenance, and inspection, while the article by Ernesto Wiedenbrug (*Wiedenbrug 2009*) discusses the use of instantaneous torque measurements to diagnose motor failures. The white paper by Artesis (*Artesis 2008*: www.artesis.com) and the presentations by Joe Barnes (*Barnes 2009a, 2009b*) describe the use of advanced on-board statistical learning algorithms to monitor three-phase electric motors, while the article by John Hutchinson (*Hutchinson 2013*) explains the latest techniques available for testing, diagnosing, and monitoring defects in a high-voltage plant.

Omega Engineering Ltd. (www.omega.co.uk) produces a catalogue (*Omega Engineering 2008*) which describes a large range of sensors and loggers. They also produce publications describing techniques for noncontact temperature measurement (*Omega Engineering 1998a*), data acquisition (*Omega Engineering 1998b*), force-related measurement (*Omega Engineering 1998c*), and flow and level measurement (*Omega Engineering 2001*). Fluke Corporation (www.fluke.com) and the Snell Group have produced an excellent book (*Fluke/Snell 2009*) going into the details of thermography, while FLIR® Systems Inc. (www.flir.com) produced short primers on thermography (*FLIR 2008*, *FLIR 2014*). The article by Paula Bowle (*Bowle 2008*) aims to help the buyer of an infrared camera understand the meaning of the specifications and determine which camera and options are suitable for a given application, while the article by Michael Stuart (*Stuart 2011*) presents an "infrared photo safari" of reliability and maintenance issues, and the article by John Snell (*Snell 2015*) discusses how to implement an infrared thermography maintenance program.

There are a large number of books, papers, and courses discussing vibration analysis, including very readable introductions to the subject by Victor Wowk (*Wowk 1991*), Simon Mills of AV Technology (*Mills 2010*: www.avtechnology.co.uk), and Commtest, owned by GE (*Commtest 2006*: www.commtest.com). The articles by Colin Sanders (*Sanders 2010a, 2010b, 2011*) give an appreciation of some of the mainstream vibration and other CM techniques, the basic premise on which they work and where and how they might be usefully employed, as well as discussing what is required when taking vibration readings, how transducers work, what they can measure, and how that can be used to determine the condition of a machine. The article by Nick Williams (*Williams 2010*) discusses the basic principles underlying the analysis of vibration time waveforms and gives several examples of its application in practice to the monitoring of rotating equipment. The presentation by John Raby (*Raby 2014*) gives an overview of vibration monitoring principles, as well as discussing available technology. John Prentis' book relates vibration theory to system dynamics and control theory (*Prentis 1970*), and there is a short course run by the University of Manchester (*Ball 2004*).

The presentations by Andrew Starr (*Starr 2013a, 2013b*) provide high-level overviews of vibration, thermography, and oil and debris analysis, while the book by Cornelius Scheffer and Paresh Girdhar (*Scheffer and Girdhar 2004*) covers these subjects in more detail. Karl Dalton's presentation (*Dalton 2014*) discusses the use of wireless sensors for vibration monitoring (pointing out the data rate compromises that result from potential power and bandwidth limitations) and how the advent of web-based remote diagnostics makes world-class CM much more widely available. The presentation by Ieuan Mogridge (*Mogridge 2012*) gives a case study in the use of vibration monitoring in the nuclear power generation industry, while the article by Jason Tranter (*Tranter 2009*) looks at the use of vibration analysis to address the challenges of keeping wind turbines operational. The articles by Steve Lacey (*Lacey 2009, 2010a, 2010b*) discuss the different sources of bearing vibration and some of the characteristic defect frequencies that may be present, giving some examples of how vibration analysis can be used to detect deterioration in machine condition and how vibration-based CM can be used to detect and diagnose machine faults and form the basis of a predictive maintenance strategy.

The article by John Atkinson (*Atkinson 2009*) discusses several of the techniques employed in vibration-based CM of bearings, pointing out the differences between them and comments on their relative usefulness and implementation. The article by Akilu Yunusa-Kaltungo and Jyoti Sinha (*Yunusa-Kaltungo and Sinha 2014*) describes a simplified and systematic vibration-based diagnosis for enabling the early detection of faults in a cement plant, which is expected to enhance its safety and availability. The article by Lin Liu and Suri Ganeriwala (*Liu and Ganeriwala 2010*) looks at using vibration analysis to detect and diagnose cavitation in a centrifugal pump. Finally, the papers by Roy Freeland (*Freeland 2014*) and Perpetuum Ltd. (*Perpetuum 2014*: www.perpetuum.com) discuss a novel technology for vibration monitoring called energy harvesting, in which the energy in the vibration being monitored is used to power the sensor and data transmission, removing the need for batteries and other power sources.

The articles by Tom Murphy (*Murphy 2011a, 2011b*) explore the range of application of both simple and sophisticated airborne and contact ultrasound technologies, while the article by Andrew Chater (*Chater 2010*) explains the basic concept of stress wave analysis and demonstrates the use of the technique for investigating typical defects encountered with diesel engines. The article by Ray Beebe (*Beebe 2009*) discusses how steam turbine performance analysis can show conditions that reduce machine efficiency and output, such as blade deposits and erosion, as well as determining whether maintenance and overhaul has been effective, while his book (*Beebe 2012*) also discusses vibration analysis of steam turbines.

The paper by Drew Troyer and Jim Fitch (*Troyer and Fitch 1999*) discusses oil analysis in some detail, while the article by TestOil (*TestOil 2015*) discusses the use of analytical ferrography to diagnose wear conditions and prevent machine failure.

Considering the huge number of diesel engines in service worldwide, details of how to apply CM techniques to these prime movers are comparatively scarce. Clifford Power Systems (http://cliffordpower.com) has produced some very useful information sheets (*Clifford Power Systems 2013*). Mohammad Abdulqader's book (*Abdulqader 2006*) has two chapters detailing the sensors and indicators that can be fitted to a diesel engine, while the report by Binh Pham and others (*Pham et al. 2012*) looks in detail at the approach needed to monitor emergency diesel generators. A presentation by Vadrevu Pramodkumar (*Pramodkumar 2010*: www.wartsila.com) discusses the technologies and sensor choices for monitoring large diesel generators, while a paper by Harold Jarrett (*Jarrett 2013*) looks at what can be gained from trending data transmitted from smaller units.

A brochure by Asentria (www.asentria.com) describes the monitoring required for the power-producing equipment for mobile phone mast sites (*Asentria 2013* and Figure 10.3) which also covers site safety and security aspects. Full monitoring solutions for such sites are also available from AIO Systems (www.aiosystems.com), Galooli Group (www.galooli.com), HMS Industrial Networks (Netbiter: www.netbiter.com), PowerOasis Ltd. (www.power-oasis.com), Ramboll Group (www.ramboll.com), MastMinder (www.mastminder.com), Infozech Software Pvt. Ltd. (www.infozech.com), Invendis Technologies Pvt. Ltd. (www.invendis.com), Nimble Wireless Inc. (www.nimblewireless.com), WebNMS (part of ZOHO Corp. Pvt. Ltd.: www.webnms.com), and many others.

Azura Engineering Ltd. (www.azura-engineering.com), Controllis Ltd. (www.controllis.com), Gemini Data Loggers (UK) Ltd. (www.geminidataloggers.com), Trimble Navigation Ltd. (www.trimble.com), and DBR and Associates (www.dbr.co.uk) are just a small selection of firms that sell data gathering and logging units.

The IMechE Tribology Group seminar papers on wind turbine CM (*IMechE 2009*) discuss a range of wind turbine monitoring techniques in some detail, while the MSc thesis by Larry Juang (*Juang 2010*) and the paper by John O'Connor and others (*O'Connor et al. 2013*) discuss methods for the CM of batteries.

FIGURE 10.3 Typical mobile phone site monitoring (Asentria Corp.: reproduced with permission).

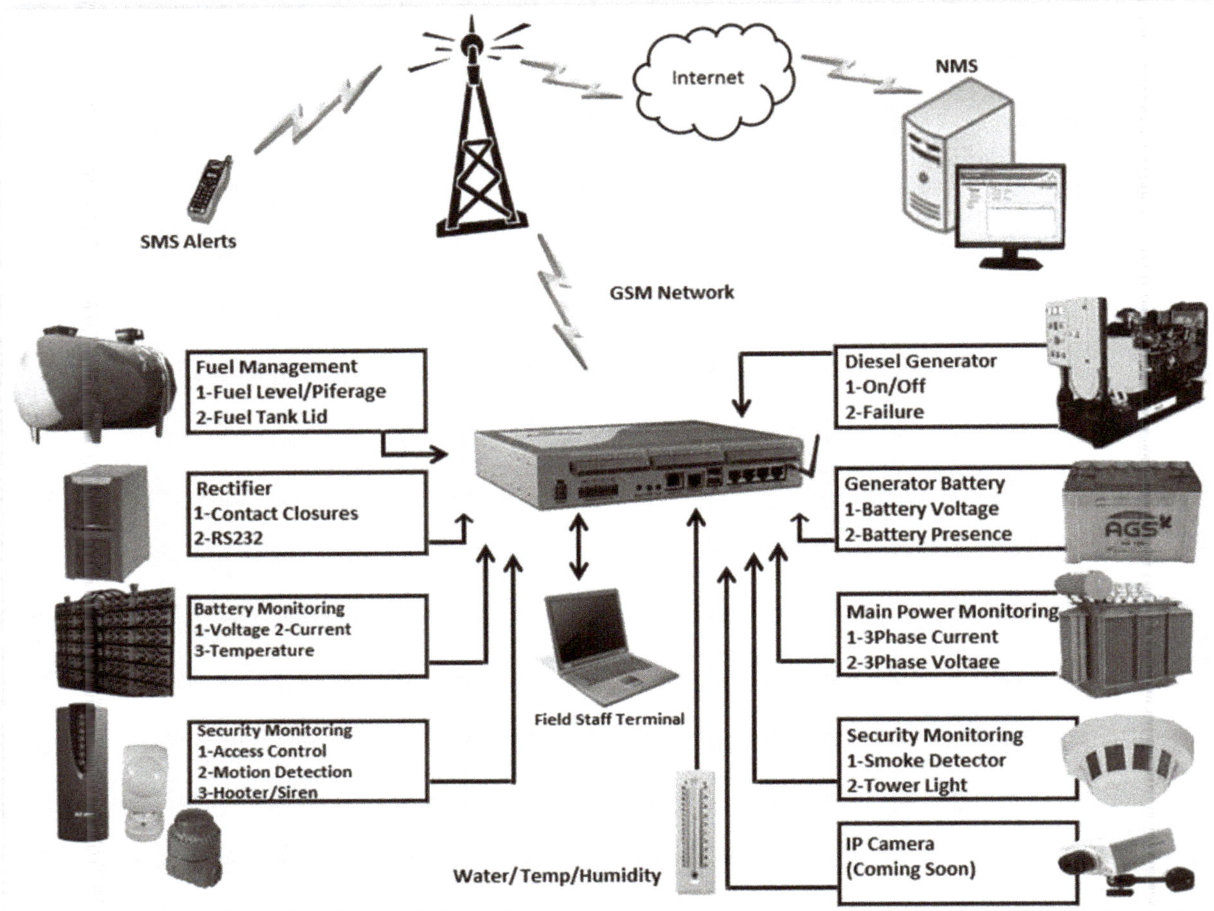

The article by Kam Chana and Donald Lyon (*Chana and Lyon 2009*) discusses the use of noncontact tip-timing eddy current probes that have made it possible to assess turbine blade health and implement condition-based predictive maintenance routines.

The presentation by Mark Simmons (*Simmons 2012*) discusses the use of image data to monitor the condition of high-voltage electricity transmission towers, while the presentation by Phil Haywood (*Haywood 2013*) describes a very innovative robotic inspection technique for high-voltage power lines.

The presentation by Hafiz Wasif and Dave Lawton (*Wasif and Lawton 2012*) describes a CM system for a food processing machine, showing how complex signals can be intelligently analyzed to produce simple red-amber-green outputs that can be easily understood by machine operators and maintainers.

The article by Martin Pilling and Les Wilkinson (*Pilling and Wilkinson 2003*) discusses the application of CM to RCM in the railway industry, while the article by Peng Guo and Nan Bai (*Guo and Bai 2011*) discusses the application of a sophisticated CM method (Auto Associative Kernel Regression (AAKR)) to temperature trend analysis of a wind turbine gearbox.

Mike Sondalini's paper (*Sondalini 2013c*) argues that, while finding a problem before it becomes a failure is good, companies can end up with so much work that the maintenance costs and backlog increase, and people become overloaded: understanding when a predictive maintenance strategy can cause uneconomic maintenance, or why CM can produce unending failures, is vital to achieve reliable equipment with low maintenance cost.

The article by Terrence O'Hanlon (*O'Hanlon 2015*) states that drones are a game changer that hold vast potential for streamlining and reducing the cost of inspection and monitoring tasks associated with reliability and asset performance.

The article by Carlos Gamez (*Gamez 2014*) discusses how CM can be used to determine asset health, current CM technical developments, and what systems and processes can be used to establish and manage a successful CM program, while the article by David Stevens (*Stevens 2014*) discusses how to switch to a CBM strategy.

The paper by Europump (*Europump 2012*) highlights the importance of in-service monitoring of pumping systems, describes the main parameters to monitor and the ways to sense them, and briefly illustrates how they can be effectively processed and analyzed to retrieve valuable information on the status of the equipment under surveillance.

Condition monitoring of ships is growing in importance, given the very arduous conditions in which many ocean-going vessels operate. Bonita Nightingale's article (*Nightingale 2009*) and *Goorangai 2010* give short overviews, while Anders Sundberg's paper (*Sundberg 2003*) reviews the economics of ship CM, other important considerations, and the links to the new DNV classification, showing that a correct maintenance strategy can create better economic returns than are shown by traditional economic models: modern management should treat such a strategy as a means of realizing the profit potential inherent in each vessel. The paper by Gabriele Manno and others (*Manno et al. 2014*) looks at the integration of online health assessment of individual components with system level failure models (such as a fault tree) in order to predict system level reliability and then develops importance measures that can be used to evaluate the criticality of components. Daniel Shorten's paper (*Shorten 2015*) looks at the reasons behind the slow adoption of CM by the shipping industry. Standards produced by the British Institute of Non-Destructive Testing (*BINDT 2015*: www.bindt.org), Lloyd's Register (*Lloyd's Register 2013*: www.lr.org), Det Norsk Veritas (*Det Norsk Veritas 2008*: www.dnvgl.com), and Germanischer Lloyd (*Germanischer Lloyd 2008*: www.dnvgl.com) cover the qualification and assessment of CM personnel and give guidance for ship machinery CM. The paper by Kongsberg Maritime (*Kongsberg Maritime 2014*: www.km.kongsberg.com) gives an overview of ship engine diagnostics, performance, and emissions monitoring, while the paper by Kevin Logan (*Logan 2011*) discusses the use of a ship's propeller as a dynamometer for ship hull CM, and the MSc thesis by Mads Aas-Hansen (*Aas-Hansen 2010*) discusses the determination of ship hull fouling resistance in some detail. The article by Christian Cabos (*Cabos 2011*) looks at the use of the latest software for ship hull monitoring that allows for 3-D imaging, integrated reporting, and the highest levels of accuracy in examining the integrity of a vessel's structure, while the paper by Geoff Walker (*Walker 2011*) looks at the monitoring of Liquid Natural Gas (LNG) carrier seawater pumps.

Finally, the article by David Manning-Ohren (*Manning-Ohren 2015*) gives ten key rules for CM, while the article by Jason Tranter (*Tranter 2015*) states that, in the majority of cases, CM techniques are used to detect fault conditions that should not exist (because they have arisen due to poor procurement, storage, work management, installation, maintenance, and other operating practices) and that, while CM is vitally important, more must be done in order to maximize plant utilization through an active defect elimination program.

Field Service

An excellent magazine devoted to the increasingly complex subject of field service is *Field Service News*, a bimonthly publication edited by Kris Oldland of 1927 Media Ltd. (http://fieldservicenews.com) that covers topics such as software, hardware, fleet operations, logistics, technology, and management and documents many of the conferences and events that are relevant to field service. Both print and pdf format versions are available free to subscribers.

The article by Mark Forrest (*Forrest 2014*) discusses the link between customer satisfaction and business performance, and the strategic role customer service plays in this, while the articles by Nick Frank (*Frank 2014a, 2014b*) draw on real-life experience to help identify a blueprint for companies looking to establish a profitable service division and give a series of real-life examples that outline how service companies can operate profitably. Kris Oldland's interview with Martin Summerhayes (*Oldland and Summerhayes 2014*) looks at the areas where many service companies can go wrong, while Aly Pinder of the Aberdeen Group (*Pinder 2014*: www.aberdeen.com) looks at four ways to excel at service in the "age of the customer," and Jim Rapoza and Ali Pinder look at how new technologies, trends in mobile, and live interactions are transforming support (*Rapoza and Pinder 2015*).

A flyer by Jen Montgomery (*Montgomery 2015*) summarizes the key challenges facing service organizations, and three truths that they need to understand, while the booklet produced for Field Service Europe by Worldwide Business Research and Pegasystems Inc. (*WBR/Pegasystems 2015*) gives the results of a survey of 115 field service executives that looked at the trends and business impacts of field service. Finally, the excellent e-book written by Advanced Field Service Solutions Ltd. (*Advanced Field Service 2015*: www.advancedfieldservice.com) discusses how it is possible to use existing, readily available technology to profitably deliver service excellence and explores future developments that could transform the field service industry in the future.

References

* denotes sources that the author found particularly helpful and/or inspiring.

Aas-Hansen, M., "Monitoring of Hull Condition of Ships," M.Sc. thesis, Department of Marine Technology, Norwegian University of Science and Technology, Trondheim, Norway, 2010. Pg. 203

Abdulqader, M., *Diesel Generator Auxiliary Systems and Instruments* (Alkhobar, Saudi Arabia: MRA Engineering, 2006). Pg. 201

Ackert, S., *Basics of Aircraft Maintenance Programs for Financiers: Evaluation and Insights of Commercial Aircraft Maintenance Programs* (San Francisco, CA: Jackson Square Aviation, 2010). Pg. 194

Ackert, S., *Engine Maintenance Concepts for Financiers: Elements of Turbofan Shop Maintenance Costs*, 2nd ed. (San Francisco, CA: Jackson Square Aviation, 2011). Pg. 194

* Advanced Field Service, "The Service Manager Handbook: 2015/16 Edition," Wilmslow, UK: Advanced Field Service Solutions Ltd., 2015. Pg. 204

Al Shaalane, A., "Improving Asset Care Plans in Mining: Applying Developments from Aviation Maintenance," MSc thesis, Department of Industrial Engineering, Stellenbosch University, Stellenbosch, South Africa, 2012. Pg. 192

Aladon, "Maintenance Management," Lutterworth, UK: Aladon Ltd., 1999. Pg. 191

Allied Reliability, "What Every Senior Manager Must Know about Maintenance and Reliability," Tulsa, OK: Allied Reliability Inc., 2006. Pg. 191

Allied Reliability, PdM Secrets Revealed: How to Improve Your PdM Program or Start One from Scratch, 5th ed. (Charleston, SC: Allied Reliability Inc., 2010). Pg. 191

Anand, G., Kodali, R., and Mishra, R., "The Evolution of Maintenance," *Asset Management & Maintenance Journal* 21, no. 3 (July 2008): 46-53. Pg. 190

Andrews, J. and Moss, T., *Reliability and Risk Assessment*, 2nd ed. (London, UK: Professional Engineering Publishing Ltd., 2002). Pg. 187

ARMS Reliability, "Why RCA Doesn't Work," Ocean Grove, Victoria, Australia: ARMS Reliability, 2012. Pg. 188

ARMS Reliability, "101 Tips and Tricks to Improve Your Root Cause Analysis," Ocean Grove, Victoria, Australia: ARMS Reliability, 2014a. Pg. 188

ARMS Reliability, "4 Simple Steps to a 5-Star Cause and Effect Chart," Ocean Grove, Victoria, Australia: ARMS Reliability, 2014b. Pg. 189

Artesis, "Artesis - Predictive Maintenance Revolution," Cambridge, UK: Artesis LLP, 2008. Pg. 199

Arthasartsri, S. and Ren, H., "Inventory Control Strategy for Airlines," *Asset Management & Maintenance Journal* 21, no. 4 (October 2008): 58-63. Pg. 195

Asentria, "Passive Monitoring Solutions," Seattle, WA: Asentria Corp., 2013. Pg. 201

Assetivity, "5 Keys to Lean Maintenance and Improving Maintenance Productivity," *Asset Management & Maintenance Journal* 27, no. 4 (July 2014): 9-11. Pg. 193

Atkinson, J., "Diagnosing Machine Bearing Faults," *Maintenance & Engineering* 9, no. 6 (November/December 2009): 43-46. Pg. 200

Atkinson, J., "Proactive Maintenance," *Maintenance & Engineering* 12, no. 5 (September/October 2012): 8-12. Pg. 190

Atout, T., "The Planner: The Heart of the Maintenance Process," *Uptime* 6, no. 47 (June/July 2011): 44-47. Pg. 195

* Bagadia, K., *Computerized Maintenance Management Systems Made Easy: How to Evaluate, Select and Manage CMMS* (New York: McGraw-Hill, 2006). Pg. 196

Ball, A., "Machinery Vibration Monitoring and Analysis: Short Course," Manchester, UK: Manchester School of Engineering, Manchester University, 2004. Pg. 200

Banerji, R. and Chakraborty, D., "Lifecycle FMECA," *Maintenance & Engineering* 11, no. 2 (March/April 2011): 53-54. Pg. 189

Barnes, J., "Simplifying Predictive Maintenance," Cambridge, UK: Artesis LLP, 2009a. Pg. 199

Barnes, J., "Simplifying Predictive Maintenance for Rotating Machinery - Technical Presentation," Cambridge, UK: Artesis LLP, 2009b. Pg. 199

Beebe, R., "Condition Monitoring of Steam Turbines by Performance Analysis," *Maintenance & Engineering* 9, no. 5 (September/October 2009): 40-45. Pg. 201

Beebe, R., *Steam Turbine Performance and Condition Monitoring*, (Fort Myers, FL: Reliabilityweb.com, 2012). Pg. 201

Belanger, D., "Top Five Maintenance and Reliability Enablers for Improved Operational Performance," *Maintenance & Engineering* 13, no. 1 (January/February 2013): 42-46. Pg. 191

Bennett, R., "Machines Don't Die… They're Murdered," *Uptime* 3, no. 29 (June/July 2008): 8-17. Pg. 191

Berneski, B., "Delivering Task Periodicities with Reliability Centered Maintenance (RCM)," *Uptime* 6, no. 44 (December/January 2011): 50-53. Pg. 190

Bernet, J., "From Near or Far: An Overview of Remote Condition Monitoring Assessment," *Uptime* 4, no. 36 (August/September 2009): 54-59. Pg. 199

Bhachu, U., "Reliability Centered Maintenance: An Introduction for the New Reliability Engineer," *Uptime* 4, no. 35 (June/July 2009): 44-47. Pg. 192

BINDT, "General Requirements for the Qualification and Assessment of Marine Machinery Condition Monitoring Personnel – Issue 1, Rev B," Northampton, UK: British Institute of Non-Destructive Testing, 2015. Pg. 203

Blache, K., "Benchmarking a Better Understanding: Benchmarks Shed Light on Maintenance and Reliability Perceptions," *Uptime* 5, no. 38 (December/January 2010): 44-49. Pg. 187

Blischke, W. and Murthy, D., *Case Studies in Reliability and Maintenance* (Hoboken, NJ: John Wiley & Sons Inc., 2003). Pg. 192

Bofillos, J., "Can a CMMS/EAM Solution Really Solve All Your Problems?," *Uptime* 7, no. 50 (December/January 2012): 24-25. Pg. 187

Bowle, P., "What You Should Know before You Buy: A Guide to Buying an Infrared Camera," *Uptime* 4, no. 31 (October/November 2008): 32-36. Pg. 200

Bradshaw, L., "AMMJ Knowledge Centre: Past Issues," Mornington, Victoria, Australia: Engineering Information Transfer Pty Ltd., 2014. Pg. 186

BS ISO 17359:2011, "Condition Monitoring and Diagnostics of Machines - General Guidelines," London, UK: British Standards Institution, 2011. Pg. 197,198

Burnett, S., "A Simplified Numerical Decision Making Toolbox for Physical Asset Management Decisions," MSc thesis, Department of Industrial Engineering, Stellenbosch University, Stellenbosch, South Africa, 2013. Pg. 192

Butters, T., Güttel, S., Shapiro, J., and Sharpe, T., "Statistical Cluster Analysis and Visualisation for Alarm Management Configuration," *IET/IAM 2014 Asset Management Conference*, Millennium Gloucester Hotel, London, UK, November 27-28, 2014 (Stevenage, UK: Institution of Engineering and Technology). Pg. 190

Cabos, C., "Condition Monitoring Support for the Hull," *Digital Ship*, May 2011, 50. Pg. 203

* Campbell, J. and Reyes-Picknell, J., *Uptime: Strategies for Excellence in Maintenance Management*, 3rd ed. (Boca Raton, FL: CRC Press/Taylor & Francis, 2015) Pg. 190

Castro, P., "Reaching for the Top: Creating a New Partner with Reliability Centered Operations," *Uptime* 5, no. 40 (April/May 2010): 8-17. Pg. 192

Castro, P., "9 Leadership Principles for a Successful Maintenance and Reliability Program," *Asset Management & Maintenance Journal* 26, no. 4 (July 2013): 3-5. Pg. 193

Chafai, M. and Refou, L., "A Failure Mode Effects and Criticality Analysis of a Cement Plant Rotary Kiln Drive System," *Asset Management & Maintenance Journal* 21, no. 1 (January 2008): 44-51. Pg. 189

Chana, K. and Lyon, D., "Turbo-Machinery Tip Timing Comes of Age," *Maintenance & Engineering* 9, no. 1 (January/February 2009): 35-40. Pg. 202

Chater, A., "Locomotive Diesel Stress Wave Analysis," *Maintenance & Engineering* 10, no. 3 (May/June 2010): 42-45. Pg. 201

Chikezie, N., "When the Bubbles Burst: Managing Pump Cavitation," *Asset Management & Maintenance Journal* 23, no. 1 (February 2010): 13-14. Pg. 188

Chikezie, N., "When the Bubbles Burst – Managing Pump Cavitation via Troubleshooting," *Maintenance & Engineering* 11, no. 1 (January/February 2011): 22-24. Pg. 188

Clifford Power Systems, "Information Sheets," Tulsa, OK: Clifford Power Systems, Inc., 2013. Pg. 201

Commtest, "Beginners Guide to Machine Vibration," Christchurch, New Zealand: Commtest Instruments Ltd., 2006. Pg. 200

Cox, S. and Tait, R., *Safety, Reliability and Risk Management: An Integrated Approach*, 2nd ed. (Oxford, UK: Butterworth-Heinemann, 1998). Pg. 187

Crotty, R., "Lean Thinking Maintenance, Repair and Overhaul Jobs," *Uptime* 10, no. 71 (June/July 2015): 18-21. Pg. 193

Cyient, "Engineering Solutions for Predictive Maintenance in Rail," Hyderabad, India: Cyient Ltd., 2015. Pg. 194

Dabbs, T. and Pereira, D., 10 Steps to Pump Reliability: Part 1," *Uptime* 8, no. 55 (October/November 2012): 48-51. Pg. 188

Dabbs, T. and Pereira, D., "10 Steps to Pump Reliability: Part 2," *Uptime* 8, no. 56 (December/January 2013): 46-47. Pg. 188

Dalton, K., "The Balance between Power and Bandwidth," *Condition Monitoring and Analysis Seminar*, Dexter House, London, UK, February 5-6, 2014 (London, UK: IBC Energy/Informa UK Ltd.). Pg. 200

Daniel, S., "The Do's and Don'ts of Effective Maintenance Planning," *Asset Management & Maintenance Journal* 25, no. 4 (August 2012): 8-10. Pg. 193

Davis, J., "'Big M' and the Performance Culture: Managing Maintenance for Production Reliability," *Uptime* 5, no. 40 (April/May 2010): 42-45. Pg. 190

Dean, P., "Performance Indicators: What You Need to Know," *Maintenance & Engineering* 10, no. 2 (March/April 2010): 45-48. Pg. 193

del Carmen Garcia Lizarraga, M. and Sinha, J., "A Quantified Approach for a Reliability-Based Maintenance Model," *Maintenance & Engineering* 15, no. 3 (May/June 2015): 41-44. Pg. 194

Det Norsk Veritas, "Guidance for Condition Monitoring – Classification Notes No. 10.2," Høvik, Norway: Foundation Det Norsk Veritas, 2008. Pg. 203

DeWald, D., "Key Performance Indicators for Stores and MRO," *Uptime* 6, no. 47 (June/July 2011): 36-39. Pg. 195

Dezfuli, H., Benjamin, A., Everett, C., Maggio, G. et al., "NASA Risk Management Handbook: Version 1.0," Washington DC: National Aeronautics and Space Administration, 2011. Pg. 187

Dezfuli, H., Stamatelatos, M., Maggio, G., Everett, C. et al., "NASA Risk-Informed Decision Making Handbook: Version 1.0," Washington DC: National Aeronautics and Space Administration, 2010. Pg. 187

Dhillon, B., *Engineering Maintenance: A Modern Approach* (Boca Raton, FL: CRC Press, 2002). Pg. 192

Dick, J., "Requirements Workshop," Bath, UK: Integrate Systems Engineering Ltd., 2015. Pg. 196

DiStefano, R., "Unlocking Big Benefits: Predictive Technologies Increase Bottom Line," Fort Myers, FL: Reliabilityweb.com/GE Energy, 2005. Pg. 185

DiStefano, R. and Covino, L., "Enterprise Reliability: Changing the Game," *Asset Management & Maintenance Journal* 20, no. 2 (April 2007): 22-30. Pg. 186

Dufresne, P., "How to Proactively Run What You Have Now," *Uptime* 10, no. 69 (February/March 2015): 6-9. Pg. 188

Dunn, S., "5 Keys to Lean Maintenance and Improving Maintenance Productivity: Part 1," *Asset Management & Maintenance Journal* 27, no. 5 (September 2014a): 8-10. Pg. 193

Dunn, S., "5 Keys to Lean Maintenance and Improving Maintenance Productivity: Parts 2 and 3," *Asset Management & Maintenance Journal* 27, no. 6 (November 2014b): 12-16. Pg. 193

Dunn, S., "5 Keys to Lean Maintenance and Improving Maintenance Productivity: Parts 4 and 5," *Asset Management & Maintenance Journal* 28, no. 1 (January 2015a): 20-24. Pg. 193

Dunn, S., "RCM, PMO and Preventative Maintenance Programs: 4 Key Concepts," *Asset Management & Maintenance Journal* 28, no. 3 (May 2015b): 8-10. Pg. 193

* Emmett, S. and Wheelhouse, P., *Excellence in Maintenance Management: A Cross-Functional Approach* (Cambridge, UK: Cambridge Academic, 2011). Pg. 190

Engineered Software, "Using Statistics to Schedule Maintenance," Lacey, WA: Engineered Software Inc., 1999. Pg. 193

Europump, "Improvement of Reliability of Pumps by Condition Monitoring: Consequences for MTBR/MTBF," Brussels, Belgium: Europump, 2012. Pg. 203

Evans, J. and Hunt, T., *Oil Analysis Handbook* (Chipping Norton, UK: Coxmoor, 2003). Pg. 199

Evans, R., "Too Small for CMMS? Think Again," *Maintenance & Engineering* 11, no. 3 (May/June 2011): 8-10. Pg. 196

Fallaize, G., "Risky Business!," *Maintenance & Engineering* 10, no. 3 (May/June 2010): 54. Pg. 185

* Flaus, J., *Risk Analysis: Socio-Technical and Industrial Systems* (London, UK: ISTE Ltd., 2013). Pg. 187

FLIR, "IR Thermography Primer," Hong Kong, PRC: FLIR® Systems Co. Ltd., 2008. Pg. 200

FLIR, "Thermal Imaging for Electrical and Mechanical Diagnostics," Wilsonville, OR: FLIR® Systems Inc., 2014. Pg. 200

Fluke/Snell, *Introduction to Thermography Principles* (Orland Park, IL: American Technical Publishers, Fluke Corp. and the Snell Group, 2009). Pg. 200

Forrest, M., "The Changing Landscape of Customer Service," *Field Service News*, no. 1 (March 2014): 12. Pg. 204

Francis, R., "Downtime More Important Than Uptime – Striving for Successful Shutdowns," *Asset Management & Maintenance Journal* 21, no. 3 (July 2008): 54-62. Pg. 191

Frank, N., "Creating Value from Services: Where to Start?," *Field Service News*, no. 1 (March 2014a): 36-37. Pg. 204

Frank, N., "Making Your Services Profitable," *Field Service News*, no. 2 (July 2014b): 44-45. Pg. 204

Fraser, A., "A Reliable Plant Is a Cost-Effective Plant - And Also a Safe One," *Maintenance & Engineering* 14, no. 3 (May/June 2014a): 17-19. Pg. 186

Fraser, A., "Using Reliability to Unlock Sustainable Success," Warrington, UK: Reliable Manufacturing Ltd., 2014b. Pg. 186

Freeland, R., "Energy Harvesting - A Practical Reality for Wireless Sensing," Southampton, UK: Perpetuum Ltd., 2014. Pg. 201

Friedman, A., "Avoiding the Pitfalls: Why Do Predictive Maintenance Programs Fail?," *Uptime* 4, no. 34 (April/May 2009a): 34-37. Pg. 190

Friedman, A., "Why PdM Programs Fail: Misuse of Technology," *Uptime* 4, no. 36 (August/September 2009b): 32-35. Pg. 190

Friedman, A., "Choose Your Partners Wisely: Or Discover Another Reason Why PdM Programs Can Fail," *Uptime* 5, no. 38 (December/January 2010): 32-35. Pg. 190

Gallimore, J., "Cutting Maintenance Costs – Safely," *Maintenance & Engineering* 9, no. 1 (January/February 2009): 25-26. Pg. 194

Gamez, C., "Asset Health and Condition Monitoring: How to Establish a Successful Program," *Asset Management & Maintenance Journal* 27, no. 2 (March 2014): 10-15. Pg. 203

Garvey, R., "Why Do Machines Go Bump in the Night?," *Uptime* 8, no. 60 (August/September 2013): 18-22. Pg. 188

Gauthier, S., "Decision Analysis to Support Condition-Based Maintenance Plus," MSc thesis, Naval Postgraduate School, Monterey, CA, 2006. Pg. 194

Germanischer Lloyd, "Rules for Classification and Construction: 1 Ship Technology - Seagoing Ships: 17 Guidelines for Machinery Condition Monitoring," Hamburg, Germany: Germanischer Lloyd AG, 2008. Pg. 203

Gluzman, D., "Cause vs Contributing Factor: An Important Aspect in Root Cause Analysis," *Uptime* 8, no. 56 (December/January 2013): 18-20. Pg. 188

GoalArt, "A Small Book on Alarms," Lund, Sweden: GoalArt AB, 2008. Pg. 190

Goly, K., "The Anatomy of a PdM Program: A Business-Based Approach to Developing an Effective Program," *Uptime* 5, no. 39 (February/March 2010): 36-39. Pg. 190

Goly, K., "How to Develop a Spare Parts Stocking Strategy," *Uptime* 8, no. 55 (October/November 2012): 36-38. Pg. 195

Goorangai, "Condition Monitoring Maintenance," Occasional Papers of the Royal Australian Naval Reserve Professional Studies Program, Volume 4, Number 3, March 2010 (Canberra, Australia: RANR Professional Studies Program, Office of the Director General Reserves (Navy)). Pg. 203

* Gulati, R., *Maintenance and Reliability Best Practices*, 2nd ed. (New York: Industrial Press Inc, 2013). Pg. 190

Gulati, R. and Mears, C., *Workbook to Accompany Maintenance and Reliability Best Practices*, 2nd ed. (New York: Industrial Press Inc., 2014). Pg. 190

Guo, P. and Bai, N., "Wind Turbine Gearbox Condition Monitoring with AAKR and Moving Window Statistic Methods," *Energies* 4, no. 11 (2011): 2077-2093. Pg. 202

Gutiérrez, E., "Integral Asset Care," *Uptime* 10, no. 70 (April/May 2015): 58-61. Pg. 194

Haarman, M., *Value Driven Maintenance®: Creating Shareholder Value with Maintenance* (Dordrecht, the Netherlands: Mainnovation, 2002). Pg. 192

Haarman, M., *Discover the Hidden Value of Maintenance* (Aberdeen, UK: Mainnovation, 2011). Pg. 192

* Haarman, M. and Delahay, G., *Value Driven Maintenance®: New Faith in Maintenance* (Dordrecht, the Netherlands: Mainnovation, 2004). Pg. 192

Haarman, M. and Delahay, G., "Value Driven Maintenance® - Discover the Hidden Value in Your Maintenance Organization," *Maintenance & Engineering* 13, no. 3 (May/June 2013): 12-17. Pg. 192

Harper, J., "Lessons from a Successful National CMMS Implementation," *Asset Management & Maintenance Journal* 23, no. 3 (July 2010): 14-19. Pg. 196

Harper, J., "Strategic Maintenance Reporting to Enable Sustained Improvement" *Asset Management & Maintenance Journal* 24, no. 1 (January 2011): 14-19. Pg. 19?

Hartigan, K., "Condition Monitoring – An Insurance Policy against Unforeseen Plant Downtime," *Maintenance & Engineering* 10, no. 5 (September/October 2010): 8-11. Pg. 199

Hawkins, B., "Maintenance Management 101: Basic Rules to Live By," *Uptime* 2, no. 25 (October/November 2007). Pg. 191

Hawkins, B., "Maintenance Management 201: More Basic Rules to Live By," *Uptime* 3, no. 30 (August/September 2008): 40-43. Pg. 191

Hawkins, B., "How Reliability Impacts Shareholder Value," Sandy Hook, CT: Management Resources Group, Inc., 2011. Pg. 186

Haywood, P., "Robotic Inspection and Repair of 400kV Overhead Conductor Using LineScout™ Technology," *3rd Annual Predictive Asset Management Seminar*, Dexter House, London, UK, December 4-5, 2013 (London, UK: IBC Energy/ Informa UK Ltd.). Pg. 202

Hendriks, S., Zaal, T., and van der Sanden, B., "Maintenance Criticality Reassessment: A FMECA Based Method for Reducing Maintenance Costs as a Result of Residual Risk Optimization," *Asset Management & Maintenance Journal* 28, no. 2 (March 2015): 14-18. Pg. 189

Herring, M., "Electric Motor Testing… the Missing Link of Condition Monitoring?," *Maintenance & Engineering* 11, no. 6 (November/December 2011): 41-43. Pg. 199

Herring, M. and Ruane, T., "The Monitoring of Electric Motors," *Maintenance & Engineering* 13, no. 5 (September/October 2013): 25-27. Pg. 199

Hide, M., "Back to the Basics: Developing and Delivering a Maintenance Plan," *Uptime* 6, no. 43 (October/November 2010): 34-37. Pg. 190

Hines, J. and Siebert, R., "Technical Review of On-Line Monitoring Techniques for Performance Assessment - Volume 1: State-of-the-Art," Knoxville, TN: University of Tennessee-Knoxville, Department of Nuclear Engineering, 2006. Pg. 199

Hines, J., Garvey, D., Siebert, R., and Usynin, A., "Technical Review of On-Line Monitoring Techniques for Performance Assessment - Volume 2: Theoretical Issues," Knoxville, TN: University of Tennessee-Knoxville, Department of Nuclear Engineering, 2008a. Pg. 199

Hines, J., Garvey, J., Garvey, D., and Siebert, R., "Technical Review of On-Line Monitoring Techniques for Performance Assessment - Volume 3: Limiting Case Studies," Knoxville, TN: University of Tennessee-Knoxville, Department of Nuclear Engineering, 2008b. Pg. 199

Holroyd, T., *Acoustic Emission and Ultrasonic Monitoring Handbook* (Oxford, UK: Coxmoor, 2000). Pg. 199

Honkanen, T., "Modelling Industrial Maintenance Systems and the Effects of Automatic Condition Monitoring," Ph.D. thesis, Helsinki University of Technology, Helsinki, Finland, 2004. Pg. 192

Hunt, T., *Level, Leakage & Flow Monitoring* (Oxford, UK: Coxmoor, 2001). Pg. 199

Hunt, T., *The Concise Encyclopaedia of Condition Monitoring* (Chipping Norton, UK: Coxmoor, 2006). Pg. 199

Hutchinson, J., "Testing, Diagnostics and Monitoring of High Voltage Plant and Equipment," *Maintenance & Engineering* 13, no. 4 (July/August 2013): 56-59. Pg. 200

IEEE, IEEE Standard Definitions for Use in Reporting Electric Generating Unit Reliability, Availability and Productivity, 2nd ed. (New York: Institute of Electrical and Electronics Engineers Power Engineering Society, 2006). Pg. 197

IMechE, "Wind Turbine Condition Monitoring," *IMechE Tribology Group Seminar*, Douglas Knoop Centre, Sheffield, UK, November 18, 2009 (London, UK: Institution of Mechanical Engineers). Pg. 201

Infante, F. and Raben, N., "IEC TS 61400-26-1: Wind Turbines - Part 26-1: Time Based Availability for Wind Turbines," Geneva, Switzerland: International Electrotechnical Commission, 2010. Pg. 197

ITEM Software, "Reliability Block Diagram (RBD)," Whiteley, UK: ITEM Software Inc., 2007. Pg. 187

Jaramillo, C., "Myths of RCM Implementation: Part 1," *Uptime* 9, no. 64 (April/May 2014a): 16-19. Pg. 193

Jaramillo, C., "Myths of RCM Implementation: Part 2," *Uptime* 9, no. 65 (June/July 2014b): 18-21. Pg. 193

* Jardine, A. and Tsang, A., *Maintenance, Replacement and Reliability: Theory and Applications* (Boca Raton, FL: CRC/Taylor & Francis, 2006). Pg. 187

Jarrett, H., "Advances in Failure Prediction through Remote Monitoring," Buford, GA: Omnimetrix LLC., 2013. Pg. 201

Jonker, R. and Haarman, M., "Value Driven Maintenance," *Uptime* 1, no. 19 (October/November 2006): 8-14. Pg. 192

Juang, L., "Online Battery Monitoring for State-of-Charge and Power Capability Prediction," MSc thesis, University of Wisconsin-Madison, Madison, WI, 2010. Pg. 201

Kalocai, J., "Weibull-Based Method for Failure Mode Characterization and Remaining Life Expectancy Estimation," *Uptime* 7, no. 54 (August/September 2012): 40-46. Pg. 188

Kamalapurka, D., Houshyar, A., White, B., and Lamberson, L., "Selecting a Computerised Maintenance Management System," *Asset Management & Maintenance Journal* 20, no. 2 (April 2007): 60-67. Pg. 196

Karalexis, G. and Smith, R., "Advanced Maintenance and Reliability Best Management Practices," *Asset Management & Maintenance Journal* 23, no. 2 (April 2010): 6-16. Pg. 188

Kelly, A., *Maintenance Management – Its Auditing and Benchmarking* (Farnham, UK: Conference Communications, 2002). Pg. 193

Kelly, T., "Some Thoughts on Maintenance Budgeting," *Maintenance & Engineering* 11, no. 6 (November/December 2011): 13-17. Pg. 194

Kennedy, P., Bapat, V., and Kurchina, P., *In Pursuit of the Perfect Plant* (New York: Evolved Technologist Press, 2008). Pg. 185

Killick, M. and Thomas, G., "Best Practice Maintenance Planning," *Asset Management & Maintenance Journal* 21, no. 4 (October 2008): 8-13. Pg. 193

Knapp, M., "The Power of Intelligent Assets," *Assets*, May 2014, 19-21. Pg. 185

Knezevic, J., *Systems Maintainability: Analysis, Engineering and Management* (London, UK: Chapman & Hall, 1997). Pg. 192

Koelzer, D., "Control Your CMMS to Cut Costs: Optimizing Planning and Scheduling," *Uptime* 5, no. 37 (October/November 2009): 18-20. Pg. 190

Kongsberg Maritime, "Engine Performance, Engine Diagnostics & Emission Monitoring: Maximizing Performance by Providing the Full Picture," Kongsberg, Norway: Kongsberg Maritime, 2014a. Pg. 203

Krüppel, K. and Lawton, T., "OMDEC Overview," Unpublished presentation, Toronto, Ontario: Optimal Maintenance Decisions Inc., 2008. Pg. 187

Lacey, S., "An Overview of Vibration Analysis," *Asset Management & Maintenance Journal* 22, no. 2 (April 2009): 42-53. Pg. 200

Lacey, S., "The Role of Vibration Monitoring in Predictive Maintenance – Part 1," *Maintenance & Engineering* 10, no. 1 (January/February 2010a): 36-45. Pg. 200

Lacey, S., "The Role of Vibration Monitoring in Predictive Maintenance – Part 2," *Maintenance & Engineering* 10, no. 2 (March/April 2010b): 38-44. Pg. 200

Lanthier, P., "Quantifying Financial Benefits from an Asset Performance Initiative," *Uptime* 6, no. 46 (April/May 2011): 46-51. Pg. 186

Larsson, J., "Condition Monitoring and Alarm Management," *Condition Monitoring and Analysis Seminar*, Dexter House, London, UK, February 5-6, 2014 (London, UK: IBC Energy/Informa UK Ltd). Pg. 190

Latino, R., "Root Cause Analysis: The Top 10 Reasons Why It Will Never Work in My Company," *Asset Management & Maintenance Journal* 28, no. 1 (January 2015): 34-35. Pg. 189

LCE, "Preventative and Predictive Maintenance," Charleston, SC: Life Cycle Engineering, 2006. Pg. 194

Lee, J. and AbuAli, M., "Intelligent Maintenance Systems: The Next Five Years and Beyond," *Asset Management & Maintenance Journal* 23, no. 3 (July 2010): 6-13. Pg. 194

* Levitt, J., *Complete Guide to Preventive and Predictive Maintenance* (New York: Industrial Press Inc., 2003). Pg. 191

Levitt, J., "CMMS: 9 Plus 50 Questions," *Asset Management & Maintenance Journal* 20, no. 2 (April 2007): 8-13. Pg. 196

* Levitt, J., *The Handbook of Maintenance Management* 2nd ed. (New York: Industrial Press Inc., 2009). Pg. 191

Liu, L. and Ganeriwala, S., "Diagnosing Tiny Bubbles: Vibration Signatures of Cavitation in a Centrifugal Pump," *Uptime* 5, no. 40 (April/May 2010): 50-59. Pg. 201

Lloyd's Register, "ShipRight Design & Construction: Machinery Planned Maintenance and Condition Monitoring," London, UK: Lloyd's Register Group, 2013. Pg. 203

Logan, K., "Using a Ship's Propeller for Hull Condition Monitoring," *ASNE Intelligent Ships Symposium IX*, Bossone Research Center, Drexel University, Philadelphia, PA, May 25-26, 2011 (Alexandria, VA: American Society of Naval Engineers). Pg. 203

Manning-Ohren, D., "Ten Key Rules for Condition Monitoring," *Maintenance & Engineering* 15, no. 3 (May/June 2015): 23-24. Pg. 203

Manno, G., Knutsen, K., and Vartdal, B., "An Importance Measure Approach to System Level Condition Monitoring of Ship Machinery Systems," *MFPT: Technology Solutions for Affordable Sustainment Conference 2014*, Sheraton Oceanfront Hotel, Virginia Beach, VA, May 21-22, 2014 (Dayton, OH: Society for Machinery Failure Prevention Technology). [Pg. 203]

McDermott, R., Mikulak, R., and Beauregard, M., *The Basics of FMEA*, 2nd ed. (New York: Productivity Press/Taylor & Francis Group LLC., 2009). [Pg. 189]

Mellor, A., "Back to Basics: How to Save More Money with Condition Monitoring," *Maintenance & Engineering* 13, no. 1 (January/February 2013): 36-41. [Pg. 199]

Miller, D., "Reliability through Optimized Setup and Changeovers - Part 1," *Uptime* 10, no. 70 (April/May 2015a): 42-44. [Pg. 188]

Miller, D., "Reliability through Optimized Setup and Changeovers - Part 2," *Uptime* 10, no. 71 (June/July 2015b): 42-45. [Pg. 188]

Mills, S., *Cost Effective Maintenance* (Handforth, UK: AV Technology Ltd., 2007). [Pg. 191]

* Mills, S., *Vibration Monitoring and Analysis Handbook* (Northampton, UK: British Institute of Non-Destructive Testing, 2010). [Pg. 200]

Mills, S., "Successful Failure Mode and Effect Analysis (FMEA)," Handforth, UK: AV Technology Ltd., 2012. [Pg. 189]

Mobley, K., "What Is Risk Management?," *Uptime* 6, no. 47 (June/July 2011): 40-41. [Pg. 187]

Mogridge, I., "Online Vibration Monitoring of Large Rotating Plant," *2nd Annual Predictive Asset Management Seminar*, Royal Garden Hotel, London, UK, September 11-13, 2012 (London, UK: IBC Energy/Informa UK Ltd.). [Pg. 200]

Montgomery, J., "Driving Differentiation and Growth through Service: The Director's Report," *Field Service Fall 2015*, 2015 (London, UK: Worldwide Business Research). [Pg. 204]

* Moore, R., *Making Common Sense Common Practice: Models for Manufacturing Excellence*, 3rd ed. (Oxford, UK: Elsevier Butterworth-Heinemann, 2004). [Pg. 185]

Moore, R., *Selecting the Right Manufacturing Improvement Tools: What Tool? When?* (Oxford, UK: Elsevier Butterworth-Heinemann, 2006). [Pg. 185]

Moore, R., "Reliability Leadership and Manufacturing Excellence - A Common Sense Strategy for Business Excellence," Knoxville, TN: RM Group, Inc., 2007. [Pg. 186]

Moore, R., "What Tool When? Some Thoughts," *Uptime* 7, no. 50 (December/January 2012): 13-16. [Pg. 193]

Moore, R., "Reliability: Essential for a Safe Cost Effective Environmentally Friendly Operation," *Society for Maintenance and Reliability Professionals Annual Conference*, Indiana Convention Center and Westin Indianapolis, Indianapolis, IN, October 14-16, 2013 (Atlanta, GA: Society for Maintenance and Reliability Professionals). [Pg. 186]

Moore, R., "Is Your Plant Reliable? It's Good for Personal and Process Safety," *Uptime* 10, no. 69 (February/March 2015): 24-27. [Pg. 186]

Moubray, J., "The Case against Streamlined RCM," 2000. [Pg. 192]

* Moubray, J., *Reliability-Centered Maintenance*, 2nd ed. (New York: Industrial Press Inc., 2001). [Pg. 192]

Murphy, T., "There's Something in the Air: Understanding Ultrasound - Part 1," *Maintenance & Engineering* 11, no. 3 (May/June 2011a): 44-48. [Pg. 201]

Murphy, T., "There's Something in the Air: Understanding Ultrasound – Part 2," *Maintenance & Engineering* 11, no. 4 (July/August 2011b): 54-57. Pg. 201

Narayan, V., *Effective Maintenance Management* (New York: Industrial Press Inc., 2004). Pg. 192

* Narayan, V., Wardhaugh, J., and Das, M., *Case Studies in Maintenance and Reliability: A Wealth of Best Practices* (New York: Industrial Press Inc., 2012). Pg. 190

NASA, "Reliability Centered Maintenance Guide for Facilities and Collateral Equipment," Washington DC: National Aeronautics & Space Administration, 2008. Pg. 192

NASA, "Tools of Reliability Analysis - Introduction and FMEAs," Cleveland, OH: Lewis Research Center, National Aeronautics & Space Administration, 2012. Pg. 189

Neale, M., "A Review of Plant Maintenance Methods and Economics," in: Pusie, H. and Pusie, S. eds., *Technology Showcase: Integrated Monitoring, Diagnostics and Failure Prevention. Proceedings of a Joint Conference*, Mobile, AL, April 22-26, 1996 (Dayton, OH: Society for Machinery Failure Prevention Technology). Pg. 191

Nelson, T., "Risk and Criticality: Understanding Potential Failure," *Uptime* 7, no. 49 (October/November 2011): 56-58. Pg. 187

Nelson, T., "Risk Calculation Methodology," *Uptime* 7, no. 51 (February/March 2012): 48. Pg. 187

Nightingale, B., "Condition Monitoring: A Stitch in Time…," *Marine Engineers Review*, May 2009, 33-36. Pg. 203

Nohynek, P., "Realising the Benefits of Condition Monitoring," *Maintenance & Engineering* 11, no. 3 (May/June 2011): 40-43. Pg. 197

Nowlan, F. and Heap, H., *Reliability-Centred Maintenance* (San Francisco, CA: Dolby Access Press, 1978). Pg. 191

O'Hanlon, T., "Surveying the Field: Computerized Maintenance Management and Enterprise Asset Management Best Practices," *Uptime* 3, no. 30 (August/September 2008): 22-25. Pg. 196

O'Hanlon, T., "Why Drones Are the Next Internet," *Uptime* 10, no. 71 (June/July 2015): 6-11. Pg. 203

O'Connor, J., Leonard, T., and Bohensky, G., "Comparing Methods to Determine the Health of Battery Systems," Rockaway, NJ: BTECH Inc., 2013. Pg. 201

O'Connor, P., Newton, D., and Bromley, R., *Practical Reliability Engineering*, 3rd ed. (Chichester, UK: John Wiley & Sons Ltd., 1995). Pg. 187

Oldland, K. and Summerhayes, M., "Talking Strategy with the Billion Dollar Man…," *Field Service News*, no. 2 (July 2014): 12-13. Pg. 204

OMDEC, "The Elusive P-F Interval," Toronto, Ontario: Optimal Maintenance Decisions Inc., 2010. Pg. 187

Omega Engineering, "Transactions Volume 1: Non-Contact Temperature Measurement, 2nd ed. (Itasca, IL: Putman Publishing Company/Omega Press LLC, 1998a). Pg. 200

Omega Engineering, "Transactions Volume 2: Data Acquisition," (Itasca, IL: Putman Publishing Company/Omega Press LLC, 1998b). Pg. 200

Omega Engineering, " Transactions Volume 3: Force-Related Measurements," (Itasca, IL: Putman Publishing Company/Omega Press LLC, 1998c). Pg. 200

Omega Engineering, "Transactions Volume 4: Flow and Level Measurement," (Itasca, IL: Putman Publishing Company/Omega Press LLC, 2001). Pg. 200

Omega Engineering, "Complete Measurement, Control and Automation Handbook and Encyclopedia," Manchester, UK: Omega Engineering, 2008. Pg. 200

Overman, R., "Core Principles of Reliability Centered Maintenance," *Asset Management & Maintenance Journal* 22, no. 1 (February 2009): 44-48. Pg. 192

Page, A. and Karalexis, G., "Are You Doing Too Much PM? 16 Ways to Save Time and Money on Preventive Maintenance," *Asset Management & Maintenance Journal* 22, no. 3 (July 2009): 22-27. Pg. 193

Page, A. and Repasz, C., "Preventative Maintenance," *Asset Management & Maintenance Journal* 28, no. 1 (January 2015): 25-33. Pg. 193

Palmer, R., *Maintenance Planning and Scheduling Handbook*, 3rd ed. (New York: McGraw-Hill, 2012). Pg. 194

Penrose, H., *Physical Asset Management for the Executive: Don't Read This if You Are on an Airplane* (Saybrook, CT: Success by Design Publishing, 2008). Pg. 185

Pernett, C., "Root Cause Analysis: The 10 Most Common Implementation Errors," *Uptime* 3, no. 29 (June/July 2008): 52-57. Pg. 188

Perpetuum, "Getting Started with Vibration Energy Harvesting," Southampton, UK: Perpetuum Ltd., 2014. Pg. 201

Peters, R., *Noise & Acoustics Monitoring Handbook* (Oxford, UK: Coxmoor, 2002). Pg. 199

Peterson, B., "Creating the Performance Culture," *Uptime* 7, no. 52 (April/May 2012): 9-14. Pg. 190

Pham, B., Lybeck, N., and Agarwal, V., "Online Monitoring Technical Basis and Analysis Framework for Emergency Diesel Generators—Interim Report for FY 2013," Idaho Falls, ID: Idaho National Laboratory, 2012. Pg. 201

* Pierce, K., Gopalkrishnan, G., Crosby, C., Thomason, D. et al., *A Guidebook to Implementing Condition-Based Maintenance (CBM) Using Real-Time Data* (San Leandro, CA: OSIsoft LLC, 2015). Pg. 194

Pilling, M. and Wilkinson, L., "Reliability-Based Maintenance and Condition Monitoring," London, UK: Asset Management Consulting Ltd., 2003. Pg. 202

Pinder, A., "Four Ways to Excel at Service in the 'Age of the Customer'," *Field Service News*, no. 2 (July 2014): 30. Pg. 204

Plucknette, D., "Shedding Light on RCM Facilitation," *Uptime* 6, no. 43 (October/November 2010): 17-24. Pg. 192

Plucknette, D. and Castro, P., "Missing a Key Player? The Role of Operations in Reliability Centered Maintenance," *Uptime* 6, no. 43 (October/November 2010): 8-15. Pg. 192

Plucknette, D. and Colson, C., "Clean, Green and Reliable," *Uptime* 6, no. 47 (June/July 2011): 12-17. Pg. 187

Plucknette, D., Mears, C., and Gulati, R., "RCM vs FMEA: There Is a Distinct Difference!," *Uptime* 8, no. 56 (December/January 2013): 52-55. Pg. 189

Posey, A. and Slater, P., "The Missing Link in Reliability: Many Equipment Failures Could Be Linked to Materials Management," *Uptime* 5, no. 40 (April/May 2010): 38-41. Pg. 195

Pramodkumar, V., "Condition Monitoring Diesel Generators," Navi Mumbai, India: Wärtsilä India, 2010. Pg. 201

Prentis, J., *Dynamics of Mechanical Systems* (London, UK: Longman, 1970). Pg. 200

Raby, J., "Dynamix 1444 Monitoring and Protection," Milwaukee, WI: Rockwell Automation Inc., 2014. Pg. 200

Rama, P., Chen, R., and Andrews, J., "A Review of Performance Degradation and Failure Modes for Hydrogen-Fuelled PEM Fuel Cells," *Proceedings of the IMechE, Part A: Journal of Power and Energy* 222 (April 22, 2008a): 421-441. Pg. 188

Rama, P., Chen, R., and Andrews, J., "Failure Analysis of Polymer Electrolyte Fuel Cells," *Proceedings of the SAE 2008 World Congress*, Cobo Center, Detroit, MI, April 14-17, 2008b (Warrendale, PA: SAE International). Pg.188

Rapoza, J. and Pinder, A., "Getting a Clear Picture of the Future of Support: 10 Ways Support Will Change in the Future," Boston, MA: Aberdeen Group, 2015. Pg. 204

Real Asset Management, "A Guide to Selecting a CMMS," *Maintenance & Engineering* 13, no. 4 (July/August 2013): 31-32. Pg. 196

Reeve, J., "CMMS – Who Is at Fault," *Asset Management & Maintenance Journal* 22, no. 2 (April 2009): 40-41. Pg. 196

Reeves, C., *Vibration Monitoring Handbook* (Oxford, UK: Coxmoor, 1998). Pg. 199

ReliaSoft, "RS475 FRACAS Principles and Applications: Course Notes," Tucson, AZ: ReliaSoft Corp., 2008. Pg. 189

ReliaSoft, "FRACAS: From Data Collecting to Problem Solving," *Asset Management & Maintenance Journal* 28, no. 3 (May 2015): 11-14. Pg. 189

Renshaw, S., "Maintenance Plan Development: Template at Your Peril," *Maintenance & Engineering* 15, no. 1 (January/February 2015): 38-40. Pg. 193

Rezendes, R., "Developing the RCM2 Mindset: The Building Blocks of a Maintenance Philosophy," *Uptime* 4, no. 36 (August/September 2009): 46-49. Pg. 192

Riccetti, S., "Product Safety and Equipment Reliability in the Food Industry through Maintenance Engineering," *Maintenance & Engineering* 11, no. 1 (January/February 2011): 36-41. Pg. 193

Riccetti, S., "How to Monitor Food Equipment Critical Parts to Design Reliable Maintenance Tasks," *Maintenance & Engineering* 14, no. 4 (July/August 2014): 42-48. Pg. 193

Rio, R., "MRO Inventory Rationalization and Optimization," *Uptime* 6, no. 46 (April/May 2011): 32-35. Pg. 196

Roider, G., "Optimised Maintenance with Transparent Maintenance Planning," *Maintenance & Engineering* 10, no. 4 (July/August 2010): 19-20. Pg. 196

Ross, J., "Establishing Proactive Maintenance," *Uptime* 4, no. 35 (June/July 2009): 32-35. Pg. 190

Rothwell, N. and Tullmin, M., *Corrosion Monitoring Handbook* (Oxford, UK: Coxmoor, 1999). Pg. 199

Roylance, B. and Hunt, T., *Wear Debris Analysis Handbook* (Oxford, UK: Coxmoor, 1999). Pg. 199

Sanders, C., "A Maintenance Manager's Guide to Vibration Analysis and Associated Techniques in Condition Monitoring – Part 1," *Maintenance & Engineering* 10, no. 5 (September/October 2010a): 42-45. Pg. 200

Sanders, C., "A Maintenance Manager's Guide to Vibration Analysis and Associated Techniques in Condition Monitoring – Part 2," *Maintenance & Engineering* 10, no. 6 (November/December 2010b): 34-37. Pg. 200

Sanders, C., "A Maintenance Manager's Guide to Vibration Analysis and Associated Techniques in Condition Monitoring – Part 3," *Maintenance & Engineering* 11, no. 1 (January/February 2011): 42-46. Pg. 200

Scheffer, C. and Girdhar, P., *Practical Machinery Vibration Analysis and Predictive Maintenance* (Oxford, UK: Newnes/Elsevier/IDC Technologies, 2004). Pg. 200

Schenkelberg, F., "The Perils of MTBF – Part 1," *Asset Management & Maintenance Journal* 26, no. 4 (July 2013a): 32-34. Pg. 188

Schenkelberg, F., "The Perils of MTBF – Part 2," *Asset Management & Maintenance Journal* 26, no. 5 (September 2013b): 23. Pg. 188

Schlumberger Sema, "Reliability Centered Maintenance (RCM) Short Course: Introductory Training," Manchester, UK: Schlumberger Sema, 2004. Pg. 192

Scott, M., *Load Monitoring Handbook (Force, Strain, Pressure & Torque)* (Chipping Norton, UK: Coxmoor, 2003). Pg. 199

Sexton, M., "Building Condition Monitoring into Design Solutions to Improve Overall System Performance," *Condition Monitoring and Analysis Seminar*, Dexter House, London, UK, February 5-6, 2014 (London, UK: IBC Energy/Informa UK Ltd.). Pg. 197

Shorten, D., "Marine Machinery Condition Monitoring: Why Has the Shipping Industry Been Slow to Adopt?," London, UK: Lloyd's Register Group, 2015. Pg. 203

Shreve, D., "Transitions: When to Move from a Walk-Around to Online Systems for PdM," *Uptime* 5, no. 37 (October/November 2009): 52-59. Pg. 199

Simmons, M., "Predictive Asset Management - OHL Tower Steelwork," *2nd Annual Predictive Asset Management Seminar*, Royal Garden Hotel, Kensington, London, UK, September 11-13, 2012 (London, UK: IBC Energy/Informa UK Ltd.). Pg. 202

Slater, P., "The Spares Stocking Policy," *Uptime* 3, no. 30 (August/September 2008a): 50-53. Pg. 196

Slater, P., "When Is Critical Really Critical? Turning a Critical Eye to Your Inventory," *Uptime* 3, no. 28 (April/May 2008b): 50-52. Pg. 196

Slater, P., "Why You Would Be Nuts to Keep Squirrel Stores," *Asset Management & Maintenance Journal* 22, no. 2 (April 2009): 8-10. Pg. 195

* Slater, P., *Smart Inventory Solutions: Improving the Management of Engineering Materials and Spare Parts*, 2nd ed. (South Norwalk, CT: Industrial Press Inc., 2010a). Pg. 195

Slater, P., "The Truth about Inventory Management Theory," *Asset Management & Maintenance Journal* 23, no. 2 (April 2010b): 24-25. Pg. 195

Slater, P., "Six Tips to Improve Spare Parts Management," *Maintenance & Engineering* 12, no. 5 (September/October 2012): 40-42. Pg. 195

Slater, P., "The Elephant in Your Storeroom," *Maintenance & Engineering* 13, no. 1 (January/February 2013): 24. Pg. 195

Slater, P. and Levitt, J., "Spare Parts Inventory: An Exercise in Risk Management," *Maintenance & Engineering* 14, no. 1 (January/February 2014): 38-42. Pg. 195

Smith, A. and Allen, T., "Risks of Using PM Templates," *Uptime* 6, no. 48 (August/September 2011): 48-53. Pg. 192

Smith, R., "7 Days to Better Equipment Reliability," *Uptime* 8, no. 55 (October/November 2012): 32-35. Pg. 187

Smith, R., "The Top 10 Reasons Why Maintenance Planning Is not Effective," *Asset Management & Maintenance Journal* 26, no. 1 (January 2013a): 18-24. Pg. 193

Smith, R., "Why People Do not Understand the P-F Curve," *Uptime* 8, no. 60 (August/September 2013b): 46-50. Pg. 187

Smith, R. and Keeter, B., "What's the FRACAS: Failure Elimination Made Simple," *Asset Management & Maintenance Journal* 24, no. 1 (January 2011): 20-25. Pg. 189

Smith, S., "What Shutdown and When?," Daresbury, UK: ICI Eutech Ltd., 2000. Pg. 194

Smith, T. and Bush, C., "Waiting for Godot? Quality Data Is Key to Finding Promising Results," *Uptime* 5, no. 38 (December/January 2010): 22-25. Pg. 196

Snell, J., "Implementing an Infrared Thermography Maintenance Program," *Asset Management & Maintenance Journal* 28, no. 3 (May 2015): 15-19. [Pg. 200]

Sondalini, M., "Ageless Maintenance and Reliability Success Secrets," Rossmoyne, WA, Australia: Lifetime Reliability Solutions, 2009a. [Pg. 191]

* Sondalini, M., *Plant and Equipment Wellness: A Process for Exceptional Equipment Reliability and Maximum Life Cycle Profits* (Crows Nest, NSW, Australia: Engineers Media, 2009b). [Pg. 191]

Sondalini, M., "The Connection between Equipment Risk and Equipment Reliability and Its Effect on Maintenance Strategy," *Uptime* 6, no. 48 (August/September 2011): 24-27. [Pg. 186]

Sondalini, M., "Stop Committing Industrial Suicide on Reliability Cliffs," Rossmoyne, WA, Australia: Lifetime Reliability Solutions, 2012. [Pg. 186]

Sondalini, M., "31 Sure Ways to Lower Operating Asset Maintenance Costs and Improve Reliability," Rossmoyne, WA, Australia: Lifetime Reliability Solutions, 2013a. [Pg. 191]

Sondalini, M., "A Common Misunderstanding about Reliability Centered Maintenance," Rossmoyne, WA, Australia: Lifetime Reliability Solutions, 2013b. [Pg. 192]

Sondalini, M., "Don't Waste Your Time and Money with Condition Monitoring," Rossmoyne, WA, Australia: Lifetime Reliability Solutions, 2013c. [Pg. 202]

Sondalini, M., "Getting High Equipment Reliability," Rossmoyne, WA, Australia: Lifetime Reliability Solutions, 2013d. [Pg. 186]

Sondalini, M., "How Plant Wellness Reduces Failures and Maintenance Costs," Rossmoyne, WA, Australia: Lifetime Reliability Solutions, 2013e. [Pg. 191]

Sondalini, M., "Machines Only Stop Working after Their Parts Fail," Rossmoyne, WA, Australia: Lifetime Reliability Solutions, 2013f. [Pg. 186]

Sondalini, M., "Maintenance Best Practices for Outstanding Equipment Reliability and Maintenance Results," Rossmoyne, WA, Australia: Lifetime Reliability Solutions, 2013g. [Pg. 191]

Sondalini, M., "Show Me the Money in Maintenance," Rossmoyne, WA, Australia: Lifetime Reliability Solutions, 2013h. [Pg. 191]

Sondalini, M., "Top Four Machine Faults," Rossmoyne, WA, Australia: Lifetime Reliability Solutions, 2013i. [Pg. 186]

Sondalini, M., "Why Do Machines and Equipment Continue to Fail in Companies?," Rossmoyne, WA, Australia: Lifetime Reliability Solutions, 2013j. [Pg. 186]

Sondalini, M., "How to See the Future of Your Business, Plant and Equipment," *Maintenance & Engineering* 14, no. 6 (November/December 2014): 32-35. [Pg. 191]

Sondalini, M. and Witt, H., "What Is Equipment Reliability and How Do You Get It," Rossmoyne, WA, Australia: Lifetime Reliability Solutions, 2013. [Pg. 186]

Spurlock, M. and Keen, J., "Understanding the Rate of Change Dangers with Alarms," *Uptime* 8, no. 55 (October/November 2012): 56-58. [Pg. 190]

Stamatelatos, M., Vesely, W., Dugan, J., Fragola, J. et al., "*Fault Tree Handbook with Aerospace Applications*, 2nd ed. (Washington, DC: National Aeronautics & Space Administration, 2002). [Pg. 188]

Starr, A., "Condition Monitoring Techniques - Overview," Unpublished presentation, Cranfield, UK: Cranfield University, 2013a. [Pg. 200]

Starr, A., "PdM Tools Overview - Condition Monitoring Techniques," *3rd Annual Predictive Asset Management Seminar*, Dexter House, London, UK, December 4-5, 2013b (London, UK: IBC Energy/Informa UK Ltd.). Pg. 200

Stevens, B., "OMDEC Solution Overview," Godfrey, Ontario: Optimal Maintenance Decisions Inc., 2008. Pg. 187

Stevens, B., "Using Business Common Sense to Improve Maintenance Practices," *Asset Management & Maintenance Journal* 23, no. 4 (October 2010): 6-11. Pg. 193

Stevens, D., "Switching Strategies: Moving to Condition-Based Maintenance," *Maintenance & Engineering* 14, no. 6 (November/December 2014): 14-16. Pg. 203

Stuart, M., "Infrared Safari: A Photo Essay on the Hunt for Reliability Problems with Thermography," *Uptime* 6, no. 44 (December/January 2011): 33-36. Pg. 200

Sundberg, A., "Management Aspects on Condition Based Maintenance - The New Opportunity for Maritime Industry," *9th International Conference on Marine Engineering Systems*, Helsinki University of Technology, Helsinki, Finland, May 19-21, 2003 (London, UK: Marine Management Organisation). Pg. 203

Swatkowski, P., "Maintenance and Steven Covey: The 7 Habits of Highly Effective Maintenance Organizations," *Uptime* 5, no. 37 (October/November 2009): 28-31. Pg. 195

Tansley, M., "Changing the Perception of Condition Monitoring," *Asset Management & Maintenance Journal* 23, no. 1 (February 2010): 7-9. Pg. 199

Tate, M., *Off-the-Shelf IT Solutions: A Practitioner's Guide to Selection and Procurement* (Swindon, UK: British Computer Society, 2015). Pg. 196

Taylor, P., "Maintenance Software Strategies," *Maintenance & Engineering* 11, no. 2 (March/April 2011): 8-12. Pg. 196

TestOil, "Analytical Ferrography: The Forensic Science of Oil Analysis," *Asset Management & Maintenance Journal* 28, no. 3 (May 2015): 24-28. Pg. 201

Thomas, R., *Thermography Monitoring Handbook* (Oxford, UK: Coxmoor, 1999). Pg. 199

Todd, P., "Maintenance and Condition Monitoring Contribution to Business Improvement Initiatives," *Asset Management & Maintenance Journal* 25, no. 3 (May-July 2012): 12-18. Pg. 193

Tomlingson, P., "The Plant Maintenance Program: It's a Team Effort," *Uptime* 10, no. 71 (June/July 2015): 58-60. Pg. 193

Tranter, J., "Blowin' in the Wind: Vibration Analysis of Wind Turbines," *Uptime* 4, no. 33 (February/March 2009): 55-59. Pg. 200

Tranter, J. and Whittle, D., "Condition Improvement Instead of Just Condition Monitoring," *Maintenance & Engineering* 12, no. 1 (January/February 2012): 8-11. Pg. 198

Tranter, T., "Condition Monitoring Is Not Enough," *Maintenance & Engineering* 15, no. 3 (May/June 2015): 36-38. Pg. 203

Troyer, D. and Fitch, J., "Oil Analysis Basics," Tulsa, OK: Noria Corp., 1999. Pg. 201

Turner, S., "Russian Roulette: Loading the Barrel with Statistics and Pulling the Trigger on Reliability," *Asset Management & Maintenance Journal* 23, no. 1 (February 2010): 33-40. Pg. 188

Tyson, L., "Improving the Reliability of a Turbofan Jet Engine by Using the Weibull Distribution for Failure Mode Analysis," *Uptime* 6, no. 47 (June/July 2011): 48-51. Pg. 188

Tyson, L., "Improving the Reliability of a Turbofan Jet Engine," *Uptime* 8, no. 55 (October/November 2012): 40-41. Pg. 188

Vandeput, D., "The Forgotten Star in Condition-Based Maintenance," *Maintenance & Engineering* 11, no. 1 (January/February 2011): 14-17. Pg. 199

Vandijk, F., "The Essential Role of Condition Monitoring," *Maintenance & Engineering* 12, no. 2 (March/April 2012): 3. Pg. 199

Vicente, F., "Taking Care of Business: The Benefits of Implementing Reliability Engineering," *Uptime* 5, no. 42 (August/September 2010): 46-49. Pg. 187

Walker, G., "Maintenance Strategy – The Critical Role of Condition Monitoring," *Maintenance & Engineering* 10, no. 3 (May/June 2010): 12. Pg. 194

Walker, G., "LNG Carrier Seawater Pump Condition Monitoring," Cambridge, UK: Artesis LLP, 2011. Pg. 203

Wang, H. and Pham, H., *Reliability and Optimal Maintenance* (London, UK: Springer-Verlag London Ltd., 2006). Pg. 187

Wasif, H. and Lawton, D., "Online Predictive Condition Monitoring System for Stork Filling Machines," *2nd Annual Predictive Asset Management Seminar*, Royal Garden Hotel, Kensington, London, UK, September 11-13, 2012. Pg. 202

WBR/Pegasystems, "The 2014/2015 European Service Trends Report," London, UK/Cambridge, MA: Worldwide Business Research/Pegasystems Inc., 2015. Pg. 204

Webster, B., "Understanding and Comparing Risk," *Uptime* 7, no. 51 (February/March 2012): 44-47. Pg. 187

Welland, S., "Embracing Future Perfect: Technological Advances Streamline and Optimize Maintenance Practices," *Uptime* 3, no. 29 (June/July 2008): 18-23. Pg. 196

Wheelhouse, P., "Maintenance Key Performance Indicators," *Maintenance & Engineering* 13, no. 5 (September/October 2013): 40-43. Pg. 193

Wheelhouse, P., "Maintenance Key Performance Indicators," *Asset Management & Maintenance Journal* 28, no. 1 (January 2015): 15-18. Pg. 193

Wiedenbrug, E., "What Torque Can Tell You: PdM of Mechanical Failures Using Electrical Measurements for Instantaneous Torque," *Uptime* 4, no. 36 (August/September 2009): 36-41. Pg. 199

Williams, G., "Risk Intuition: Blurring the Lines between Procedure and Practice," *Asset Management & Maintenance Journal* 26, no. 4 (July 2013): 14-18. Pg. 188

Williams, G., "Risk Intuition: Blurring the Lines between Procedure and Practice," *Maintenance & Engineering* 14, no. 5 (September/October 2014): 48-50. Pg. 188

Williams, N., "Going Back to Basics – Using Vibration Time Waveform Data," *Maintenance & Engineering* 10, no. 1 (January/February 2010): 46-50. Pg. 200

Williamson, M., "Affordable Oil Condition Monitoring," *Maintenance & Engineering* 13, no. 1 (January/February 2013): 19-20. Pg. 199

Wilmott, P., "TPM – The 'Pinch-Point' Solution for Lean Manufacturing," *Maintenance & Engineering* 10, no. 6 (November/December 2010): 38-43. Pg. 193

Wireman, T., "Common Pitfalls of Planning and Scheduling Maintenance Activities," *Uptime* 6, no. 48 (August/September 2011): 44-47. Pg. 194

Wowk, V., *Machinery Vibration* (New York: McGraw-Hill, 1991). Pg. 200

Wurzbach, R., "A Web-Based Cost Benefit Analysis Method for Predictive Maintenance," Brouge, PA: Maintenance Reliability Group, 2000. Pg. 191

Yardley, E. ed., *Condition Monitoring: Engineering the Practice* (Bury St Edmunds, UK: Professional Engineering Publishing Ltd., 2002). Pg. 197

Yunusa-Kaltungo, A. and Sinha, J., "Vibration Monitoring: A Case Study," *Maintenance & Engineering* 14, no. 1 (January/February 2014): 43-48. Pg. 201

Zach, T., "Understanding Criticality: Myths and Pitfalls to Avoid," *Uptime* 10, no. 69 (February/March 2015): 16-18. Pg. 188

Zieler, J., "The Maintenance Storeroom: Keys to Efficient Maintenance Operation," *Uptime* 7, no. 50 (December/January 2012): 40-45. Pg. 198

Websites

Aberdeen Group Pg. 204	www.aberdeen.com
Advanced Field Service Solutions Ltd. Pg. 204	www.advancedfieldservice.com
AIO Systems Pg. 201	www.aiosystems.com
Allied Reliability Pg. 191	www.alliedreliability.com
ARMS Reliability Pg. 188	www.armsreliability.com
Artesis LLP Pg. 199	www.artesis.com
Asentria Corp. Pg. 201	www.asentria.com
Assetivity Pg. 193	www.assetivity.com.au
AV Technology Pg. 200	www.avtechnology.co.uk
Azura Engineering Ltd. Pg. 201	www.azura-engineering.com
British Institute of Non-Destructive Testing Pg. 203	www.bindt.org
Clifford Power Systems Pg. 201	http://cliffordpower.com
Commtest Pg. 200	www.commtest.com
Controllis Pg. 201	www.controllis.com
DBR & Associates Ltd. Pg. 201	www.dbr.co.uk
Decision Evaluation Ltd. Pg. 196	www.decisionevaluation.co.uk
Det Norsk Veritas Pg. 203	www.dnvgl.com
Engineered Software Inc. Pg. 193	www.engineeredsoftware.com
Field Service News Pg. 204	http://fieldservicenews.com
FLIR Systems Inc. Pg. 200	www.flir.com
Fluke Corp. Pg. 200	www.fluke.com
Galooli Group Pg. 201	www.galooli.com
Gemini Data Loggers (UK) Ltd. Pg. 201	www.geminidataloggers.com
Germanischer Lloyd Pg. 203	www.dnvgl.com
GoalArt® Pg. 189	www.goalart.com
HMS Industrial Networks (Netbiter) Pg. 201	www.netbiter.com

Infozech Software Pvt. Ltd. Pg. 201	www.infozech.com
Integrate Systems Engineering Pg. 196	www.integrate.biz
International Electrotechnical Commission Pg. 197	www.iec.ch
Invendis Technologies India Pvt. Ltd. Pg. 201	www.invendis.com
Kongsberg Maritime Pg. 203	www.km.kongsberg.com
Life Cycle Engineering Pg. 194	www.lce.com
Lifetime Reliability Solutions Pg. 191	www.lifetime-reliability.com
Lloyd's Register Pg. 203	www.lr.org
Mainnovation Pg. 192	www.mainnovation.com
Maintenance Online Pg. 185	www.maintenanceandengineering.com
MastMinder Pg. 201	www.mastminder.com
National Aeronautics and Space Administration Pg. 189	www.nasa.gov
Nimble Wireless Inc. Pg. 201	www.nimblewireless.com
Omega Engineering Ltd. Pg. 200	www.omega.co.uk
Optimal Maintenance Decisions Inc. Pg. 187	www.omdec.com
Paul Barringer & Associates Inc. Pg. 188	www.barringer1.com
Perpetuum Pg. 201	www.perpetuum.com
Phillip Slater Pg. 195	www.phillipslater.com
PowerOasis Ltd. Pg. 201	www.power-oasis.com
Ramboll Group Pg. 201	www.ramboll.com
Reliability Engineering Resources (Weibull Analysis) Pg. 188	www.weibull.com
ReliaSoft Corp. (xFRACAS) Pg. 186	www.reliasoft.com
Sabisu Pg. 190	www.sabisu.co
SparePartsKnowHow Pg. 195	www.sparepartsknowhow.com
Trimble Navigation Ltd. Pg. 201	www.trimble.com
Uptime Pg. 186	www.uptimemagazine.com
US Nuclear Regulatory Commission Pg. 199	www.nrc.gov
Wärtsilä Pg. 201	www.wartsila.com
WebNMS Pg. 201	www.webnms.com

Additional Material

References

* denotes sources that the author found particularly helpful and/or inspiring.

Alachmanetis, K., "LAROS: Integrated Vessel Performance and Remote Condition Monitoring System," Alexandroupolis, Greece: Prisma Electronics SA, 2015.

Andgren, N., "How Proactive and Predictive Maintenance Reduce Process Risk," Centreville, VA: Upper Occoquan Service Authority, 2016.

Ando, H., "Performance Monitoring and Analysis for Operational Improvements," *German Society for Maritime Technology International Conference on Ship Efficiency*, Hotel Hafen Hamburg, Hamburg, Germany, September 26-27, 2011 (Tokyo, Japan: Monohakobi Technology Institute).

Aschwanden, T., "Predictive Maintenance and Service: An Overview Presentation," Biel, Switzerland: SAP (Schweiz) AG, 2014.

Barringer, H., "Availability, Reliability, Maintainability, and Capability," Humble, TX: Barringer & Associates Inc., 1997.

DE&S, "Reliability & Maintainability: An Introduction for MOD Staff," Bristol, UK: Defence Equipment & Support, TLS Reliability Team, UK Ministry of Defence, 2009.

Diamond, S. and Marfatia, A., *Predictive Maintenance for Dummies*, IBM Limited Edition (Hoboken, NJ: John Wiley & Sons Inc., 2013).

* Eberline, P., *A Diamond in the Rough: Unleashing the Power of Field Service Transformation* (Pleasanton, CA: ServiceMax Inc, 2015).

Elsayed, E., "Reliability and Maintainability Engineering: An Overview," New Brunswick, NJ: Department of Industrial and Systems Engineering, Rutgers University, 2013.

Fathom Maritime Intelligence, "Fathom Focus: Driving Profit through Integrated Operations," Slough, UK: Fathom Maritime Intelligence, 2014a.

Fathom Maritime Intelligence, "Fathom Focus - Ship Performance Management," Slough, UK: Fathom Maritime Intelligence, 2014b.

FLIR, "The Ultimate Infrared Handbook for R&D Professionals," Wilsonville, OR: FLIR® Systems Inc., 2013.

FLIR, "Thermal Imaging for Science/R&D," Wilsonville, OR: FLIR® Systems Inc., 2015.

* Galar, D., Sandborn, P., and Kumar, U., *Maintenance Costs and Life Cycle Cost Analysis* (Boca Raton, FL: CRC Press/Taylor & Francis, 2017).

Hansen, S., "Vessel Performance Management," *American Bureau of Shipping Seminar*, Genoa, Italy, April 18, 2016 (Houston, TX: American Bureau of Shipping).

Hansen, S., Petersen, J., Jensen, J., and Lützen, M., "Performance Monitoring of Ships," Department of Mechanical Engineering, Technical University of Denmark, Lyngby, Denmark, 2012.

Hart, D. and Eberline, P., "Top 5 Field Service Power Metrics," Pleasanton, CA: ServiceMax Inc., 2016.

Hasselaar, T., "An Investigation into the Development of an Advanced Ship Performance Monitoring and Analysis System," School of Marine Science and Technology, Newcastle University, Newcastle upon Tyne, UK, 2010.

Hockley, C., Huby, G., James, I., Fedrick, V. et al., *No-Fault-Found (NFF): A Best Practice Guide* (Cranfield, UK: ESPRC Centre for Innovative Manufacturing - Through-life Engineering Services, 2016).

Hood, C., Woods, R., and Cooper, D., "Digitising the Aircraft Technical Log: The Move from Paper," Glasgow, UK: NVable Ltd., 2016.

* Hubbard, D., *The Failure of Risk Management: Why It's Broken and How to Fix It* (Hoboken, NJ: John Wiley & Sons Inc., 2009).

Karlskind, M., Smith, S., and Berry, A., "Service Is Hard: Turning Common Field Service Challenges into Customer Engagement Opportunities," Burlington, MA: ClickSoftware Technologies, 2016.

Khan, S., Phillips, P., Hockley, C., and Jennions, I., *No Fault Found: The Search for the Root Cause* (Warrendale, PA: SAE International, 2015).

Kongsberg Maritime, "Vessel Performance Optimizer: Cost Efficient Vessel Operation," Kongsberg, Norway: Kongsberg Maritime, 2014b.

Koo, W. and Van Hoy, T., "Determining the Economic Value of Preventive Maintenance," Chicago, IL: Jones Lang LaSalle, IP, Inc., 2003.

Kroes, M., Watkins, W., Delp, F., and Sterkenburg, R., *Aircraft Maintenance and Repair*. 7th ed. (New York: McGraw-Hill, 2013).

Lewis, J., "NFF - A Practical Perspective for Modern Rolling Stock: Dealing with NFF in Reliability Growth," *Reducing the Impact of NFF through System Design Symposium*, EPSRC TES Centre, Cranfield University, Bedford, UK, March 18, 2013 (Derby, UK: Bombardier Transportation).

NASA, "NASA-STD-8729.1: Planning, Developing and Managing an Effective Reliability and Maintainability (R&M) Program," Washington, DC: National Aeronautics & Space Administration, 1998.

Nexala, "Transport Life Cycle Management Solutions," Dublin, Ireland: Nexala, 2015.

Oldland, K. ed., "*The Handy Little Book for Field Service Managers*, 2016 Edition (London, UK: 1927 Media Ltd., 2016).

Oldland, K. ed., *The Handy Little Book for Field Service Managers*, 2017 Edition (London, UK: 1927 Media Ltd., 2017).

Oldland, K. ed., *The Handy Little Book for Field Service Managers*, 2018 Edition (London, UK: 1927 Media Ltd., 2018).

Owen, R., "Big Data Takes Off," *Professional Engineering*, December 2016, 24-28.

Pallath, P., "Using Predictive Maintenance to Approach Zero Downtime: How Predictive Analytics Makes This Possible," Walldorf, Germany: SAP SE, 2015.

Railnova, "How to Choose and Implement a Predictive Maintenance Platform," Brussels, Belgium: Railnova Sprl., 2017.

Sahay, A., *Leveraging Information Technology for Optimal Aircraft Maintenance, Repair and Overhaul (MRO)* (Sawston, UK: Woodhead Publishing Ltd., 2017).

* Savage, S., *The Flaw of Averages: Why We Underestimate Risk in the Face of Uncertainty* (Hoboken, NJ: John Wiley & Sons Inc., 2012).

Senior, R., *Train Doctor: Troubleshooting with Diesel and Electric Traction* (Barnsley, UK: Pen and Sword Books Ltd., 2016).

ServiceMax, "The Definitive Guide to Modern Field Service Management," London, UK: ServiceMax Europe, 2015.

ServicePower, "Field Service Management: The Changing Technology Landscape," McLean, VA: ServicePower Inc., 2016.

* SMRP, *SMRP Best Practices*, 5th ed., Maintenance & Reliability Body of Knowledge (Atlanta, GA: Society for Maintenance and Reliability Professionals, 2017).

Sondalini, M., "The Pocket Maintenance Advisor: A Maintenance and Operation Crew Best Practices Reference Book," St Louis, MO: Feed Forward Publications/ Business Industrial Network, 2003.

Sondalini, M., "Summary Report on Using and Introducing Precision Maintenance," Rossmoyne, WA, Australia: Lifetime Reliability Solutions, 2017.

Sullivan, P., Pugh, R., Melender, A., and Hunt, W., "Operations and Maintenance Best Practices: A Guide to Achieving Operational Efficiency - Release 3," Richland, WA: Pacific Northwest National Laboratory, US Department of Energy, 2010.

Trelleborg Marine Systems, "SeaTechnik™ Ship Performance Monitor (SPM): Level 1 Monitoring for All Ship Types," Hawarden, UK: Trelleborg Marine Systems UK Ltd., 2015.

Van Hauwermeiren, M., Vose, D., and Vanden Bossche, S., *A Compendium of Distributions Used in ModelRisk*. 2nd ed. (Sint-Amandsberg, Belgium: Vose Software, 2012).

* Vose, D., *Risk Analysis: A Quantitative Guide*, 3rd ed. (Chichester, UK: John Wiley & Sons Ltd., 2008).

Vose, D., "Fitting Distributions to Data, and Why You Are Probably Doing it Wrong," Sint-Amandsberg, Belgium: Vose Software, 2010.

Vose, D., "Probability and Uncertainty," Sint-Amandsberg, Belgium: Vose Software, 2018a.

Vose, D., "Ten Steps to Incorporating Risk Analysis into Your Business," Sint-Amandsberg, Belgium: Vose Software, 2018b.

Vose, D., "The Perplexing Math of Uncertainty," Sint-Amandsberg, Belgium: Vose Software, 2018c.

Websites

Advanced Services Group	www.advancedservicesgroup.co.uk
American Bureau of Shipping	ww2.eagle.org
Business Industrial Network	www.bin95.com
ClickSoftware Technologies	www.clicksoftware.com
Fathom Maritime Intelligence	http://fathom.world
Flaw of Averages	www.flawofaverages.com
Monohakobi Technology Institute	www.monohakobi.com
Nexala	www.nexala.com
Prisma Electronics S.A.	www.prismaelectronics.eu
Railnova Sprl	www.railnova.eu
ServiceMax Europe	www.servicemax.com
ServicePower Inc.	www.servicepower.com
Shire Systems Ltd.	www.shiresystems.com

Society for Maintenance and Reliability Professionals	http://smrp.org
Trelleborg Marine Systems UK Ltd.	www.trelleborg.com
Vose Software	www.vosesoftware.com

CHAPTER 11

Standards and Procedures

Introduction

This chapter gives a high-level overview of the generic asset management and condition monitoring standards and procedures that the author is aware of. The literature on this subject is growing rapidly, as the importance of asset management is increasingly understood: readers are urged to make themselves aware of standards that are relevant to their businesses and fields of expertise, as well as study the references and standards mentioned here.

Asset Management

Robert Davis, an ex-president of the Institute of Asset Management (IAM), has produced an excellent short introduction to the subject (*Davis 2011*). The article by Grahame Fogel (*Fogel 2012*) discusses the focus of asset management on asset value realization and states that appropriate techniques, processes, and methodologies should be used to deliver this need. Two publications by Raconteur (*Tame et al. 2013*; *Betts et al. 2014*: www.raconteur.net) give good summaries of asset management and maintenance.

The IAM (www.theiam.org) is an independent, not-for-profit organization for professionals involved in asset management. They have created a conceptual model for asset management (Figure 11.1) which shows how the acquire-operate-maintain-dispose life cycle of assets is driven by asset management strategy and planning (which is itself driven by an organization's strategic plan, asset management decision-making, and people and processes within an organization), asset knowledge and risk assessment, and review processes. Full details of this model are given in the *PAS 55 Specification for the Optimized Management of Physical Assets* (*PAS 55-1:2008 2008*) and the accompanying

FIGURE 11.1 IAM asset management conceptual model (Institute of Asset Management: reproduced with permission).

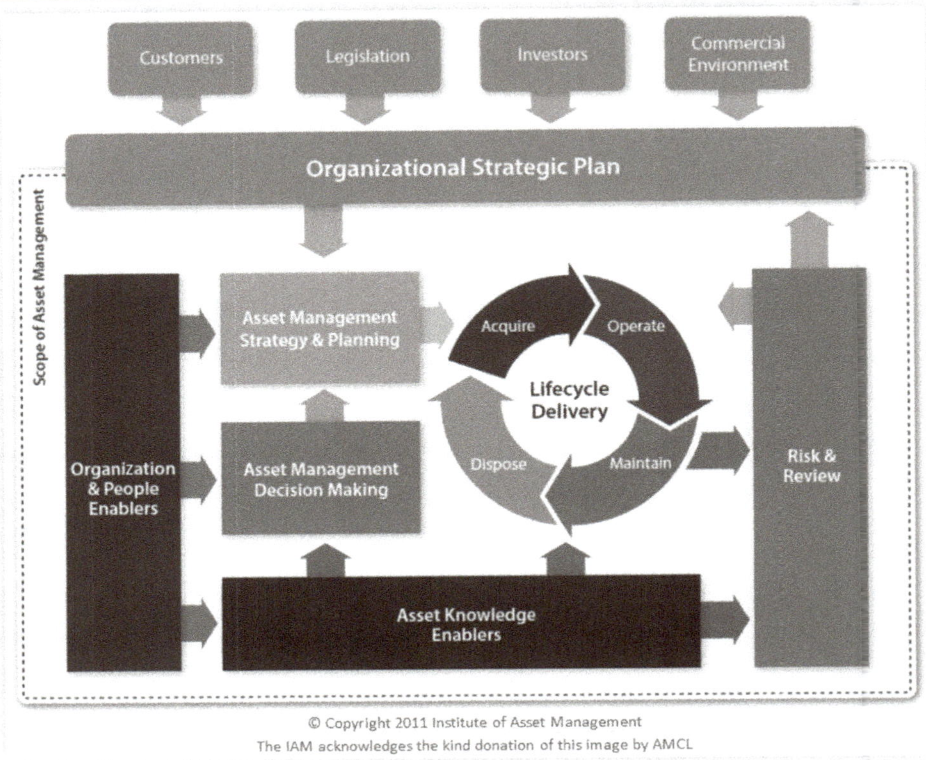

guidelines for its use (*PAS 55-2:2008 2008*). Other helpful publications by the IAM include a broad overview of the subject (*IAM 2015*) and detailed description of each of the elements, summarized in Figure 11.2, that make up asset management (*IAM 2011*).

The figure shows three capabilities that the IAM has developed that emerge from the conceptual model in Figure 11.1:

- A competencies framework, leading to asset management training and qualifications.
- A list of asset management subjects, feeding the IAM knowledge framework.
- An assessment methodology for asset management capability and maturity.

Detailed listings of the IAM asset management competencies and how to apply them are given in several other IAM publications (*IAM 2008a, 2008b, 2009*). The Global Forum on Maintenance & Asset Management (GFMAM: www.gfmam.org) has produced a document (*GFMAM 2014*) that describes the asset management landscape, lists the basic principles, and defines the various core subjects that impact asset management. The excellent large-format posters produced by the IAM (*IAM 2014a, 2014b*) provide broad overviews of the key asset management concepts in a highly readable and engaging form that shows how the important ideas interact.

PAS 55 has now been superseded by BS ISO 55000 (*BS ISO 55000:2014 2014*, available from the British Standards Institution (BSI): www.bsigroup.com) which provides an overview of asset management summarized in Figure 11.3 and discusses principles and terminology.

FIGURE 11.2 IAM asset management framework (Institute of Asset Management: reproduced with permission).

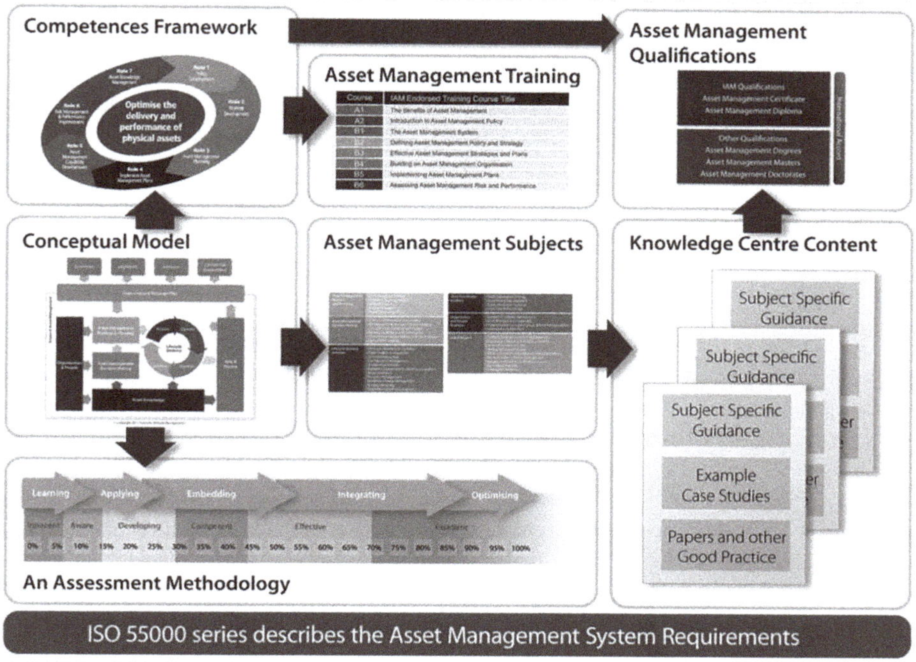

FIGURE 11.3 Overview of an asset management system (*BS ISO 55000:2014*: reproduced with permission).

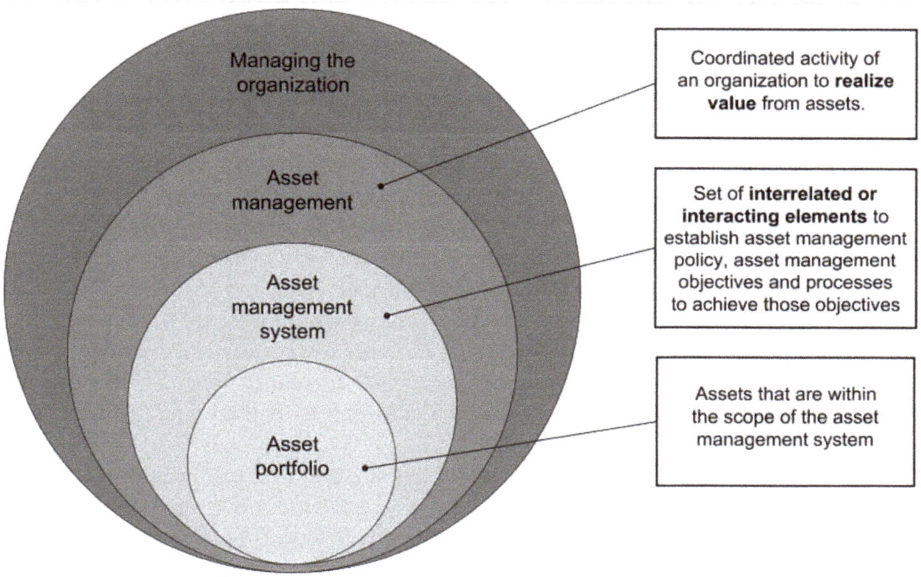

BS ISO 55000 is about more than management of assets. Its underlying principles are that:

- Asset management enables an organization to realize value (tangible or intangible, financial or nonfinancial) from assets in the achievement of its organizational objectives.
- Asset management enables an organization to examine the need for, and examine the performance of, assets and asset systems at different levels.
- Asset management enables the application of analytical approaches toward managing assets over the different stages of their life cycles.
- Asset management does not focus on assets themselves but on the value assets can provide to the organization.

BS ISO 55001 states the requirements for the establishment and maintenance of an asset management system (*BS ISO 55001:2014 2014*), while BS ISO 55002 provides more detailed guidance (*BS ISO 55002:2014 2014*). An asset management system directs, coordinates, and controls asset management activities, provides improved risk control, and gives assurance that asset management objectives will be achieved consistently. Critical asset management activities beyond the system include leadership, culture, motivation, and behavior. Guidance on how to make the transition from PAS 55 to BS ISO 55000 is available from the BSI (*BSI 2014*): a presentation by David McKeown (*McKeown 2013*) also describes the transition from PAS 55 to BS ISO 55000. Figure 11.4 shows the elements of an asset management system and how they link together: the numbers refer to sections in BS ISO 55001.

The book edited by Terrence O'Hanlon (*O'Hanlon 2014*) is an excellent reference that provides a good deal of knowledge and guidance on BS ISO 55000, written by many of its most respected developers and thought leaders. The articles by Terry Wireman (*Wireman 2011, 2012*) discuss the background to and production of ISO 55000 for asset management. The article by Ron Moore (*Moore 2014*) discusses the philosophies behind ISO 55000, while the article by Rhys Davies (*Davies 2014*) makes the point that the processes defined in ISO 55000 should allow value to be delivered more consistently. The paper by David Gazda (*Gazda 2014*) gives an overview of the BS ISO 55000 process, illustrated with case studies, while the article by Karen Conneely (*Conneely 2014*) shows how the introduction of the ISO 55000 standard for asset management is the beginning of a significant sea-change in corporate attitudes to asset value, and that it places assets under the corporate spotlight. The article argues that it is time for all organizations to recognize the new strategic imperatives of asset management.

The article by Terrence O'Hanlon (*O'Hanlon 2015*) gives an accessible overview of ISO 55000 and its relationship to risk and reliability. Assetivity (www.assetivity.com.au) have produced a strategic asset management plan (SAMP) template (*Assetivity 2015*) that ensures that senior executives and stakeholders can quickly understand the key elements of an organization's asset management strategy and why these are appropriate. The article by Scott Yates (*Yates 2015*) summarizes the factors that make up a good SAMP, while the article by Mark Ruby (*Ruby 2015*) covers the development of asset management plans in more detail.

The book edited by Madeleine Berenyi (*Berenyi 2014*) and produced by the Asset Management Council of Engineers Australia (www.amcouncil.com.au) provides a concise picture of the principles, concepts, and processes of asset management, including emphasis on the key roles of stakeholders, leadership, culture, and asset management maturity. It presents an intellectual framework and context through which asset management information can be developed and universally understood, while

FIGURE 11.4 Elements of an asset management system (*BSI 2014*: reproduced with permission).

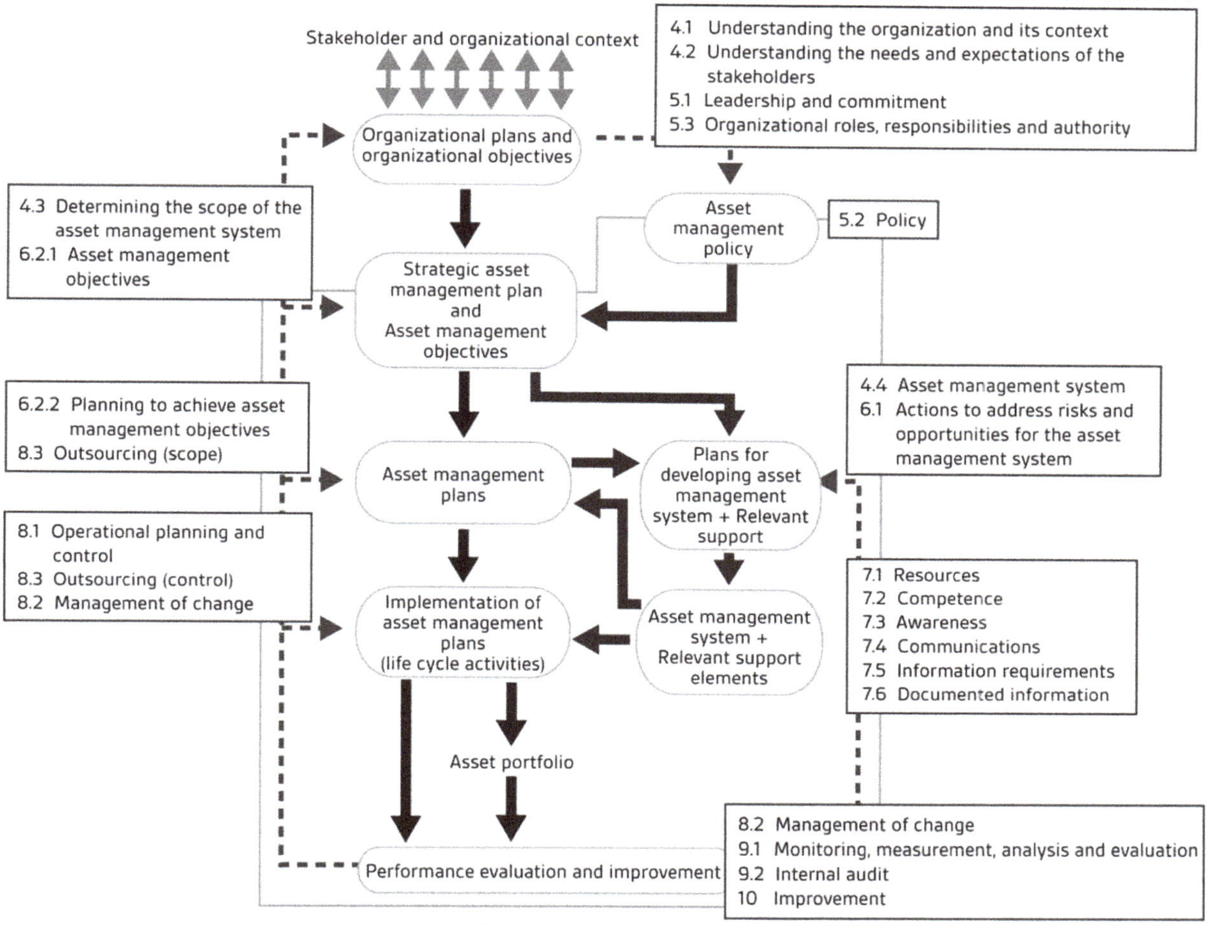

providing opportunities for both individuals and organizations to build their asset management capabilities.

The reports by Reliabilityweb.com (*Reliabilityweb.com 2014a, 2014b*: www.reliabilityweb.com) discuss the results of research carried out to discover the asset management practices, investments, and challenges faced by approximately 1,000 asset managers from a wide variety of industries worldwide. One of the main conclusions is that organizational culture is cited as the top challenge faced by over 40% of respondents.

The excellent book by João Ricardo Barusso Lafraia and John Hardwick (*Lafraia and Hardwick 2013*: www.livingassetmanagement.com) highlights the abundant potential to develop and change leadership, culture, and behavior so that asset management will produce the desired outcomes. The authors make the point that physical assets and management systems are visible and tangible, but leadership, emotions, culture, and behaviors, despite being invisible and intangible, are essential to an organization. Without the right leadership, culture, and behaviors, an organization cannot produce its desired outcomes. The presentation by João Ricardo Barusso Lafraia (*Lafraia 2012*) summarizes many of these issues.

Two documents by the Infrastructure Asset Management Exchange (*IAM Exchange 2014, 2015*) discuss the need to develop and strengthen infrastructure asset management strategies and secure senior leadership buy-in, as well as what strategies are being implemented and where investments need to be made to overcome the challenges of managing infrastructure assets.

The UK-based Asset Management Academy was set up in 2012 to provide training leading to internationally recognized asset management qualifications. A flyer (*Asset Management Academy 2015*) gives an overview of how to build a career in asset management, while the website (www.am-academy.com) gives full details of all the courses, which last between two and eight days and are run in several worldwide locations, that are currently available. Similar asset management courses are also run by The Woodhouse Partnership Ltd. (www.twpl.com, www.assetmanagementacademy.com).

The paper by Mike Dixon (*Dixon 2014*) argues that the way to sell the concept and value of asset management is to create compelling stories that interpret iconic achievements and events in history in asset management terms. There is huge potential to develop a series of illustrative studies to highlight the power of the asset management approach in a way which should appeal to a general business readership and thus engage existing and developing business leaders across a wide range of enterprises.

Books and Papers

The books edited by Joe Amadi-Echendu and others (*Amadi-Echendu et al. 2010, 2012*), published under the auspices of the International Society of Engineering Asset Management (ISEAM: www.iseam.org), contain a number of articles that address a wide range of asset management issues in some detail.

The article by Wayne Reed (*Reed 2013*) makes the point that attention should be focused not just on the conduct of asset management certification and/or gap analysis, but more crucially on the subsequent ordered preparation and execution of asset management improvement plan(s), proving what worked and to what extent, examining this and leveraging learning. The presentation by Nick Waller (*Waller 2014*), which builds on the papers by Joe Peppard and others (*Peppard et al. 2007*; *Ward et al. 2008*), explains the necessity of putting business needs before technology drivers.

John Woodhouse, Managing Director of The Woodhouse Partnership Ltd. (TWPL: www.twpl.com), is one of the driving forces behind the development of asset management. He has written many articles that describe the basic principles. His earlier papers (*Woodhouse 2000, 2001, 2004, 2008a, 2008b*) focus on describing asset management and its relationship to risk, while his later articles and papers (*Woodhouse 2011a, 2011b, 2012a, 2012b*) explore some of the core concepts that need to be considered when interpreting and implementing good asset management practices, discuss the difficulties of defining an asset life cycle (where the life stages may not be clear-cut; have physical existence periods that span multiple cycles of acquisition, usage, and disposal by various organizations; or could have an infinite life through maintenance and renewal of individual elements), and highlight how many organizations struggle with the practical challenges of determining and demonstrating the business case for what is worth doing and when, particularly when faced with uncertain assumptions and the need to determine the right compromise between competing priorities, such as costs versus risks, short-term versus long-term benefits, or tangible versus intangible goals. The SALVO and MACRO projects that were managed by TWPL are discussed in Chapter 2.

Several books are available that give wide-ranging overviews of asset management. The books written by Nicholas Hastings (*Hastings 2010*) and edited by Alan Wilson

(*Wilson 2013*) are very comprehensive manuals, touching on almost all aspects of asset management, while the book edited by John Mitchell and John Hickman (*Mitchell and Hickman 2012*) describes the processes that make up a successful asset management program. The book by John Campbell and others (*Campbell et al. 2011*) goes into more detail on maintenance, reliability, and optimizing asset life cycle costs. Chris Lloyd's books (*Lloyd 2010, 2012*) bring together a number of articles and case studies by many of the leading authorities in asset management, while Clive Deadman's book (*Deadman 2010*) looks at the use of asset management to the utility sector. The presentation by Paul Gibbons and Tom Sharp (*Gibbons and Sharp 2013*) presents a rigorous application of PAS 55 to the management of airport assets.

The article by Mark Brunner (*Brunner 2010*) discusses the most critical elements of asset management, covering selection of equipment and design for operability, reliability, and maintainability, control of operating parameters, as well as maintenance and processes for root cause analysis of failures. The tutorial edited by Lina Bertling (*Bertling 2007*) covers asset management, maintenance, and replacement strategies and shows how maintenance can be turned into a strategic tool for asset management. It reviews maintenance policies, shows links to probabilistic approaches and reliability-centered maintenance methods, and shows how condition monitoring can be used to optimize maintenance decisions. The articles by Rohit Banerji (*Banerji 2008a, 2008b*) discuss the evolution, philosophy, and deployment of world-class asset management from the enterprise perspective, as well as standard measures of the effects of its deployment and indicative industry benchmarks, while the paper by Nicholas Clarke (*Clarke 2011*: www.tessella.com) discusses how asset monitoring in conjunction with an asset register can assist with asset maintenance and optimization.

The article by Jim Davis (*Davis 2009*) looks at the pitfalls to avoid when implementing an Enterprise Asset Management (EAM) system, while the article by John Mitchell (*Mitchell 2008*) makes the point that an Asset Performance Management (APM) program must be a top-down process: management must establish and insist upon overall strategies, enforce standards of performance, drive behaviors, and assure that tasks are completed on time with the right tools.

Many utility companies are now embracing asset management, given their diverse, extensive, and aging infrastructure bases (*ICE 2014*) where failures can cause widespread disruption. The presentations by Carl Johnstone (*Johnstone 2011*), Damien Culley (*Culley 2012*), and Derrick Dunkley (*Dunkley 2013*) discuss the application of risk-based asset management to the UK electricity grid, while the presentation by Jonathan Booth (*Booth 2013*) discusses a risk analysis of a UK electricity distribution network and presentations by Matt Wheeldon (*Wheeldon 2011, 2012*) describe the condition monitoring and risk analysis of a UK network of drains and sewers. The presentations by Chris Watts (*Watts 2013*) and Mark Worsfold (*Worsfold 2011*) show the interest UK utility regulators are now taking in asset management. The article by Don Angell and others (*Angell et al. 2013*) shows how utility power transformer data can be used to manage asset health, while the presentation by Neil Gregory (*Gregory 2011*) discusses the asset management of a hydroelectric plant in New Zealand.

The AeroSpace and Defence Industries Association of Europe (ASD: www.asd-europe.org) has produced a series of specifications for materials management (*ASD 2012*), logistics support analysis procedures (*ASD 2010*), and the development and continuous improvement of predictive maintenance (*ASD 2014*) in the military and civil aviation sectors.

Mike Sondalini of Lifetime Reliability Solutions (www.lifetime-reliability.com) has produced several papers on the management of risk in asset management

(*Sondalini 2013b, 2013c*) as well as course on asset management for CEOs and other senior managers (*Sondalini 2013a*).

Another well-respected consultancy in the field of infrastructure and railway asset management is Asset Management Consulting Ltd. (AMCL: www.amcl.com): they run a number of online and off-line courses on the subject.

The article by Paul Wheelhouse (*Wheelhouse 2009*) discusses Plant Asset Care (PAC), a program that allows a business to plan, repair, and replace its plant and equipment to suit its real needs: the prize is the optimum balance of safety, cost, performance, and availability, taking into account the short-term constraints and the longer-term needs of the business.

Identifying assets accurately is one of the cornerstones of asset management. The article by Andrew Davies (*Davies 2012*) compares the benefits of Radio-Frequency IDentification (RFID) with barcoding. The article by Thomas Carroll (*Carroll 2009*) discusses the risk of "rogue" components (those with short in-service periods and recurring faults) entering an asset management program, and how this risk should be controlled.

The article by Richard Jones (*Jones 2015*) looks at the characteristics of world-class asset management, while the paper by Amy Attwater and others (*Attwater et al. 2014*) reviews the state of play of performance measurement for asset management systems: the paper concludes that organizations may not be clear about how to measure their performance even if they have well-defined asset management systems.

The articles by Ross Dentten (*Dentten 2014, 2015*) discuss the application of asset management techniques to Crossrail, the largest construction project in Europe, involving 42 km of tunnels, 10 new stations, 50 main subprojects, 200 contracts given to 100 contractors, and with approximately one million assets recorded in its asset management information system. The paper by Tim Kersley and Andrew Sharp (*Kersley and Sharp 2014*) describes how Network Rail and the UK rail regulator approach asset management, and how this has been used to support improvements in Network Rail's asset management capabilities over the last decade.

The paper by Yvonne Power (*Power 2014a*) presents asset health management strategies which are being successfully implemented within large-scale resource organizations, discusses the challenges of implementing these strategies at all levels of an organization from those in the field to upper level management, and presents the benefits of implementing methods which are sustainable and suitable for the long-term benefit of the organization and for the resources industry as a whole. Another paper (*Power 2014b*) explores the importance of intelligent Asset Performance Management (iAPM), an automated system that integrates disparate data and deploys advanced monitoring and diagnosis algorithms to continuously evaluate asset performance (equipment, process, control, and infrastructure) so that every aspect of an organization's asset condition is completely visible and available to all users online in near real time across the entire supply chain. Results are targeted at each stakeholder group (executive, analytical, and operational) and integrated into daily operational workflows.

The article by Alan Wilson (*Wilson 2015*) states that the integration of maintenance within physical asset management should begin at the conceptual phase of a capital project: as the detailed design is being progressed, developments can and should be made in risk evaluation, asset access, and the proposed maintenance and spare parts programs.

The paper by Andrew Crossley and Steven Male (*Crossley and Male 2014*) describes the planning, design, and structure of a new undergraduate module in asset management to be delivered from January 2015 to students at the University of Bristol in the United Kingdom. The paper by Ali Zuashkiani and others (*Zuashkiani et al. 2014*) presents the results of a preliminary examination of the state of education programs in asset management in universities in North America, Europe, and Australia, comparing and contrasting various graduate level programs focused on areas related to asset management against the 39 subjects listed by the IAM anatomy and competences framework and finding that none of the programs have a complete coverage of all the subjects but have a reasonable coverage of all the subject groups.

Finally, the U.S. Environmental Protection Agency (US EPA: www.epa.gov) has produced a very simple and accessible asset management best practices guide (*US EPA 2008*), while Steve Albee and Duncan Rose have produced a set of course notes (*Albee and Rose 2012*) explaining the basics via a fictional case study.

Condition Monitoring

The BSI has published several international standards for condition monitoring:

- BS ISO 13372 (*BS ISO 13372:2012 2012*) provides a vocabulary of condition monitoring terms.
- BS ISO 13374 (*BS ISO 13374-1:2003 2003, BS ISO 13374-2:2007 2007, BS ISO 13374-3:2012 2012*) discusses the details of data processing and communication.
- BS ISO 13379 (*BS ISO 13379-1:2012 2012*) provides general guidelines for data interpretation and diagnosis of faults.
- BS ISO 13381 (*BS ISO 13381-1:2004 2004*) provides general guidelines for prognosis of future fault progression.
- BS ISO 17359 (*BS ISO 17359:2011 2011*) provides some general guidelines and suggested parameter lists for different machinery types.

The article by Simon Mills (*Mills 2011*) provides an overview of ISO 17359:2011.

The International Association of Marine Aids to Navigation and Lighthouse Authorities/Association Internationale de Signalisation Maritime (IALA/AISM: www.iala-aism.org) has produced a very comprehensive guide covering all aspects of remote control and monitoring of aids to navigation (*IALA/AISM 2009*). The guide looks at system goals and objectives, monitoring methods, choice of sensors and data communication, display and storage, as well as maintenance, documentation, and training needs. The principles in this guide can be easily generalized to cover the monitoring of any physical asset (*Provost 2012*).

Figure 11.5 summarizes the processes involved in condition monitoring, based on a diagram in *BS ISO 17359:2011* but modified by the author in the light of experience: note that business development, engineering, operations, and other skills not shown for simplicity (e.g., IT, communications) are also needed, as well as asset management capabilities. The reader is referred to previous chapters for coverage of the various techniques mentioned in Figure 11.5.

An Observation

Standards, rulebooks, and company processes rarely, if ever, keep pace with developments in technology and new ideas created within an organization, or brought in from academia or other industries, so the reader should not let the business, organizational,

FIGURE 11.5 Condition monitoring procedure flowchart (*BS ISO 17359:2011*, modified by the author: reproduced with permission).

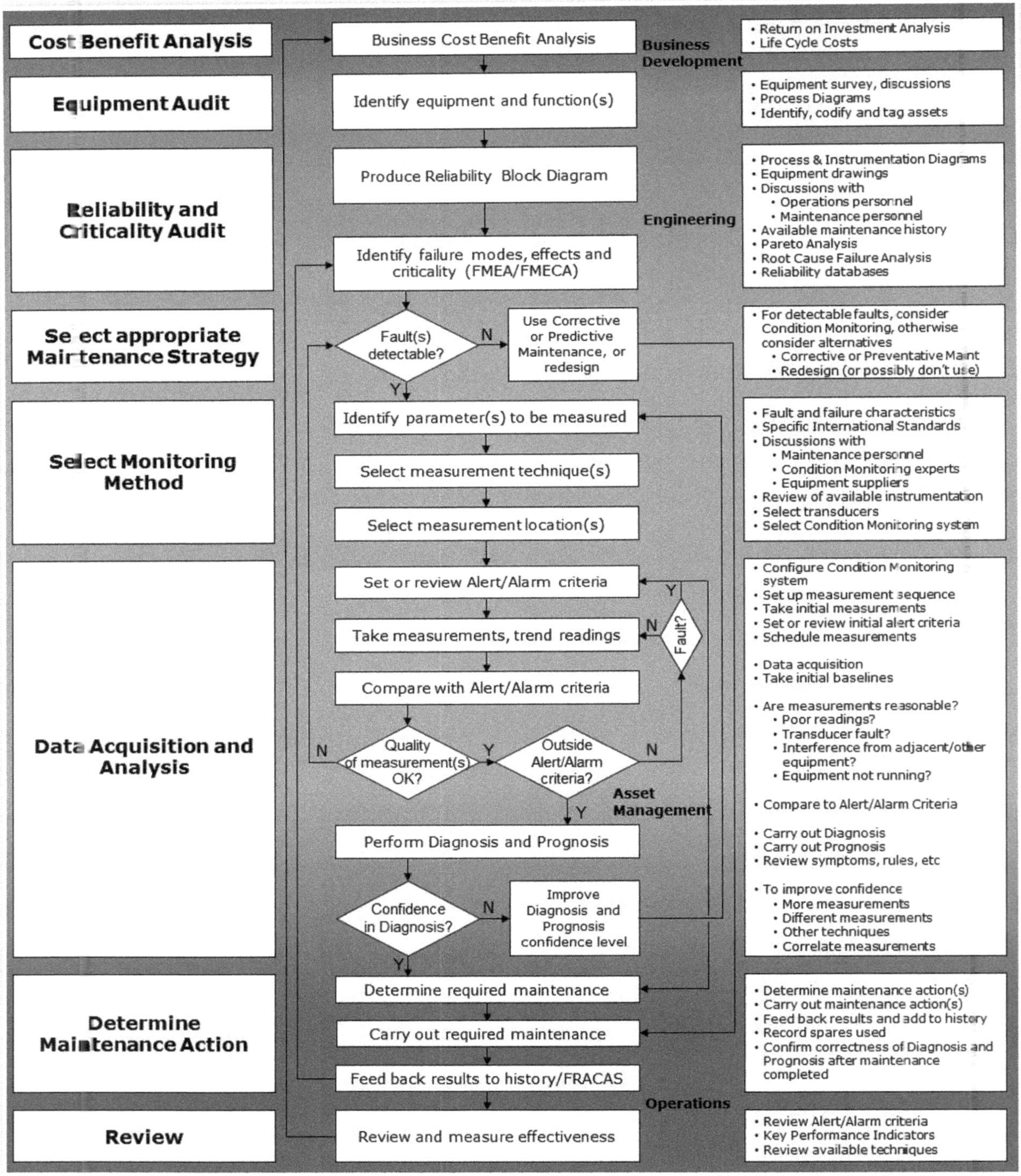

and technical creativity that underpins good asset management practice suffer because the rule-makers have not caught up.

References

* denotes sources that the author found particularly helpful and/or inspiring.

Albee, S. and Rose, D., *The Fundamentals of Asset Management: Course Notes* (Washington, DC: US Environmental Protection Agency Office of Water, 2012). Pg. 235

Amadi-Echendu, J., Brown, K., Willett, R., and Mathew, J. eds., *Definitions, Concepts and Scope of Engineering Asset Management* (London, UK: Springer-Verlag London Ltd, 2010). Pg. 232

Amadi-Echendu, J., Brown, K., Willett, R. and Mathew, J. eds., *Asset Condition, Information Systems and Decision Models* (London, UK: Springer-Verlag London Ltd, 2012). Pg. 232

Angell, D., McGrail, T., and Elkinson, K., "Asset Management—Data, Analysis and Decisions," *Power Grid International* (July 2013): 26-30. Pg. 233

ASD, "S2000M Issue 5: International Specification for Materiel Management," Brussels, Belgium: AeroSpace and Defence Industries Association of Europe, 2012. Pg. 233

ASD, "S3000L Issue 1: International Procedure Specification for Logistics Support Analysis LSA," Brussels, Belgium: AeroSpace and Defence Industries Association of Europe, 2010. Pg. 233

ASD, "S4000P Issue 1: International Specification for Developing and Continuously Improving Preventative Maintenance," Brussels, Belgium: AeroSpace and Defence Industries Association of Europe, 2014. Pg. 233

Asset Management Academy, *Seven Top Tips for Building a Successful Career in Asset Management* (London, UK: Asset Management Academy Ltd, 2015). Pg. 232

Assetivity, *Strategic Asset Management Plan (SAMP) Template* (Burswood, Western Australia: Assetivity Pty, 2015). Pg. 230

Attwater, A., Wang, J., Parlikad, A., and Russell, P., "Measuring the Performance of Asset Management Systems," *IET/IAM 2014 Asset Management Conference*, Millennium Gloucester Hotel, London, UK, November 27-28, 2014 (Stevenage, UK: Institution of Engineering and Technology). Pg. 234

Banerji, R., "A World Class Approach to Asset Management—Part 1," *Asset Management & Maintenance Journal* 21, no. 3 (July 2008a): 22-27. Pg. 233

Banerji, R., "A World Class Approach to Asset Management—Parts 2 and 3," *Asset Management & Maintenance Journal* 21, no. 4 (October 2008b): 46-56. Pg. 233

Berenyi, M. eds., *Framework for Asset Management*, 2nd ed. (Hawthorn, Victoria, Australia: Asset Management Council Ltd/Engineers Australia, 2014). Pg. 230

Bertling, L. eds., "Tutorial Book on Asset Management—Maintenance and Replacement Strategies," *IEEE Tutorial on Asset Management—Maintenance and Replacement Strategies*, Tampa, FL, June 24-28, 2007 (New York, NY: Institute of Electrical and Electronics Engineers Power Engineering Society). Pg. 233

Betts, B., Dean, J., Gagan, O., Leroux, M. et al., *Asset Management and Maintenance* (London, UK: Raconteur/IAM/Times Newspapers Ltd., May 7, 2014). Pg. 227

Booth, J., Some of my assets are more equal than others. Gas and Power Asset Management, Hilton Birmingham Metropole, Birmingham, UK, 4th July 2013. East Grinstead, UK: Faversham House Ltd, 2013. Pg. 233

Brunner, M., "What Is Asset Management and What Are the Most Critical Elements," *Asset Management & Maintenance Journal* 23, no. 3 (July 2010): 20-27. Pg. 233

BS ISO 13372:2012, *Condition Monitoring and Diagnostics of Machines—Vocabulary* (London, UK: British Standards Institution, 2012). Pg. 235

BS ISO 13374-1:2003, "Condition Monitoring and Diagnostics of Machines—Data Processing, Communication and Presentation—Part 1: General Guidelines," London, UK: British Standards Institution, 2003. Pg. 235

BS ISO 13374-2:2007, "Condition Monitoring and Diagnostics of Machines—Data Processing, Communication and Presentation—Part 2: Data Processing," London, UK: British Standards Institution, 2007. Pg. 235

BS ISO 13374-3:2012, "Condition Monitoring and Diagnostics of Machines—Data Processing, Communication and Presentation—Part 3: Communication," London, UK: British Standards Institution, 2012. Pg. 235

BS ISO 13379-1:2012, "Condition Monitoring and Diagnostics of Machines—Data Interpretation and Diagnostics Techniques—Part 1: General Guidelines," London, UK: British Standards Institution, 2012. Pg. 235

BS ISO 13381-1:2004, "Condition Monitoring and Diagnostics of Machines—Prognostics—Part 1: General Guidelines," London, UK: British Standards Institution, 2004. Pg. 235

BS ISO 17359:2011, "Condition Monitoring and Diagnostics of Machines—General Guidelines," London, UK: British Standards Institution, 2011). Pg. 236

* BS ISO 55000:2014, "Asset Management: Overview, Principles and Terminology," London, UK: British Standards Institution, 2014. Pg. 228

* BS ISO 55001:2014, "Asset Management: Management Systems—Requirements," London, UK: British Standards Institution, 2014. Pg. 230

* BS ISO 55002:2014, "Asset Management: Management Systems—Guidelines for the Application of ISO 55001," London, UK: British Standards Institution, 2014. Pg. 230

* BSI, "Moving from PAS 55 to BS ISO 55001—The New International Standard for Asset Management—Transition Guide," London, UK: British Standards Institution, 2014. Pgs. 230, 231

* Campbell, J., Jardine, A., and McGlynn, J. eds., *Asset Management Excellence: Optimizing Equipment Life-Cycle Decisions*, 2nd ed. (Boca Raton, FL: CRC Press/Taylor & Francis, 2011). Pg. 233

Carroll, T., "Going Rogue: Rogue Components—Their Effect and Control," *Uptime* 4, no. 33 (February/March 2009): 32-36. Pg. 234

Clarke, N., *Asset Monitoring, Management and Optimization* (Abingdon, UK: Tessella plc, 2011). Pg. 233

Conneely, K., "ISO 55000—Taking Asset Management into the Boardroom," *Maintenance & Engineering* 14, no. 3 (May/June 2014), 25. Pg. 230

Crossley, A. and Male, S., "Designing an Undergraduate Module in Asset Management," *IET/IAM 2014 Asset Management Conference*, Millennium Gloucester Hotel, London, UK, November 27-28, 2014 (Stevenage, UK: Institution of Engineering and Technology). Pg. 234

Culley, D., "Strategic Asset Management at National Grid—Avoiding Asset Failure," *2nd Annual Predictive Asset Management Seminar*, Royal Garden Hotel, Kensington, London, UK, September 11-13, 2012 (London, UK: IBC Energy/Informa UK Ltd). Pg. 233

Davies, A., "To RFID or Not to RFID?," *Maintenance & Engineering* 12, no. 4 (July/August 2012): 26-29. Pg. 234

Davies, R., "ISO 55000: It's All about the Value," *Uptime* 9, no. 62 (December/January 2014): 7-12. Pg. 230

Davis, J., "Successful EAM Implementation: Seven Critical Pitfalls to Avoid," *Uptime* 4, no. 36 (August/September 2009): 18-21. Pg. 233

* Davis, R., *An Introduction to Asset Management* (Chester, UK: EA Technology Ltd, 2011). Pg. 227

Deadman, C., *Strategic Asset Management: The Quest for Utility Excellence* (Leicester, UK: Matador Business/Troubador Publishing, 2010). Pg. 233

Dentten, R., "Enabling Quality Asset Information to Support the Crossrail Smart Railway," *Uptime* 9, no. 67 (October/November 2014): 6-11. Pg. 234

Dentten, R., "As Above, So Below," *Assets*, February 2015, 17-19. Pg. 234

Dixon, M., "Asset Management: Making the 'Unreasonable' Reasonable," *IET/IAM 2014 Asset Management Conference*, Millennium Gloucester Hotel, London, UK, November 27-28, 2014 (Stevenage, UK: Institution of Engineering and Technology). Pg. 232

Dunkley, D., "Predictive Techniques for Condition Based Risk Management," *3rd Annual Predictive Asset Management Seminar*, Dexter House, London, UK, December 4-5, 2013 (London, UK: IBC Energy/Informa UK Ltd). Pg. 233

Fogel, G., "Top 10 Reasons Why We Should Rethink the Use and Purpose of Asset Management Standards," *Uptime* 8, no. 55 (October/November 2012): 15-17. Pg. 227

Gazda, D., "How Can ISO55000 Help with Condition Monitoring?," *Condition Monitoring and Analysis Seminar*, Dexter House, London, UK, February 5-6, 2014 (London, UK: IBC Energy/Informa UK Ltd). Pg. 230

GFMAM, *The Asset Management Landscape: English Version*, 2nd ed. (Zürich: Global Forum on Maintenance & Asset Management, 2014). Pg. 228

Gibbons, P. and Sharp, S., "How We Do Predictive Asset Management at Gatwick," *3rd Annual Predictive Asset Management Seminar*, Dexter House, London, UK, December 4-5, 2013 (London, UK: IBC Energy/Informa UK Ltd). Pg. 233

Gregory, N., Plant Asset Management—Median Energy, New Zealand," *1st Annual Predictive Asset Management Seminar*, Mayfair Conference Centre, London, UK, October 19-20, 2011 (London, UK: IBC Energy/Informa UK Ltd). Pg. 233

* Hastings, N., *Physical Asset Management* (London, UK: Springer-Verlag London Ltd, 2010). Pg. 232

IALA/AISM, *Guideline 1008—Remote Control and Monitoring of Aids to Navigation*, 2nd ed. (Saint Germain en Laye, France: International Association of Marine Aids to Navigation and Lighthouse Authorities/Association Internationale de Signalisation Maritime (IALA/AISM), 2009). Pg. 235

IAM Exchange, *Aligning Asset Management & Investment Strategy with Wider Organisational Objectives* (London, UK: IQPC Ltd, 2014). Pg. 231

IAM Exchange, *Infrastructure Asset Management Report: From Inspiration to Practical Application* (London, UK: IQPC Ltd, 2015). Pg. 231

IAM, *The IAM Competences Framework: Part 1: Asset Management Competence Requirements Framework* (Bristol, UK: Institute of Asset Management, 2008a). Pg. 228

IAM, *The IAM Competences Framework: Part 2: Guidance on Using the 2008 Asset Management Requirements Framework* (Bristol, UK: Institute of Asset Management, 2008b). Pg. 228

IAM, "The PAS55 Assessment Methodology: General Guidance Notes on the Use of the IAM Assessment Methodology for Use with BSI PAS 55:2008," 2009. Institute of Asset Management, Bristol, UK. Pg. 228

IAM, *Asset Management Landscape: Draft A* (Bristol, UK: Institute of Asset Management, 2011). Pg. 228

* IAM, *The Asset Management Journey: Creating Value through Good Asset Management* (Bristol, UK: Institute of Asset Management, 2014a). Pg. 228

* IAM, *What is Asset Management? Asset Management: The Coordinated Activity of an Organisation to Realise Value from Assets* (Bristol, UK: Institute of Asset Management, 2014b). Pg. 228

* IAM, *Asset Management—An Anatomy*, 3rd ed. (Bristol, UK: Institute of Asset Management, 2015). Pg. 228

ICE, *State of the Nation—Infrastructure 2014* (London, UK: Institution of Civil Engineers, 2014). Pg. 233

Johnstone, C., Strategic Asset Management within a Utility," *1st Annual Predictive Asset Management Seminar*, Mayfair Conference Centre, London, UK, October 19-20, 2011 (London, UK: IBC Energy/Informa UK Ltd). Pg. 233

Jones, R., "What Does World Class Asset Management Look Like?" *Maintenance & Engineering* 15, no. 1 (January/February 2015): 6-11. Pg. 234

Kersley, T. and Sharp, A., "The Asset Management Journey: A Case Study of Network Rail's Journey Supported by An Excellence Model," *IET/IAM 2014 Asset Management Conference*, Millennium Gloucester Hotel, London, UK, November 27-28, 2014 (Stevenage, UK: Institution of Engineering and Technology). Pg. 234

Lafraia, J., "Asset Management Leadership and Culture," *Institute of Asset Management 2012 Conference*, Radcliffe Conference Centre, Warwick University, Coventry, UK, June 18-20, 2012 (Bristol, UK: Institute of Asset Management). Pg. 231

* Lafraia, J. and Hardwick, J., *Living Asset Management* (Hawthorn, Victoria, Australia: Asset Management Council Ltd, 2013). Pg. 231

* Lloyd, C. eds., *Asset Management: Whole-Life Management of Physical Assets* (London, UK: Thomas Telford, 2010). Pg. 233

Lloyd, C. eds., *International Case Studies in Asset Management* (London, UK: Institution of Civil Engineers, 2012). Pg. 233

McKeown, D., "ISO 55000—The IAM Perspective," *7th Annual Canadian Network of Asset Managers Workshop*, Westin Edmonton, Edmonton, Alberta, May 5-8 2013 (Calgary, Alberta: Canadian Network of Asset Managers). Pg. 230

Mills, S., "A New Standard for Condition Monitoring," *Maintenance & Engineering* 11, no. 3 (May/June 2011): 13-14. Pg. 235

Mitchell, J., "Develop, Implement and Measure: Raising an Asset Management Performance Management Program," *Uptime* 3, no. 27 (February/March 2008): 22-29. Pg. 233

* Mitchell, J. and Hickman, J. eds., *Physical Asset Management Handbook*, 4th ed. (Fort Myers, FL: Reliabilityweb.com, 2012). Pg. 233

Moore, R., "Asset Management, or Maintenance Management Rebranded?" *Uptime* 9, no. 63 (February/March 2014): 12-15. Pg. 230

O'Hanlon, T., "Asset Management PAS 55/ISO 55000—The Sustainable Business Strategy for Operational Excellence," *Maintenance & Engineering* 15, no. 2 (March/April 2015a): 45-50. Pg. 230

* O'Hanlon, T. eds., *The (New) Asset Management Handbook—A Guide to ISO 55000* (Fort Myers, FL: Reliabilityweb.com, 2014). Pg. 230

PAS 55-1:2008, "Asset Management Part 1: Specification for the Optimized Management of Physical Assets," London, UK: Institute of Asset Management/British Standards Institution, 2008. Pg. 227

PAS 55-2:2008, "Asset Management Part 2: Guidelines for the Application of PAS 55-1," London, UK: Institute of Asset Management/British Standards Institution, 2008. Pg. 228

Peppard, J., Ward, J., and Daniel, E., "Managing the Realization of Business Benefits from IT Investments," Cranfield, UK: Cranfield School of Management, Cranfield University, 2007. Pg. 232

Power, Y., "Asset Health Management: A Strategic Perspective," *AMPEAK Asset Management Conference 2014*, Crown Convention Centre, Crown Perth, Perth, Australia, June 2-5, 2014a (Canberra, Australia: Engineers Australia). Pg. 234

Power, Y., "iAPM for Asset Health Management: New Benchmarks for Resource Industries," *Australian Journal of Multi-Disciplinary Engineering* 11, no. 1 (April 2014b): 71-80. Pg. 234

Provost, M., "Guide to Remote Control and Monitoring of Assets—Based on IALA Guideline no 1008 3rd ed, Internal presentation, Loughborough, UK: Intelligent Energy Ltd, June 2012. Pg. 235

Reed, W., "Proving Asset Management Delivers," *Uptime* 8, no. 57 (February/March 2013): 18-22. Pg. 232

Reliabilityweb.com, "Asset Management Practices, Investments and Challenges 2014-2019," Fort Myers, FL: Reliabilityweb.com, 2014a. Pg. 231

Reliabilityweb.com, "Asset Management Practices, Investments and Challenges 2014-2019—Charts and Slides," Fort Myers, FL: Reliabilityweb.com, 2014b. Pg. 231

Ruby, M., "Developing Asset Management Plans: Creating Value from Physical Assets," *Uptime* 10, no. 71 (June/July 2015): 46-49. Pg. 230

Sondalini, M., *Enterprise Physical Asset Management Success—The Plant Wellness Way for CEOs, Executives and Senior Managers* (Rossmoyne, WA, Australia: Lifetime Reliability Solutions, 2013a). Pg. 234

Sondalini, M., *Fortunes Realised with Risk Reduction* (Rossmoyne, WA, Australia: Lifetime Reliability Solutions, 2013b). Pg. 234

Sondalini, M., *Risk Management Wisdom* (Rossmoyne, WA, Australia: Lifetime Reliability Solutions, 2013c). Pg. 234

Tame, S., Dean, J., Gagan, O., Hopkins, K. et al., *Asset Management and Maintenance* (London, UK: Raconteur/ IAM/Times Newspapers Ltd., June 6, 2013). Pg. 227

US EPA, "Asset Management: A Best Practices Guide," Washington, DC: US Environmental Protection Agency Office of Water, 2008. Pg. 235

Waller, N., *Uncovering the Hidden Value of Asset and Facilities Management Systems* (Portsmouth, UK: IBM Corp., 2014). Pg. 232

Ward, J., Daniel, E., and Peppard, J. "Building Better Business Cases for IT Investments," *MIS Quarterly Executive* 7, no. 1 (March 2008): 1-15. Pg. 232

Watts, C., "Regulatory Perspective on Asset Management," *3rd Annual Predictive Asset Management Seminar*, Dexter House, London, UK, December 4-5, 2013 (London, UK: IBC Energy/Informa UK Ltd). Pg. 233

Wheeldon, M., "Understanding the Deterioration of Long Life Assets," *1st Annual Predictive Asset Management Seminar*, Mayfair Conference Centre, London, UK, October 19-20, 2011 (London, UK: IBC Energy/Informa UK Ltd). Pg. 233

Wheeldon, M., "Achieving and Sustaining PAS55 and the Role of Predictive Maintenance," *2nd Annual Predictive Asset Management Seminar*, Royal Garden Hotel, Kensington, London, UK, September 11-13, 2012 (London, UK: IBC Energy/Informa UK Ltd). Pg. 233

Wheelhouse, P., "Creating Value from Plant Asset Care," *Maintenance & Engineering* 9, no. 6 (November/December 2009): 15-16. Pg. 234

Wilson, A., "Early Asset Management—Two Examples to Support This Approach," *Maintenance & Engineering* 15, no. 3 (May/June 2015): 6-10. Pg. 234

* Wilson, A. eds., *Asset Management, Focusing on Developing Maintenance Strategies and Improving Performance* (Farnham, UK: Conference Communications, 2013). Pg. 232

Wireman, T., "The Business of Asset Management—Part 1," *Uptime* 7, no. 49 (October/November 2011): 22-25. Pg. 230

Wireman, T., "The Business of Asset Management—Part 2," *Uptime* 7, no. 50 (December/January 2012): 20-23. Pg. 230

Woodhouse, J. *Risk-Based Decision-Making in Maintenance, Inspection, Spares and Asset Renewal* (Newbury, UK: The Woodhouse Partnership Ltd, 2000). Pg. 232

Woodhouse, J., *Asset Management* (Newbury, UK: The Woodhouse Partnership Ltd, 2001). Pg. 232

Woodhouse, J., *Asset Management: Concepts and Practices* (Kingsclere, UK: The Woodhouse Partnership Ltd, 2004). Pg. 232

Woodhouse, J., *Asset Management: A Science Emerging* (Kingsclere, UK: The Woodhouse Partnership Ltd, 2008a). Pg. 232

Woodhouse, J., *Decision-Support: Technology and People in Solving Problems and Making Better Asset Management Decisions* (Kingsclere, UK: The Woodhouse Partnership Ltd, 2008b). Pg. 232

Woodhouse, J., "Setting the Right Standards in Asset Management," *Maintenance & Engineering* 11, no. 5 (September/October 2011a): 8-10. Pg. 232

Woodhouse, J., "Value, Risks and Decision-Making," *Maintenance & Engineering* 11, no. 6 (November/December 2011b): 26-28. Pg. 232

Woodhouse, J., "Asset Life Cycles and other Horizons," *Maintenance & Engineering* 12, no. 1 (January/February 2012a): 15-16. Pg. 232

Woodhouse, J., *Setting a Good Standard in Asset Management* (Kingsclere, UK: The Woodhouse Partnership Ltd, 2012b). Pg. 232

Worsfold, M., "Regulators Perspective on Asset Management," *1st Annual Predictive Asset Management Seminar*, Mayfair Conference Centre, London, UK, October 19-20, 2011 (London, UK: IBC Energy/Informa UK Ltd). Pg. 233

Yates, S., "What Does a Good Strategic Asset Management Plan Look Like?," *Asset Management & Maintenance Journal* 28, no. 2 (March 2015): 7-10. Pg. 230

Zuashkiani, A., Schoenmaker, R., Parlikad, A., and Jafari, M., "A Critical Examination of Asset Management Curriculum in Europe, North America and Australia," *IET/IAM 2014 Asset Management Conference*, Millennium Gloucester Hotel, London, UK, November 27-28, 2014 (Stevenage, UK: Institution of Engineering and Technology). Pg. 234

Websites

AeroSpace and Defence Industries Association of Europe Pg. 233	www.asd-europe.org
Asset Management Academy Pg. 232	www.am-academy.com
Asset Management Consulting Ltd. Pg. 234	www.amcl.com
Asset Management Council of Australia Pg. 230	www.amcouncil.com.au
Assetivity Pg. 230	www.assetivity.com.au
British Standards Institution Pg. 228	www.bsigroup.com
Global Forum on Maintenance & Asset Management Pg. 228	www.gfmam.org
Institute of Asset Management Pg. 227	www.theiam.org
International Association of Marine Aids to Navigation and Lighthouse Authorities Pg. 235	www.iala-aism.org
International Society of Engineering Asset Management Pg. 232	www.iseam.org
Lifetime Reliability Solutions Pg. 233	www.lifetime-reliability.com
Living Asset Management Pg. 231	www.livingassetmanagement.com
Raconteur Pg. 227	www.raconteur.net
Reliabilityweb.com Pg. 231	www.reliabilityweb.com
Tessella plc. Pg. 233	www.tessella.com
The Woodhouse Partnership Ltd. Pg. 232	www.twpl.com
TWPL Asset Management Courses Pg. 232	www.assetmanagementacademy.com
US Environmental Protection Agency Pg. 235	www.epa.gov

Additional Material

References

* denotes sources that the author found particularly helpful and/or inspiring.

* Beedles, M., *Asset Management for Directors* (Sydney, New South Wales, Australia: Australian Institute of Company Directors, 2016).

BS ISO 44001:2017, "ISO 44001 Collaborative Business Relationship Management: Your Implementation Guide," London, UK: British Standards Institution, 2017).

CHAPTER 11

* BSI Group, "PAS 280-2018: Though-Life Engineering Services: Adding Value through a Common Framework: Guide," London, UK: British Standards Institution, 2018).

Bullock, C., Gagan, O., Hillsdon, M., McClelland, J. et al., *Asset Management* (London, UK: Raconteur/IAM/Times Newspapers Ltd., May 17, 2017).

Davies, R. and Humphrey, D., "ISO55000 Asset Management—A Biography" Fort Myers, FL: Reliabilityweb.com, 2016.

Hurley, J., Jackson, F., McClelland, J., and Swan, R., *Asset Management* (London, UK: Raconteur/IAM/Times Newspapers Ltd., March 29, 2018).

Institute for Collaborative Working, "Insight into ISO 44001," London, UK: Institute for Collaborative Working, 2017.

McKeon, P. and Ramshaw, D., *Implementing Enterprise Asset Management for Dummies*, CGI Limited Edition (Chichester, UK: John Wiley & Sons Ltd, 2013).

Mitie Group, *Beyond FM… to the Connected Workspace*. London, UK: Mitie Group plc, 2018.

* Sondalini, M., *Industrial and Manufacturing Wellness: The Complete Guide to Successful Enterprise Asset Management* (South Norwalk, CT: Industrial Press Inc., 2016).

Websites

Industrial Press Inc.	http://new.industrialpress.com
Institute for Collaborative Working	www.instituteforcollaborativeworking.com
Mitie Group plc.	www.mitie.com

CHAPTER

12

Summary and Conclusions

Summary

This book has covered an enormous amount of ground, in the process referring the reader to many published books, papers, and presentations produced by experts in this and related fields that the author has found useful over the years.

The business imperative of asset management is addressed; the value in what an asset *does*, not what it *is*, is emphasized; and a willingness for customers to pay more for the former than the latter is discussed. Tools and techniques for assessing asset life-cycle costs and the need for asset monitoring are introduced, emphasizing the decision-maker's requirement for information rather than just data. Some of the excellent work done by academics and consultants on servitization is introduced, and the successes that companies have achieved from embracing this concept are also summarized.

The power that modelling and simulation capabilities can give the producers and users of assets is discussed, including references to sources of detailed information for both general modelling and simulation capabilities and modelling of particular asset types. Analysis methods that can be used to determine asset health and the component performance changes that drive overall asset performance shifts are reviewed. A technique developed for optimizing asset sensor suites is presented and methods for analyzing time series data gathered from assets are reviewed. The importance of providing compelling visualizations of data and information is discussed, before an overview of more complex analysis techniques is presented, augmented by references to more detailed works covering this wide range of ideas, tools, and techniques. The main requirements of data gathering and software architectures are reviewed.

The practicalities of asset reliability, failure mode analysis, alarms, maintenance, management of maintenance and spare parts, availability, and condition monitoring are each briefly covered by referring the reader to the extensive literature that is available on these important subjects. Standards (particularly PAS 55 and ISO 55000) that have been created recently to guide and advise asset management practitioners are presented.

As the reader can appreciate, the subjects of servitization and physical asset management are vast, with many different but equally valid approaches and perspectives being discussed by large numbers of expert practitioners in the various interconnected areas. This book aims to open the reader's eyes to what is possible, how important it is, and where to look for advice.

Servitization and Asset Management: Nine Key Points

1. Assets are used by businesses to create value that can generate profits.
    - They must be available for use, at the right time and place.
    - They must function and operate correctly, in order to deliver business benefit.
    - They must function and operate on demand, since lack of functional delivery at the right time and place could have costly repercussions.
    - Time, effort, and material resources needed to keep them functioning and operating correctly must be minimized.
    - They are items of value, on the balance sheet(s) of one or more businesses.
2. Assets need to be looked after: they are hardly ever "fit and forget" business fixtures.
    - They required time and effort to create and consumed materials during manufacture.
    - They may have to last for a long time, since replacement may be technically and/or financially difficult (or even impossible).
3. Making rational business decisions based on information about the assets used by a business can produce significant financial improvements.
    - Increased revenues.
    - Reduced costs.
    - Increased shareholder value, from more efficient and longer-lived assets
4. Proper asset management can mean the difference between business success and business failure.
5. Understanding what has happened, is happening, and may be about to happen to the assets used by a business is crucial.
    - Are they working as required?
    - How are they being treated?
    - What are their component parts and how are they behaving?
    - What has gone wrong and when do the faults/failures need to be fixed?
    - What is about to go wrong and can/should faults be fixed before failure?
    - Where are the assets?
    - When and where do they need to be available, for operation or maintenance?

    - Are substitutes needed, because of functional or operational failures?
    - Do the assets need maintaining or replacing?

6. Current technologies allow data to be gathered as often as required to understand many issues.
    - Sensors that can measure a wide variety of parameters are cheap, accurate, reliable, and usually fitted anyway as part of asset control system(s).
    - Communication is cheap and reliable, with high bandwidths across any distance.
    - Storing and processing large quantities of raw data to create and provide information a business can use is becoming easier and cheaper. The processed outputs are also becoming easier to understand.
7. Accurate information about the functional and operational performance of the assets used in a business enables better business planning and results in fewer unpleasant surprises.
8. Since most asset users want the functionality an asset provides, rather than the asset itself, servitization and asset management represent significant business opportunities for the asset manufacturers and/or asset users and/or third parties.
    - The guarantee of functionality is usually worth as much as (or more than) the asset itself.
    - Those with the knowledge and mind-set required to manage assets effectively give themselves considerable competitive advantage.
9. Servitization and asset management remove many significant business risks, such as
    - Internal: business development, cost, and balance sheet value risks.
    - External: regulatory, competition, revenue, and customer satisfaction risks.

Conclusions

The article by Chris Lloyd and Charles Johnson (*Lloyd and Johnson 2014*) of CA Solutions Ltd. (www.casolutions.co.uk) discusses seven revelations that can help businesses understand where they are on the road to asset management excellence:

- Asset management is a strategic approach, not a formula.
- The asset management system is the end of the beginning, not the beginning of the end.
- A collective shift in beliefs and attitudes is needed.
- Asset management imposes a responsibility on individuals and groups to learn from each other.
- Asset management is driven by collective learning underpinned by collectively shared knowledge.
- Asset management requires personal commitment as well as professional development.
- Asset management demands openness about past performance.

FIGURE 12.1 Snakes and ladders (*Hodkiewicz 2012*: reproduced with permission).

Melinda Hodkiewicz's paper (*Hodkiewicz 2012*) likens asset management strategy development to a game of snakes and ladders. She has developed the game into a tool (Figure 12.1) which allows executives to create their own board, identifying the "ladders" that they need to put in place to implement asset management and the "snakes" that can impact success.

Two other drawings reproduced overleaf nicely summarize the struggles of implementing an asset management strategy. Figure 12.2 overleaf was produced by

FIGURE 12.2 The realities (Delta7 Change Ltd.: reproduced with permission).

The realities of change

FIGURE 12.3 The salesman (Google Creative Commons).

Delta7 Change Ltd. (www.delta7.com) who create stories and pictures aimed at engaging employees with complex organizational issues and helping them relate to change: it shows how complex reality can be compared to the vision presented here. Figure 12.3 is a version of the classic salesman cartoon, in which the medieval king is too busy fighting a battle to see the potential of game-changing new ideas. The reader, on his or her servitization and/or asset management journeys, will meet many obstacles and

FIGURE 12.4 From measurements to asset management (*Provost 2012*).

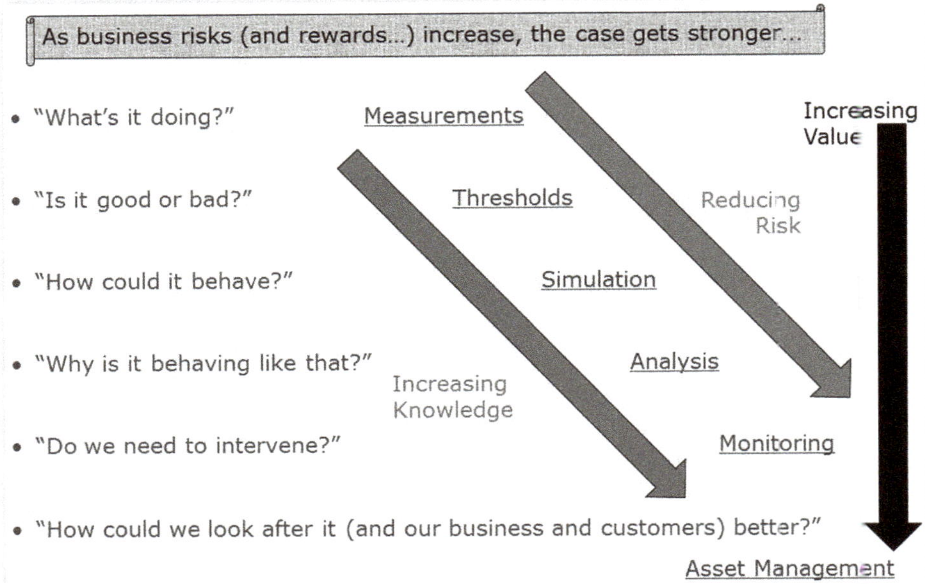

come across many people who either cannot or will not see the potential. However, he or she should not be put off, because perseverance pays and the professional and personal results that successful servitization and asset management strategies can achieve are well worth the struggle to get there.

Figure 12.4 shows how asset measurements lead to asset management.

The author sees the following as key success factors for asset management:

- Asset management has to be business focused. IT and hardware are necessary enablers, but they should not be the drivers.
- A broad view of asset management has to be taken, since the benefits may not be where anyone expects them to be.
- Integration is vital: 1 + 1 +1 = 5!
- A partnership approach, in which all interested parties share asset data, information, and knowledge, brings enormous benefits to everyone.
- Without process change no value is created, and asset management then becomes a box-ticking exercise or an engineers' playground.
- Human factors are vital. Current "heroes" will feel threatened and others will be indifferent, so a great deal of effort and perseverance is required to change the business culture.
- Remember the three reactions to innovation:
  1. "That's crazy! It will never work!"
  2. "It might work, but it's not practical…"
  3. "I told you all along that it was a great idea!"

Finally, those looking for inspiring stories on the uses of many of the principles discussed in this book (modelling, simulation, analysis, data gathering and visualization, practicalities, and procedures, all driven by the most critical of imperatives) should look no further than how NASA (www.nasa.gov) brought home the astronauts on Apollo 13 (*Baker 2013*) and managed the operation of the Voyager probes to the outer planets and beyond (*Riley et al. 2015*).

References

* denotes sources that the author found particularly helpful and/or inspiring.

Baker, D., *NASA Mission AS-508: Apollo 13 Owners' Workshop Manual* (Yeovil, UK: Haynes Publishing, 2013). Pg. 251

Hodkiewicz, M., "Snakes and Ladders—Identifying Risks to Asset Management Strategy," *Third International Engineering Systems Symposium, CESUN 2012*, Delft University of Technology, Delft, the Netherlands, June 18-20, 2012 (Delft, the Netherlands: Delft University of Technology). Pg. 248

* Lloyd, C. and Johnson, C., "The 7 Revelations," *Assets*, May 2014, 14-15. Pg. 247

* Provost, M., "Planes, Trains and Clean Energy: A Life of Simulation, Analysis, Monitoring and Asset Management," Unpublished presentation, Loughborough, UK: Intelligent Energy Ltd, 2012. Pg. 250

Riley, C., Corfield, R., and Dolling, P., *NASA Voyager 1 & 2 Owners' Workshop Manual* (Yeovil, UK: Haynes Publishing, 2015). Pg. 251

Websites

Competence Assurance Solutions Ltd. Pg. 247	www.casolutions.co.uk
Delta7 Change Ltd. Pg. 249	www.delta7.com
National Aeronautics and Space Administration Pg. 251	www.nasa.gov

Additional Material

Reference

* denotes sources that the author found particularly helpful and/or inspiring.

* Hill, P., *Mission Control Management: The Principles of High Performance and Perfect Decision Making Learned from Leading at NASA* (London, UK: Nicholas Brealey Publishing/John Murray Press, 2018).

appendix A: simplified servitization example

Customer details:

- Iron ore railways in Western Australia.
    - 1,100 km of railway connecting mines to seaport.
- Typical operation:
    - **16 trains/day.**
    - **250 wagons/train.**
    - **AU$1m profit/train.**
    - 1 in 2,000 chance of wagon having a fault.
    - 1 in 18 chance of fault causing wagon to derail.
    - No trains for one day while derailment dealt with.
    - AU$250k to deal with each derailment.

Customer requirement:

- Eliminate derailments.
- Real need: **ore throughput.**

Customer economics:

- No of derailments/year:
    - 16 trains/day @ 250 wagons/train = 4,000 wagons/day running on the railway.
    - 4,000 × 1/2,000 = 2 wagon faults/day, on average.
    - On average, every 18th fault causes a derailment.
        - There is a derailment, on average, every 9 days.
    - Average number of derailments/year = 365 ÷ 9 ≈ **40.**
- Average yearly derailment costs= 40 × (16 × AU$1m + AU$0.25m) = **AU$650m/year**.

Servitization economic opportunity:

- Set up a service to remove faulty wagons: running costs of **AU$10m/year**.
- Buy a bulldozer to knock faulty wagons detected out of train formations: cost **AU$100k**.
- Set up advanced computer system to detect faulty wagons: cost **AU$10m**.
- Charge **AU$300k/day** (30% profit from one train/day) for a zero-derailment guarantee.
- Earnings ≈ **AU$100m/year**, for as long as the ore is mined.
- Experience of detecting and dealing with faulty wagons can be used with other customers.
- Customer saves ≈ **AU$540m/year** (after costs) from avoided derailments.
- Value split: **84% customer/16% service provider**.
- Risk: for each derailment, customer demands **AU$16.25m** compensation.

(N.B. Example and all figures notional, for demonstration purposes only.)

appendix B: power system example

Introduction

This appendix discusses a fictitious but realistic electric power system consisting of five solar photovoltaic (PV) arrays, two wind turbines, and three fuel cells that are connected together to provide electric power for a remote Pacific island community. This fictitious system was inspired by the work of the Renewable Energy Development Division of the Office of the Prime Minister, Government of the Cook Islands (*Renewable Energy Development Division 2011*). However, readers should note that the simulation was created using notional data and assumptions purely to generate data that has been used to demonstrate many of the points made. This simulation does <u>not</u> claim to represent a functionally acceptable, economically viable, or physically realizable system and the inputs, outputs, and simulated faults do <u>not</u> necessarily reflect the actual functionality, performance, deterioration, failure modes, or reliability of any hardware that is in development or commercially available, either currently or in the future.

The Scenario

Palmerston Island Atoll is a small group of islands in the Pacific Ocean, just south of the Equator in Polynesia (Figure B.1). Following many years of living without electricity, the islanders have finally had an electric power system installed that meets their requirements of self-sufficiency and low environmental impact. The system supplies electricity for the islanders' own use and powers a lighthouse and marker buoys (a shipwreck would devastate the atoll's ecology and destroy the islanders' way of life), a medical center, satellite communications, and drinking water purification and supply. Any excess electricity is used to charge a battery bank and also to produce hydrogen for fuel cell-powered boats, enabling the islanders to move around the atoll without the expense, noise, and exhaust pollution that would result from using costly, imported diesel fuel. The islanders' lifestyle requires very little electricity and there are few, if any, legacy electrical items on the atoll, so power generation and distribution requirements are low because any new electrical equipment is considered to be reasonably efficient.

The HOMER package mentioned in Chapter 3 (available from www.homerenergy.com: *Lambert et al. 2006*) was used to model this notional system. Figure B.2 shows the system layout, based on that displayed by the HOMER software. Note that, by default, HOMER shows only one icon for each equipment type, so the author has modified the standard HOMER diagram in order to show more clearly the equipment configuration modelled. Note that:

- The hydrogen load (a tank providing a source of fuel for powering small boats), electrical base load, and deferrable load (water pumps, etc.) were sized for demonstration purposes only: in reality these assumed levels would not necessarily meet the islanders' needs.

FIGURE B.1 Palmerston Island Atoll (source Wikipedia Creative Commons).

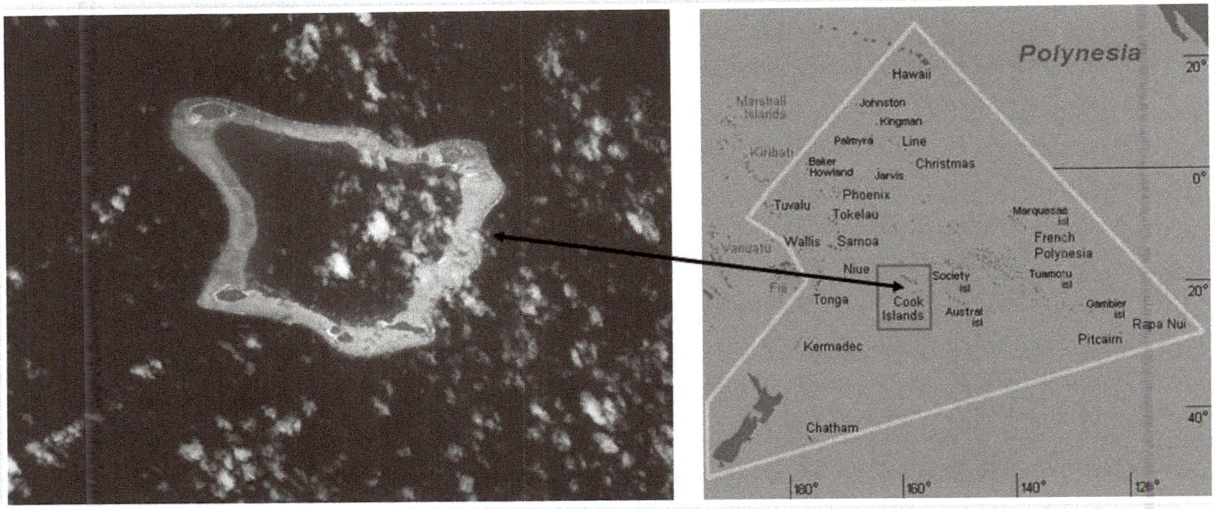

FIGURE B.2 Palmerston Island Atoll HOMER power system model.

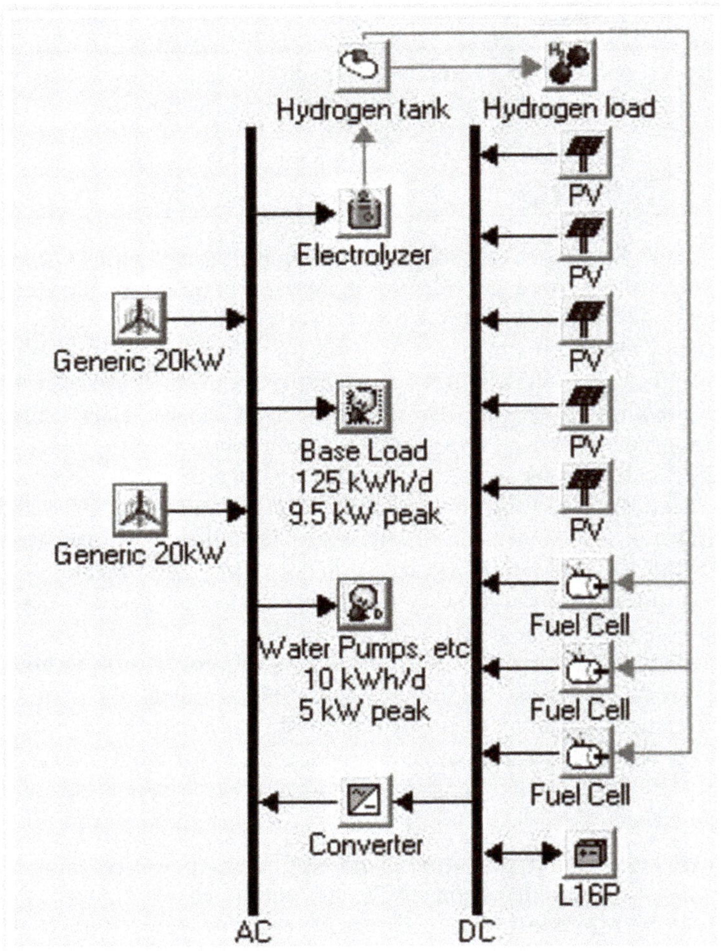

- The converter, electrolyzer, hydrogen tank, and battery (L16P) were sized for demonstration purposes.
- Solar irradiation data was sourced by HOMER from NASA's surface solar energy dataset, as described in the HOMER help file.

Fuel Cell and Wind Turbine Modelling

Figure B.3 shows an example of a type of stationary fuel cell power unit that could be used in the scenario described. A notional smaller version of this system was modelled in HOMER.

Figure B.4 shows the basic workings of a Proton Exchange Membrane (PEM) fuel cell. Gaseous hydrogen fuel is broken down into protons and electrons by platinum catalysts on the anode side of a polymer membrane, which allows the protons (but not the electrons) to move across to the cathode side. The electrons travel around the electric circuit from anode to cathode, where they react with the protons and oxygen in the air to produce water vapor. The reaction is about 50% efficient, so the airflow on the cathode side is also used to remove waste heat from the fuel cell. As discussed in the references listed in Chapter 3, fuel cells are sized to deliver the required current and arranged in stacks to deliver the required voltage and power.

For fuel cell modelling demonstration purposes, notional public domain data was used (*Larminie and Dicks 2003*; *Gou et al. 2010*) supplemented by notional modelling of the effects of ambient humidity changes on fuel cell performance added by the author. Figure B.5 shows the output from a web-based calculator of as-new fuel cell stack performance, sized to give a net system output of approximately 3 kW (after accounting for system parasitic losses).

The public domain logic was converted into a Microsoft Excel® spreadsheet (Figure B.6) by the author; then notional deterioration effects were added (Figure B.7). Finally, the deteriorated fuel cell stack model was incorporated into a fuel cell system model, which included parasitic losses from fan cooling (driven by fuel cell stack waste heat calculations and notional fan airflow, revolutions per minute (rpm), and power relationships) and output voltage conversion (Figure B.8). The spreadsheet also included the ability to analyze measurements by adjusting assumptions, as described in the Iterative Analysis section of Chapter 4. These calculations were then run at the assumed hourly environmental conditions and at one-third of each of the three-hourly fuel cell system net power outputs from the HOMER model to create simulated measurements for the three modelled fuel cell systems.

The performance of the two 20 kW wind turbines was loosely based on published data from Ampair Energy Ltd.

FIGURE B.3 Typical 5 kW stationary power fuel cell unit (source Intelligent Energy Ltd.: reproduced with permission).

FIGURE B.4 Proton Exchange Membrane (PEM) fuel cell (source Wikipedia Creative Commons).

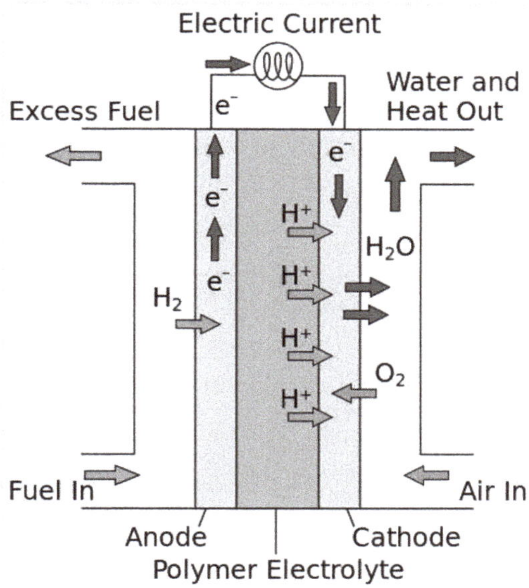

FIGURE B.5 Fuel cell stack performance calculations (source Professor Jason Keith, Department of Chemical Engineering, Michigan Technological University: see www.chem.mtu.edu/~jmkeith/fuelcellcalculator/h2fuelcellpvcalculator.swf: reproduced with permission).

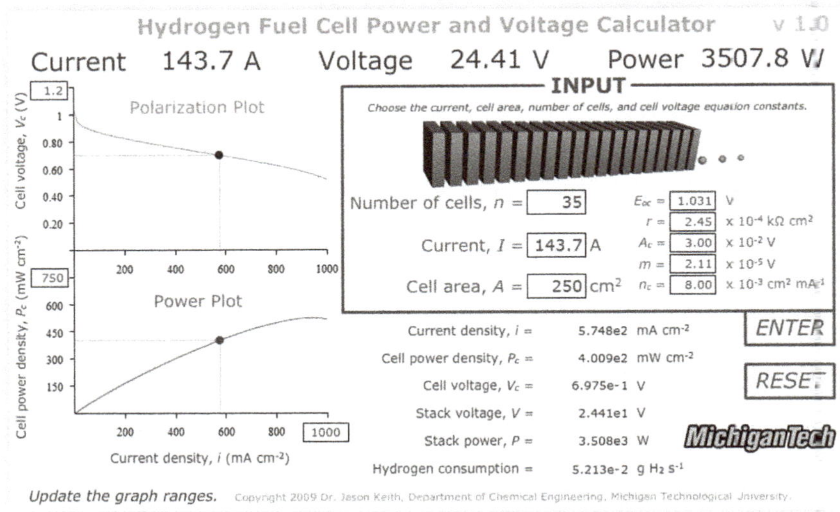

(*Ampair 2013*: www.ampair.com) and Westwind Wind Turbines (*Westwind Wind Turbines 2014*: www.westwindturbines.co.uk). The wind turbines were modelled very simply, using published curves of power output and rotor rpm variation with wind speed, notional waste heat balance calculations to simulate gearbox and electrical generator temperatures, and notional broadband vibration levels extrapolated from typical machinery vibration characteristics (*Mills 2010*; *Reeves 1998*).

Notional ambient temperature, relative humidity, wind speed, and wind direction data were estimated by the author from various sources. Notional faults in the solar PV arrays, fuel cells, and wind turbines and the hydrogen tank level sensor were simulated using engineering judgement to adjust the assumptions made in the models and the resulting sensor readings.

Simulation Outputs

Figures B.9 and B.10 show summaries of some of the time series outputs for the complete Palmerston Island Atoll power system simulation, run hourly every day for a year. The measurement list is as follows:

- Ambient conditions: temperature (°C), pressure (mBar), relative humidity (%), wind speed (m/s), and wind direction (°).
- Overall system parameters: alternating current (AC) primary load (kW), AC bus voltage (V), direct current (DC) bus voltage (V), battery state of charge (%), and stored hydrogen (kg).
- Power output from each solar PV array (kW).
- Measurements from each of the two wind turbines: net power output (kW), rotor rpm, yaw angle (°), broadband vibration (mm/s), gearbox temperature (°C), and generator temperature (°C).
- Measurements from each of the three fuel cells: net output power (kW), stack temperature (°C), stack current (A), stack voltage (V), fan current (mA), and fan rpm.

FIGURE B.6 Fuel cell stack performance model.

Simple PEM Fuel Cell Model

E0	1,031	mV
A	30	mV
Tstack	80	°C
RHamb	50%	
R	245	mΩcm^2
m	21.1	µV
n	8	cm^2/A
Acell	250	cm^2
# cells	35	

$$E = \text{Standard Potential} - \text{Activation Loss} - \text{Ohmic Loss} - \text{Concentration Loss}$$
$$= E_0 - A * \text{Log}_e(I) - I * R - m * e^{(n * I)}$$

Notional datum, for model demonstration purposes

FuelCellPro PEM/SOFC Model iPhone App by Kleiner Berliner:
Constant E0, Nernst Term off, Activation and Concentration Losses use Equation 2
Parameters are for Ballard Mark V PEM Fuel Cell, from Larminie & Dicks p60, Gou, Na & Diong p18

Icell	Ecell	PWcell	Istack	Vstack	PWstack	H_2 cons'n
mA/cm^2	mV/cell	mW/cm^2	A	V	kW	g/hour
10.0	959.4	9.6	2.5	33.6	0.084	3.3
100.0	868.3	86.8	25.0	30.4	0.760	32.9
200.0	822.9	164.6	50.0	28.8	1.440	65.8
300.0	786.2	235.8	75.0	27.5	2.064	98.7
400.0	752.7	301.1	100.0	26.3	2.635	131.6
500.0	720.9	360.5	125.0	25.2	3.154	164.5
600.0	689.5	413.7	150.0	24.1	3.620	197.4
700.0	657.3	460.1	175.0	23.0	4.026	230.4
800.0	621.8	497.4	200.0	21.8	4.352	263.3
900.0	578.2	520.3	225.0	20.2	4.553	296.2
1,000.0	515.9	515.9	250.0	18.1	4.514	329.1
1,100.0	411.4	452.6	275.0	14.4	3.960	362.0
1,200.0	212.8	255.3	300.0	7.4	2.234	394.9
1,260.0	4.7	5.9	315.0	0.2	0.051	414.6
574.6	697.5	400.8	143.7	24.4	3.507	189.1

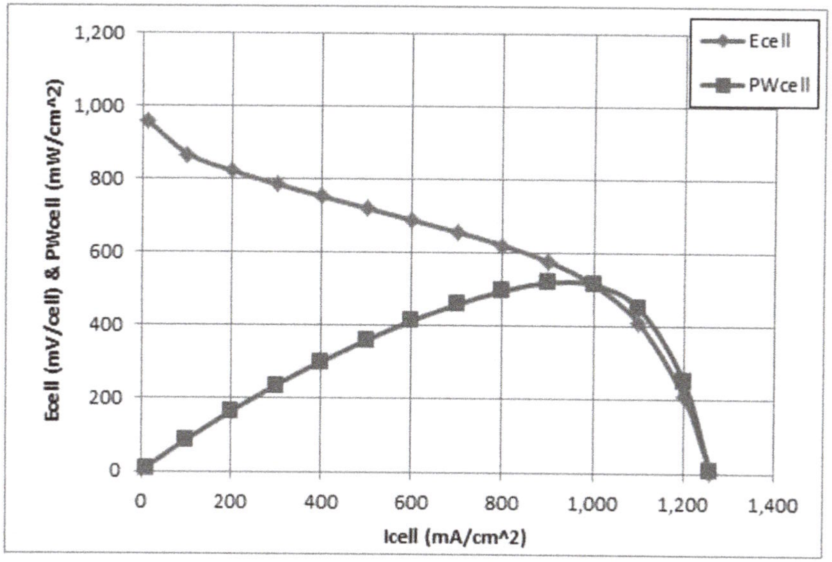

FIGURE B.7 Fuel cell stack performance model including deterioration effects.

Simple PEM Fuel Cell Model with Deterioration (Age and Cell Conditions)

E0	1,089 mV	-2 mV per 1000 hours		2 mV/% RH: Notional variations, for model demonstration purpose		
A	30 mV	Tstack >=	39 °C: ΔR =	-6 mΩcm^2/°C	Parameters are for Ballard Mark V PEM Fuel Cell, from Gou, Na & Diong p101	
Tstack	70 °C	Tstack <	39 °C: ΔR =	-13 mΩcm^2/°C		
RHamb	80%	Notional deterioration rates for E0 and R are assumed derived for constant Tstack o		80 °C, and also assumed to vary linearly to 0 at Tstack =		0 °C
R	309 mΩcm^2	5 mΩcm^2 per 1000 hours: Notional deterioration, for demonstration purposes, at Tstacl		80 °C		
m	21.1 µV					
n	8 cm^2/A					
Acell	250 cm^2					
# cells	35	1,000 hours				

Icell mA/cm^2	Ecell mV/cell	PWcell mW/cm^2	Istack A	Vstack V	PWstack kW	H_2 cons'n g/hour
10.0	1017.1	10.2	2.5	35.6	0.089	3.3
100.0	920.1	92.0	25.0	32.2	0.805	32.9
200.0	868.3	173.7	50.0	30.4	1.520	65.8
300.0	825.1	247.5	75.0	28.9	2.166	98.7
400.0	785.2	314.1	100.0	27.5	2.748	131.6
500.0	747.0	373.5	125.0	26.1	3.268	164.5
600.0	709.2	425.5	150.0	24.8	3.723	197.4
700.0	670.4	469.3	175.0	23.5	4.107	230.4
800.0	628.5	502.8	200.0	22.0	4.400	263.3
900.0	578.5	520.6	225.0	20.2	4.556	296.2
1,000.0	509.7	509.7	250.0	17.8	4.460	329.1
1,100.0	398.9	438.7	275.0	14.0	3.839	362.0
1,200.0	193.8	232.5	300.0	6.8	2.034	394.9
1,260.0	0.0	0.0	315.0	0.0	0.000	414.6
574.6	718.8	413.0	143.7	25.2	3.614	189.1

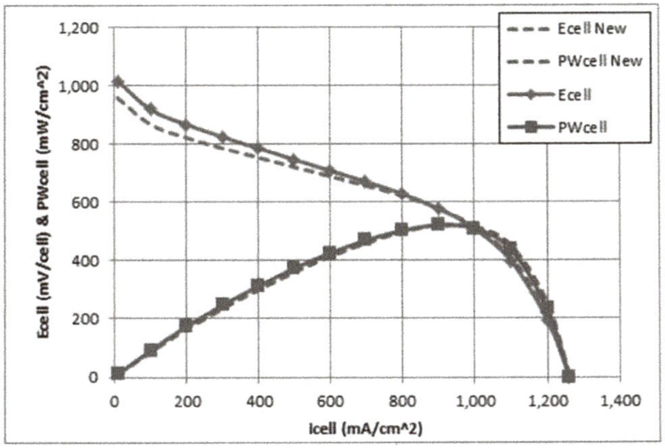

FIGURE B.8 Fuel cell system performance model—synthesis and analysis.

Simple PEM System Model & Analysis

Parameter	Value		Analysis Factors/Deltas			Associated Measurements	Measurement Differences
ηconv	90.0%	Set Istack for Desired Power: 3.200 kW	E0	0 mV		Vstack (V) 1st Goal Seek, then choose largest error	0.00%
ηfan	60.0%		R	0 mΩcm^2			
Tstack	70 °C	Fix Measurements	ηconv	0.0%	90%	Inet (A)	0.00%
RHamb	80%	Find Analysis Factors/Deltas	ηfan	0.0%	60%	Ifan (A)	0.00%
Tamb	30 °C		ΔPfan	0.0%	5.0 kPa		
Vsystem	50 V	1,000 hours	Ndesign	0.0%	1,000 rpm	Nfan (rpm)	0.00%
ΔPfan	5.00 kPa	@ Wfan 0.200 m^3/s	H2 cons'n	0.0%	(Calculated difference)		
Ndesign	1,000.0 rpm						

Notional values, for model demonstration purposes

Istack	Vstack	PWstack	H_2 cons'n	ηstack	Hstack	Wfan	PWfan	Ifan	Nfan	PWnet	Inet	ηsystem
A	V	kW	g/hour		kW	m^3/s	kW	A	rpm	kW	A	gH_2/kWh
2.5	35.6	0.089	3.3	82.7%	0.019	0.000	0.000	0.0	2	0.080	1.6	41.1
25.0	32.2	0.805	32.9	74.8%	0.271	0.007	0.000	0.0	34	0.725	14.5	45.4
50.0	30.4	1.520	65.8	70.6%	0.633	0.016	0.001	0.0	81	1.367	27.3	48.2
75.0	28.9	2.166	98.7	67.1%	1.063	0.027	0.004	0.1	135	1.946	38.9	50.7
100.0	27.5	2.748	131.6	63.8%	1.557	0.040	0.013	0.3	198	2.462	49.2	53.5
125.0	26.1	3.268	164.5	60.7%	2.113	0.054	0.032	0.6	269	2.912	58.2	56.5
150.0	24.8	3.723	197.4	57.7%	2.734	0.070	0.070	1.4	348	3.288	65.8	60.1
175.0	23.5	4.107	230.4	54.5%	3.427	0.087	0.138	2.8	436	3.572	71.4	64.5
200.0	22.0	4.400	263.3	51.1%	4.210	0.107	0.256	5.1	536	3.729	74.6	70.6
225.0	20.2	4.556	296.2	47.0%	5.131	0.131	0.463	9.3	653	3.683	73.7	80.4
250.0	17.8	4.460	329.1	41.4%	6.302	0.160	0.858	17.2	802	3.242	64.8	101.5
275.0	14.0	3.839	362.0	32.4%	8.000	0.203	1.756	35.1	1,017	1.875	37.5	193.0
300.0	6.8	2.034	394.9	15.8%	10.881	0.277	4.417	88.3	1,384	0.000	0.0	#N/A
315.0	0.0	0.000	414.6	#N/A	#N/A	#N/A	#N/A	#N/A	#N/A	#N/A	#N/A	#N/A
143.7	25.2	3.614	189.1	58.4%	2.570	0.065	0.058	1.2	327	3.200	64.0	59.1
	25.2		189.1	Measured Values				1.2	327		64.0	

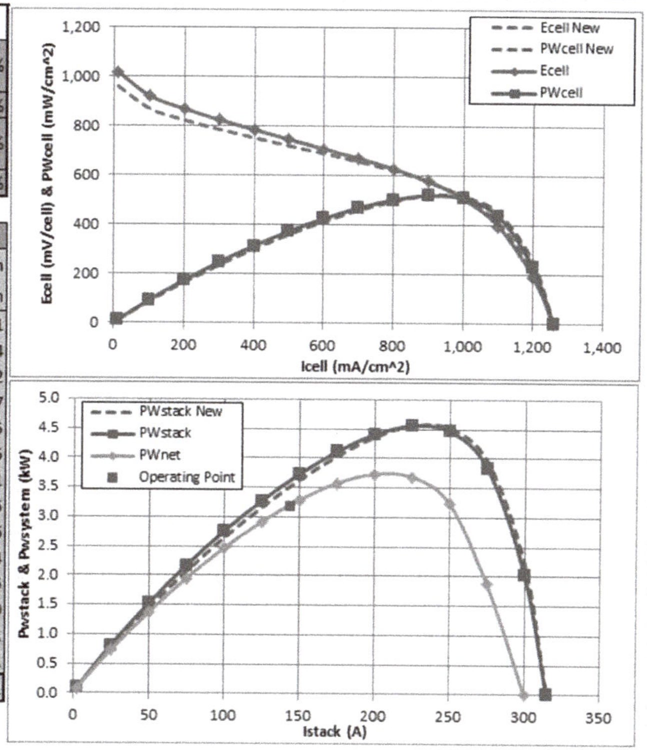

FIGURE B.9 Palmerston Island Atoll power system simulation outputs.

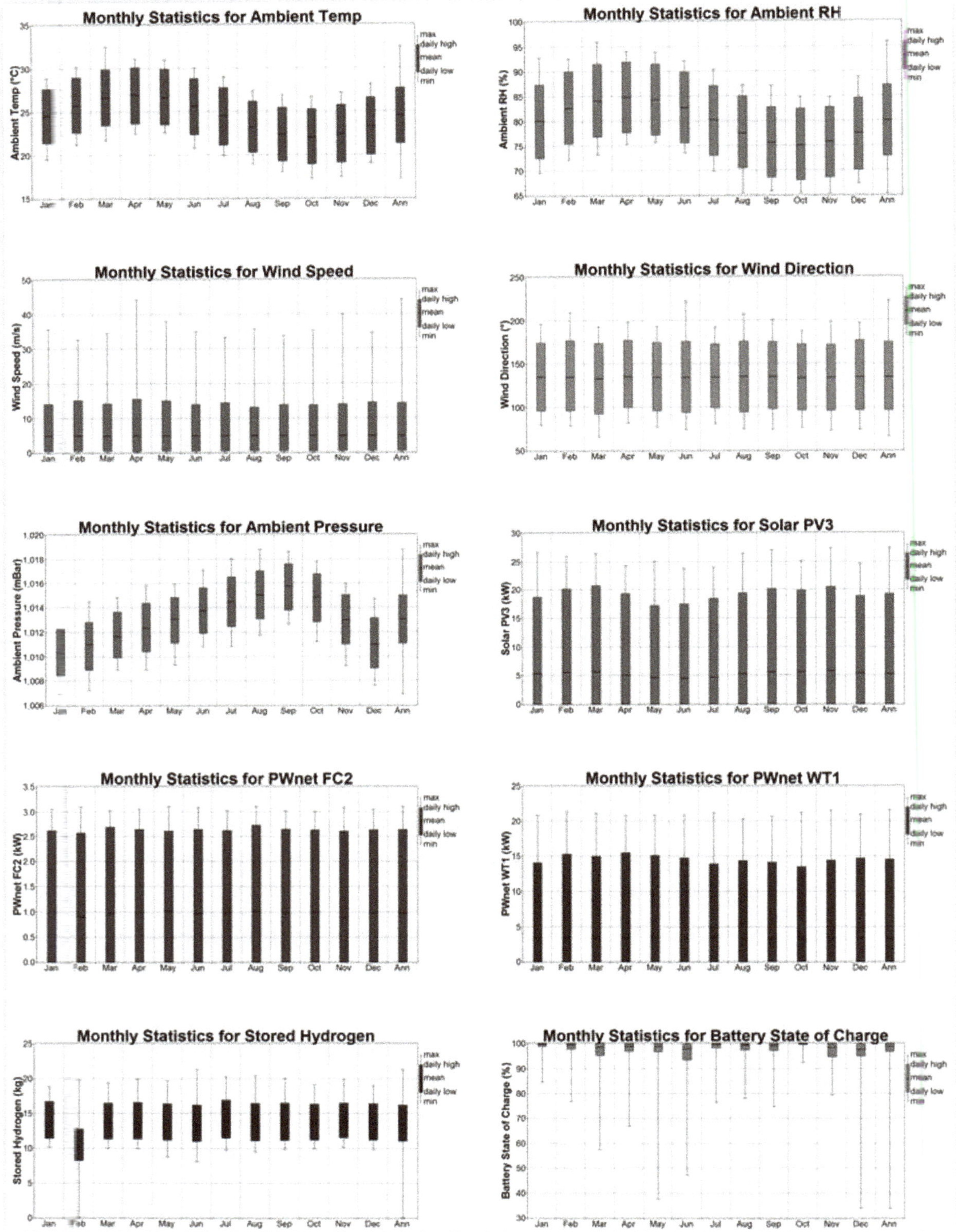

FIGURE B.10 Palmerston Island Atoll power system simulation outputs (continued).

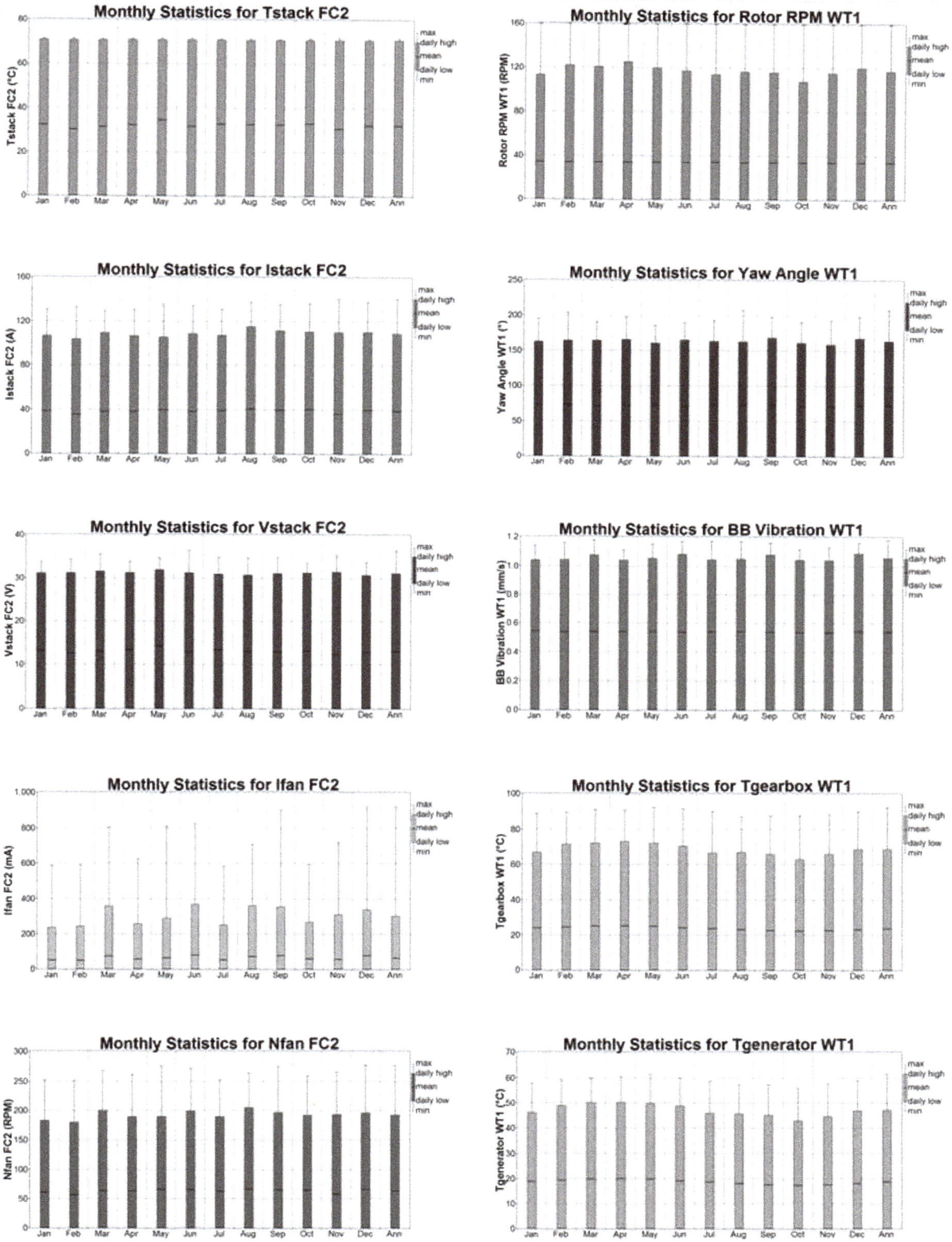

These figures were created using a program called DView, available from the U.S. National Renewable Energy Laboratory (www.nrel.gov). This program will also plot, for any parameter in the dataset, hourly, daily, or monthly time series, average daily profiles, probability distribution and cumulative distribution functions, and a view called DPlot, which is essentially a granular contour plot. Figure B.11 shows these displays for ambient temperature.

FIGURE B.11 Palmerston Island Atoll power system simulation outputs (continued).

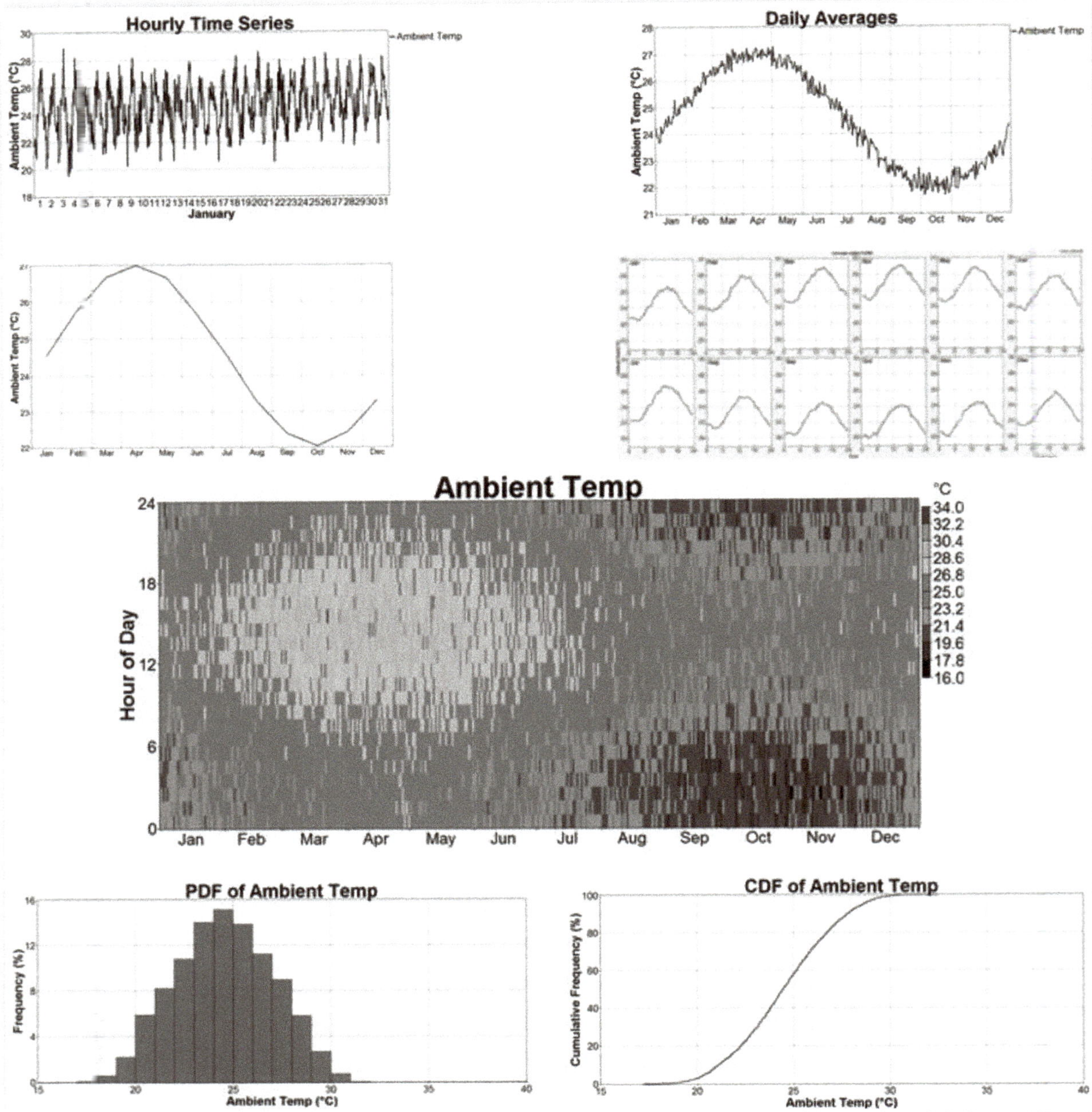

Analysis of Simulated Power System Data

Several of the analysis techniques discussed in Chapters 4 and 6 and in the references in Chapter 8 are demonstrated below. The analyses attempt to find known equipment faults that were introduced when the simulated power system data was created. The usefulness of some of the data visualization methods discussed in Chapter 7 is also shown.

Conventional Analysis

The following measurements of the performance of fuel cell system 3 were simulated:

Fuel cell system net output power (PWnet = Vsystem × Inet)	2.786 kW
Fuel cell stack power (PWstack = Vstack × Istack)	3.314 kW
Fuel cell system fan current (Ifan)	655 mA
DC system bus voltage (Vsystem)	49.969 V

Therefore

Fan power = Ifan × Vsystem = 0.655 × 49.969 = 33 W
Voltage converter efficiency (ηconv) = output power/input power
= 2.786 ÷ (3.314 – 0.033)
= 84.9%

While there is nothing incorrect about this result, the analyst is given no clue about whether the value is good or bad; the calculation also gives no indication about whether there are any other faults present in this fuel cell system that would explain the recorded sensor readings.

Iterative Analysis

Figure B.12 shows a model of fuel cell system 3, as shown in Figure B.8, run at the operational and environmental conditions at which the above measurements were taken. All the other measurements taken, as determined in the discussion on observability in Chapter 5, have also been supplied; all these inputs are outlined with black rectangles.

It is seen that increases in fan back pressure (ΔPfan) of 16.1% above nominal (due to filter blockage) and fuel consumption (H2 cons'n) of 8.3% have been calculated (shown inside the small black ovals). There are also unexplained differences between the observed measurements and the modelled values of fuel cell stack voltage (Vstack), fuel cell system net output current (Inet), fuel cell system fan current (Ifan), and fuel cell system fan rpm (shown inside the large black oval). By adjusting the assumptions made about voltage converter efficiency (ηconv) and fan design rpm (Ndesign), the parameters chosen for analysis in Chapter 5, it is seen from Figure B.13 that the measurement differences shown inside the large black oval are virtually eliminated if the voltage converter efficiency is reduced by 5% (shown inside the small black oval). The analyst is now presented with a result that shows explicitly that the voltage converter performance has fallen. He or she can also see that this analysis of filter blockage and voltage converter performance drop completely explains the observed measurements at the fuel cell system operational and environmental conditions at which they were recorded.

FIGURE B.12 Fuel cell system performance model with measurements.

Simple PEM System Model & Analysis

ηconv	90.0%	Set Istack for Desired Power	2.786 kW			
ηfan	60.0%					
Tstack	69.5 °C	Fix Measurements				
RHamb	86%	Find Analysis Factors/Deltas				
Tamb	27.3 °C					
Vsystem	49.969 V	2,514 hours				
ΔPfan	5.81 kPa	@ Wfan	0.200 m^3/s			
Ndesign	1,000.0 rpm					

Analysis Factors/Deltas			Associated Measurements	Measurement Differences
E0		0 mV	Vstack (V) 1st Goal Seek, then choose largest error	-1.84%
R		0 mΩcm^2		
ηconv	0.0%	90%	Inet (A)	-0.01%
ηfan	0.0%	60%	Ifan (A)	39.45%
ΔPfan	16.1%	5.0 kPa		
Ndesign	0.0%	1,000 rpm	Nfan (rpm)	11.87%
H2 cons'n	8.3%	Calculated difference)		

Notional values, for model demonstration purposes

Istack	Vstack	PWstack	H_2 cons'n	ηstack	Hstack	Wfan	PWfan	Ifan	Nfan	PWnet	Inet	ηsystem
A	V	kW	g/hour		kW	m^3/s	kW	A	rpm	kW	A	gH_2/kWh
2.5	35.9	0.090	3.3	83.5%	0.018	0.000	0.000	0.0	2	0.081	1.6	40.7
25.0	32.5	0.813	32.9	75.5%	0.263	0.006	0.000	0.0	32	0.732	14.6	45.0
50.0	30.7	1.534	65.8	71.3%	0.619	0.015	0.001	0.0	75	1.380	27.6	47.7
75.0	29.1	2.185	98.7	67.7%	1.044	0.025	0.004	0.1	127	1.963	39.3	50.3
100.0	27.7	2.770	131.6	64.3%	1.535	0.037	0.013	0.3	186	2.482	49.7	53.0
125.0	26.3	3.291	164.5	61.2%	2.090	0.051	0.032	0.6	254	2.933	58.7	56.1
150.0	25.0	3.745	197.4	58.0%	2.712	0.066	0.069	1.4	329	3.309	66.2	59.7
175.0	23.6	4.127	230.4	54.8%	3.407	0.083	0.137	2.7	414	3.591	71.9	64.2
200.0	22.1	4.416	263.3	51.3%	4.194	0.102	0.255	5.1	509	3.744	74.9	70.3
225.0	20.3	4.566	296.2	47.1%	5.120	0.124	0.465	9.3	622	3.692	73.9	80.2
250.0	17.9	4.464	329.1	41.5%	6.298	0.153	0.865	17.3	765	3.239	64.8	101.6
275.0	13.9	3.834	362.0	32.4%	8.005	0.194	1.776	35.5	972	1.852	37.1	195.5
300.0	6.7	2.019	394.9	15.6%	10.896	0.265	4.480	89.7	1,323	0.000	0.0	#N/A
[illegible]	0.0	0.000	414.6	#N/A	#N/A	#N/A	#N/A	#N/A	#N/A	#N/A	#N/A	#N/A
116.4	[illegible]	[illegible]	[illegible]	[illegible]	[illegible]	[illegible]	[illegible]	[illegible]	[illegible]	[illegible]	[illegible]	55.0
	26.3		165.9	Measured Values				0.7	257		55.8	

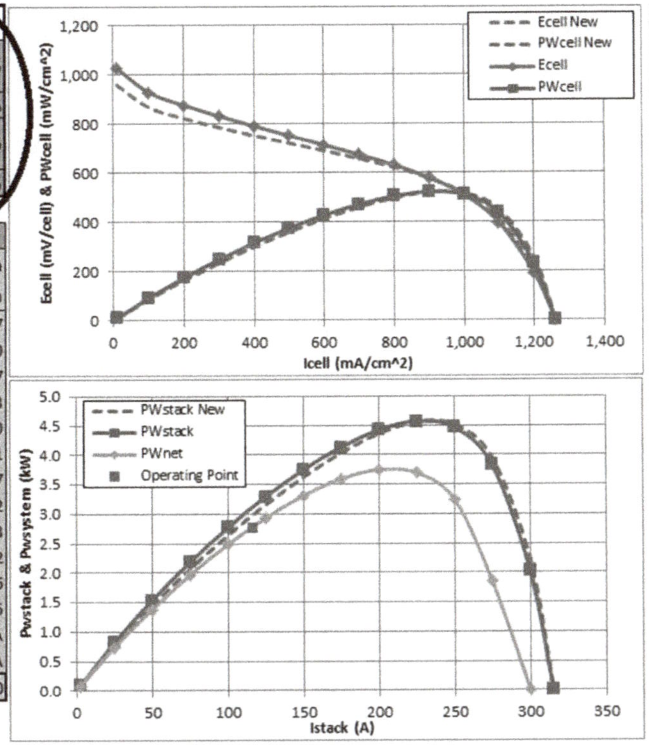

FIGURE B.13 Iterative Analysis of fuel cell system.

Simple PEM System Model & Analysis

ηconv	85.0%	Set Istack for Desired Power	2.786 kW	EO	0 mV	Vstack (V) 1st Goal Seek, then choose largest error	0.11%
ηfan	60.0%			R	0 mΩcm^2		
Tstack	69.5 °C	Fix Measurements		ηconv	-5.0% 90%	Inet (A)	0.00%
RHamb	86%	Find Analysis Factors/Deltas		ηfan	0.0% 60%	Ifan (A)	0.04%
Tamb	27.3 °C			ΔPfan	16.1% 5.0 kPa		
Vsystem	49.969 V	2,514 hours		Ndesign	0.0% 1,000 rpm	Nfan (rpm)	0.14%
ΔPfan	5.81 kPa	@ Wfan	0.200 m^3/s	H2 cons'n	0.0% (Calculated difference)		
Ndesign	1,000.0 rpm			Notional values, for model demonstration purposes			

Column headers above: Analysis Factors/Deltas; Associated Measurements; Measurement Differences.

Istack A	Vstack V	PWstack kW	H_2 cons'n g/hour	ηstack	Hstack kW	Wfan m^3/s	PWfan kW	Ifan A	Nfan rpm	PWnet kW	Inet A	ηsystem gH_2/kWh
2.5	35.9	0.090	3.3	83.5%	0.018	0.000	0.000	0.0	2	0.076	1.5	43.1
25.0	32.5	0.813	32.9	75.5%	0.263	0.006	0.000	0.0	32	0.691	13.8	47.6
50.0	30.7	1.534	65.8	71.3%	0.619	0.015	0.001	0.0	75	1.303	26.1	50.5
75.0	29.1	2.185	98.7	67.7%	1.044	0.025	0.004	0.1	127	1.854	37.1	53.3
100.0	27.7	2.770	131.6	64.3%	1.535	0.037	0.013	0.3	186	2.344	46.9	56.2
125.0	26.3	3.291	164.5	61.2%	2.090	0.051	0.032	0.6	254	2.770	55.4	59.4
150.0	25.0	3.745	197.4	58.0%	2.712	0.066	0.069	1.4	329	3.125	62.5	63.2
175.0	23.6	4.127	230.4	54.8%	3.407	0.083	0.137	2.7	414	3.391	67.9	67.9
200.0	22.1	4.416	263.3	51.3%	4.194	0.102	0.255	5.1	509	3.536	70.8	74.4
225.0	20.3	4.566	296.2	47.1%	5.120	0.124	0.465	9.3	622	3.486	69.8	84.9
250.0	17.9	4.464	329.1	41.5%	6.298	0.153	0.865	17.3	765	3.059	61.2	107.6
275.0	13.9	3.834	362.0	32.4%	8.005	0.194	1.776	35.5	972	1.749	35.0	207.0
300.0	6.7	2.019	394.9	15.6%	10.896	0.265	4.480	89.7	1,323	0.000	0.0	#N/A
315.0	0.0	0.000	414.6	#N/A	#N/A	#N/A	#N/A	#N/A	#N/A	#N/A	#N/A	#N/A
126.0	26.3	3.310	165.9	61.0%	2.114	0.051	0.033	0.7	257	2.786	55.8	59.5
	26.3		165.9	Measured Values				0.7	257		55.8	

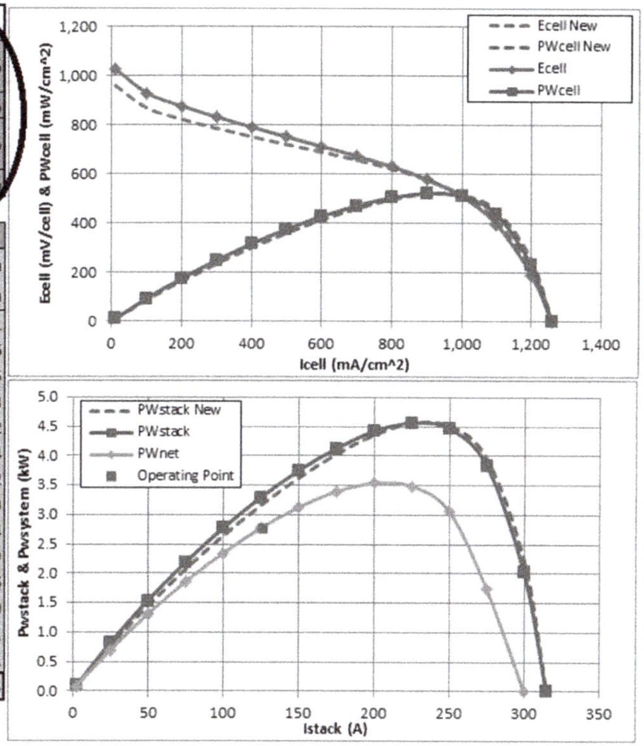

Kalman Filtering

The starting point for the Kalman Filter is the generation of measurement differences from expectation. These are taken from Figure B.12, with the additional information that the fuel cell stack current (Istack) measured at 126.0 A is 8.25% above the modelled value of 116.4 A. The other inputs (as described in Chapter 4) are:

- The system matrix **C**, derived in the observability analysis in Chapter 5 (Figure 5.11).
- The covariance matrix of component performance parameter changes and sensor biases **Q** and the measurement repeatability covariance matrix **R**, which are shown in Figure B.14.

Note the off-diagonal terms in the **R** matrix that arise because the observed measurement repeatabilities are a combination of the repeatabilities of the environmental, operational, and diagnostic measurements, which are related as shown in the **C**″ matrix in Figure B.14. The matrix **R**″ in Figure B.14 is the measurement covariance matrix for the environmental, operational, and diagnostic measurements: **R** = **C**″ × **R**″ × **C**″T (*Provost 1994*).

The Kalman Filter result is shown in Figure B.15, together with Kalman Filter analyses for 1% changes in ηconv, Ndesign, Istack sensor bias, and Vstack sensor bias, obtained simply by the matrix multiplication of the Kalman gain matrix **K** by the system matrix **C** (**K** × **C**). It is seen from inspection of the **K** × **C** matrix that the Kalman Filter underestimates the magnitudes of faults (particularly in the voltage converter efficiency (ηconv) and stack current (Istack) bias). This effect (which can be mitigated by the "concentrator" enhancement to the Kalman Filter described in *Provost (1994)*, *Provost and Nevell (1992)*, *Provost and Nevell (1994)* and the author's paper on Kalman Filtering in *Mathioudakis and Sieverding (2003)*) is due to the least-squares basis of the algorithm. However, the off-diagonal elements in the **K** × **C** matrix are not very large, so the analysis does accurately identify faults.

Figure B.15 shows that the simulated 5% deterioration in voltage converter efficiency (ηconv) is underestimated by about 0.6%, which probably is not serious since the choice of maintenance action to fix this issue would be essentially unaffected by this "error."

FIGURE B.14 Kalman Filter **Q** and **R** matrices.

Q	ηconv	Ndesign	Istack	Vstack
ηconv	2.000	0.000	0.000	0.000
Ndesign	0.000	2.000	0.000	0.000
Istack	0.000	0.000	1.000	0.000
Vstack	0.000	0.000	0.000	1.000

R	Istack	Vstack	Ifan	Nfan
Istack	0.439	-0.148	2.290	0.759
Vstack	-0.148	0.180	-1.290	-0.427
Ifan	2.290	-1.290	20.995	6.915
Nfan	0.759	-0.427	6.915	2.298

C"	PWnet	Tstack	Istack	Vstack	Ifan	Nfan
Istack	-1.567	0.683	1.000	0.000	0.000	0.000
Vstack	0.474	-0.537	0.000	1.000	0.000	0.000
Ifan	-6.725	9.048	0.000	0.000	1.000	0.000
Nfan	-2.234	2.991	0.000	0.000	0.000	1.000

R"	PWnet	Tstack	Istack	Vstack	Ifan	Nfan
PWnet	0.100	0.000	0.000	0.000	0.000	0.000
Tstack	0.000	0.200	0.000	0.000	0.000	0.000
Istack	0.000	0.000	0.100	0.000	0.000	0.000
Vstack	0.000	0.000	0.000	0.100	0.000	0.000
Ifan	0.000	0.000	0.000	0.000	0.100	0.000
Nfan	0.000	0.000	0.000	0.000	0.000	0.010

FIGURE B.15 Results of Kalman Filter analysis.

x^	
ηconv	-4.38%
Ndesign	-1.17%
Istack	-0.08%
Vstack	0.79%

=

K	Istack	Vstack	Ifan	Nfan
ηconv	-0.1986	0.0141	-0.0685	-0.0013
Ndesign	-0.0057	0.0020	-0.3261	0.9897
Istack	0.6867	0.0162	-0.1441	-0.0028
Vstack	0.0162	0.9077	0.0587	0.0010

×

y	
Istack	8.25%
Vstack	-1.84%
Ifan	39.45%
Nfan	11.87%

K×C	ηconv	Ndesign	Istack	Vstack
ηconv	0.87%	0.00%	-0.20%	0.01%
Ndesign	0.00%	0.99%	-0.01%	0.00%
Istack	-0.10%	0.00%	0.69%	0.02%
Vstack	0.01%	0.00%	0.02%	0.91%

There is also, at least with this setup of the analysis, an apparent "error" of just over 1% in the analysis of fan speed (Ndesign) change. However, the matrix algebra behind the technique allows it to be very simply applied to the analysis of bulk time series data. Replacing the vector of measurement differences $\underline{\mathbf{y}}$ in the equation $\underline{\mathbf{x}}^\wedge = \underline{\mathbf{K}} \times \underline{\mathbf{y}}$ with a matrix of measurement differences gathered over time $\underline{\mathbf{Y}}$, a matrix of estimates $\underline{\mathbf{X}}^\wedge$ can be obtained ($\underline{\mathbf{X}}^\wedge = \underline{\mathbf{K}} \times \underline{\mathbf{Y}}$). When combined with the optimal tracker (Chapter 6), this enables very simple and effective analysis of time series that brings out features that are not apparent in the time series visualization of the raw sensor measurements, as shown in a following section.

Time Series Visualization

Simple visualization of time series can easily pick up large errors or zero values in sensors, as shown in Figure B.16. However, comparisons can sometimes be more powerful; the failed yaw sensor in one of the wind turbines, which appears as a constant yaw angle value when plotted as a time series in Figure B.17, is more obvious (and much easier

FIGURE B.16 Sensor failure in solar PV array power reading.

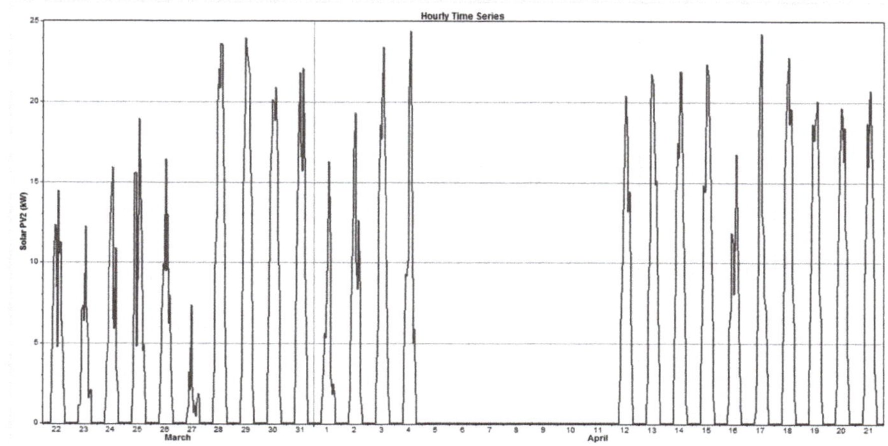

FIGURE B.17 Sensor failure in wind turbine yaw angle reading.

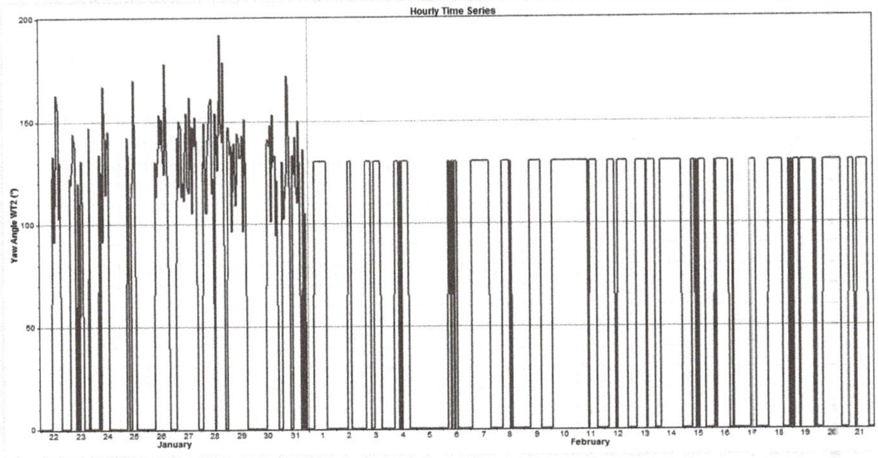

FIGURE B.18 Difference between yaw angle and wind direction.

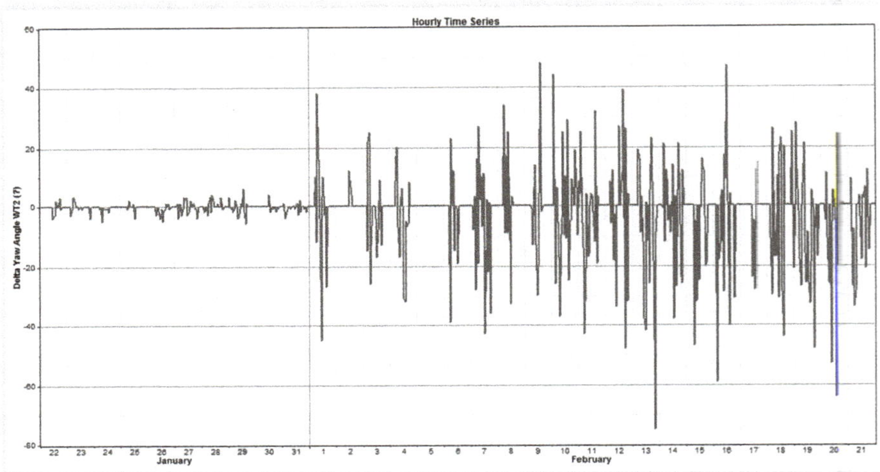

to assign warning and action limits to and therefore to find automatically) when the differences between the wind turbine yaw angle and the wind direction (which should be quite small) are plotted as a time series (Figure B.18).

Note that, while the time base is usually calendar time (either local time or UTC (universal time coordinates), which used to be known as GMT (Greenwich Mean Time)), it is sometimes better to use other measures of time, such as operating hours or operating cycles, particularly if assets are stopped and started frequently.

The Optimal Tracker

Figure B.19 shows the smoothed level estimates from the Optimal Tracker for fuel cell stack voltage (Vstack) and stack current (Istack) covering the period during which the sudden failure in the voltage inverter that has been analyzed previously occurred. Note the change of time base from calendar time to operation hours. For the purposes of demonstration, random noise has been added to the sensor readings. It is not immediately

FIGURE B.19 Fuel cell stack voltage and current observations and optimal tracker smoothed level estimates.

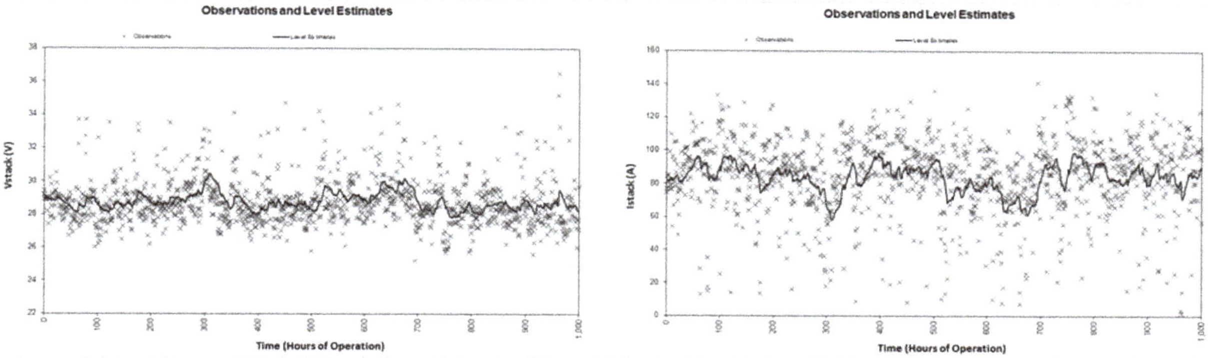

apparent from these plots that anything has changed. The measurements are dominated by changes in the environment and operation settings, which mask more subtle changes in the performance of the components making up the fuel cell system. However, by plotting the differences in these two parameters relative to their modelled values, the fact that something has happened is much more apparent, as demonstrated in Figure B.20, which shows the smoothed level and smoothed slope estimates from the Optimal Tracker for the same fuel cell stack voltage (Vstack) and stack current (Istack) readings as before, after comparison with their respective modelled values.

FIGURE B.20 Fuel cell stack voltage and current differences from model and optimal tracker smoothed level and smoothed slope estimates.

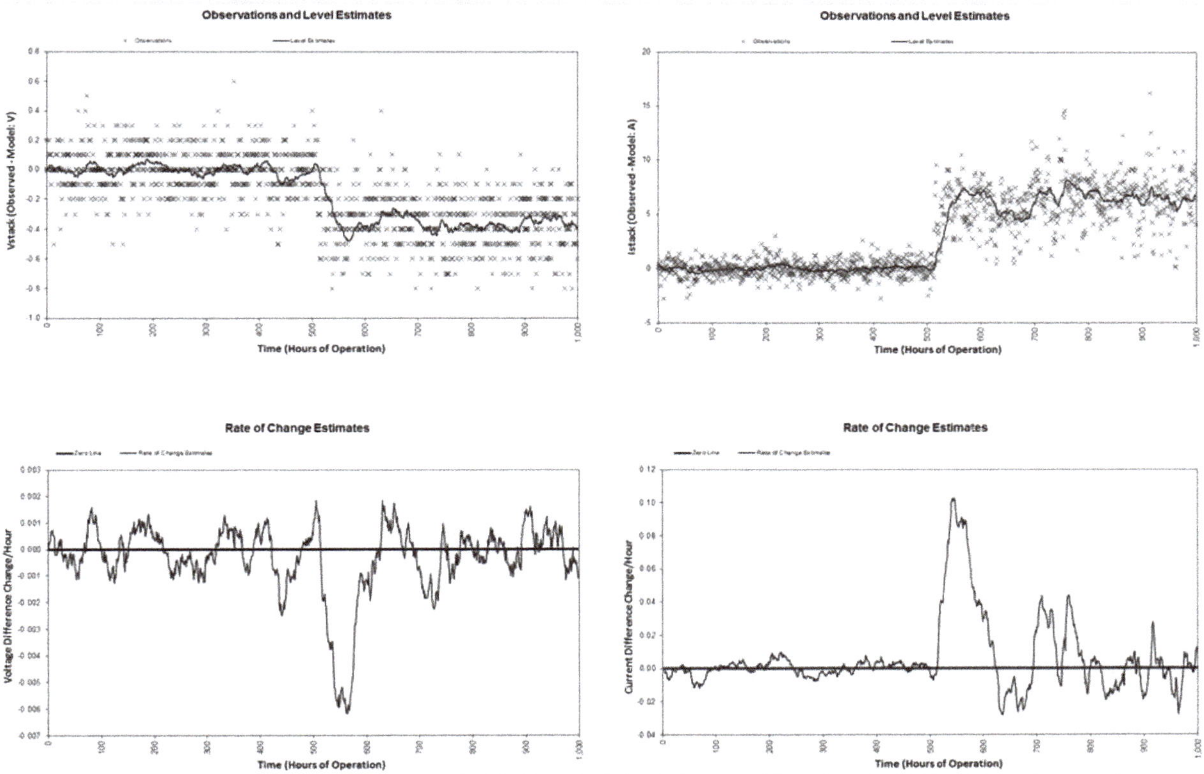

FIGURE B.21 Analysis of faulty fuel cell voltage converter using Kalman Filter and optimal tracker.

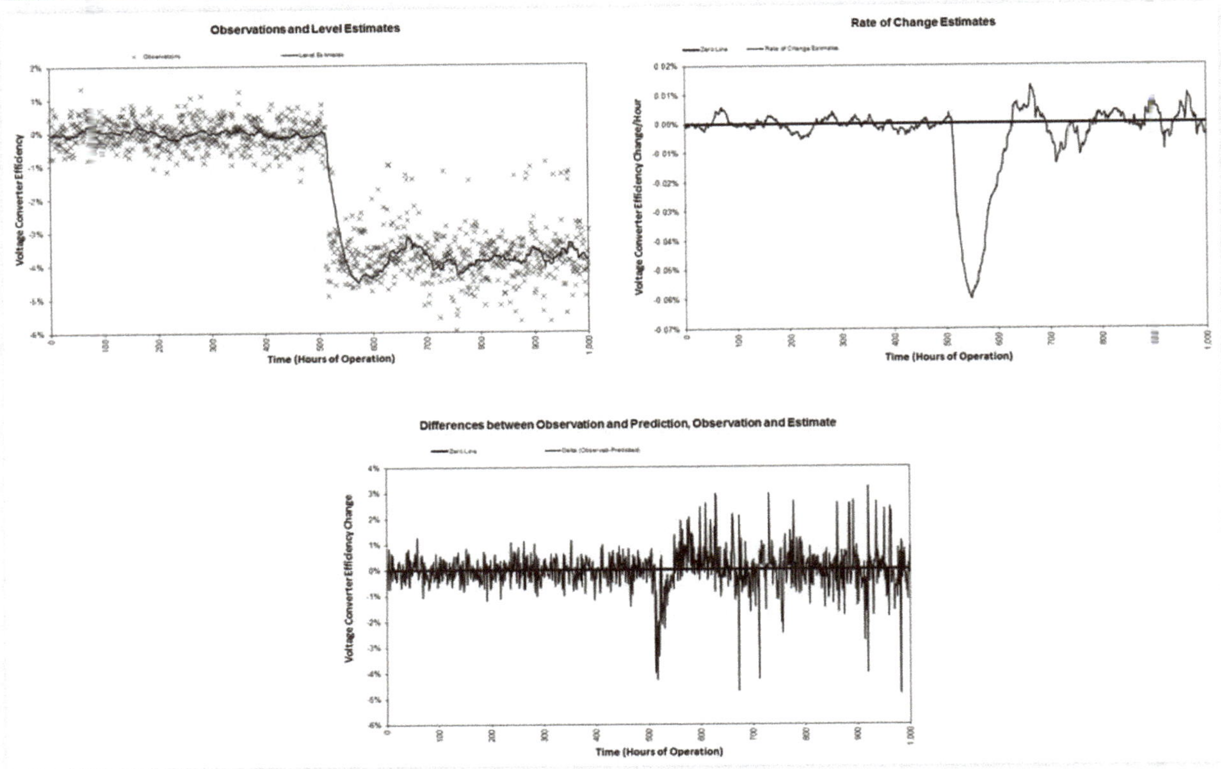

This analysis can be taken further. By analyzing the time series of the differences in all four diagnostic parameters (fuel cell stack voltage (Vstack), fuel cell stack current (Istack), fuel cell system fan current (Ifan), and fuel cell system fan rpm (Nfan)) relative to their modelled values using the Kalman Filter as described earlier, the faulty voltage converter is correctly diagnosed and the time when the fault occurred is determined, as shown in Figure B.21.

Linear Regression

As discussed in Chapter 7, x-y plots can also provide a powerful diagnostic capability; Figure 7.15 shows an example of this. Extending this further, linear regression techniques can be used to quantify apparent differences in plotted relationships. Figure B.22 shows a plot of stack voltage versus stack current for all three of the fuel cell systems simulated here. From the displayed regression equations, it is seen that the rate of change of fuel cell stack voltage with fuel cell stack current for fuel cell system 1 (−68.1 mV/A, or −68.1 mΩ) is greater than for the other two systems (−60.4 mV/A, or −60.4 mΩ) implying that this system has a greater fuel cell stack internal resistance than the others.

Statistics

Almost all data analysis and display packages can calculate and display the common measures of central tendency (mean, median, and mode) and spread (minimum, maximum,

FIGURE B.22 Linear regression of fuel cell stack voltages against stack currents.

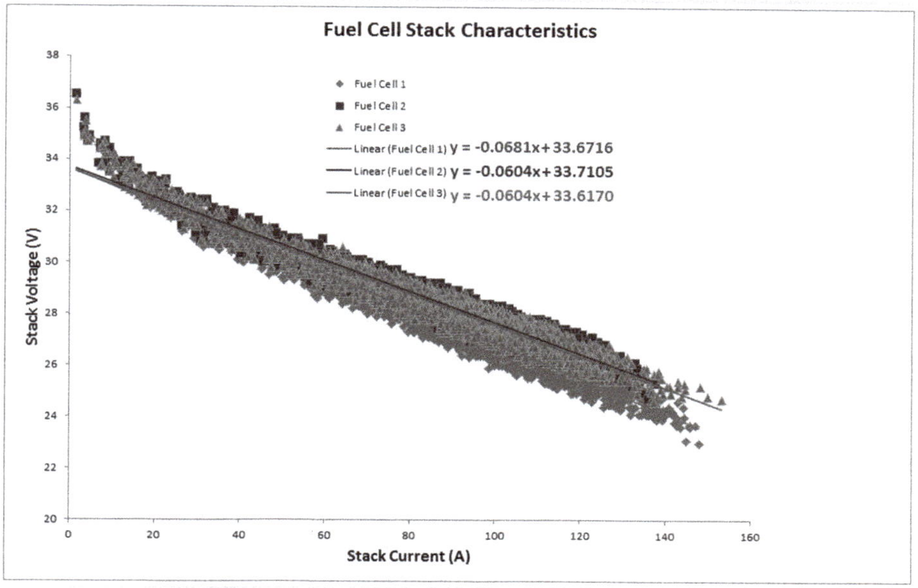

standard deviation, quartiles, deciles, etc.). Examples of statistical displays created using the data from the fictitious power system created by the author are shown above (Figures B.9 through B.11) and in Chapter 7 (Figures 7.12 and 7.13). Figure 7.11 shows how such displays can be linked together to provide "smart analytic" and visual data mining capabilities.

Parallel Coordinates Plot

An example of a parallel coordinates plot using data from one of the example wind turbines is given in Figure 7.10.

Pattern Matching

Figure B.23 shows an example of how the Cybula Signal Data Explorer software tool referenced in Chapter 8 can be used to find patterns in vibration data.

Neural Networks

In the absence of a mathematical model of an asset, neural networks (see Chapter 8) can be used on a training dataset to predict one or more outputs from one or more inputs. Figure B.24 shows the performance of four neural network models of the fault-free fuel cell system simulated in this fictitious power system example. These were set up as follows:

- Seven inputs: ambient temperature (Tamb), ambient relative humidity (RHamb), net power output (PWnet), DC system bus voltage (Vsystem), fuel cell system age, fuel cell stack temperature (Tstack), and fuel cell system filter differential pressure (ΔPfilter).

FIGURE B.23 Vibration signature detection using Cybula Signal Data Explorer (source Cybula Ltd.: reproduced with permission).

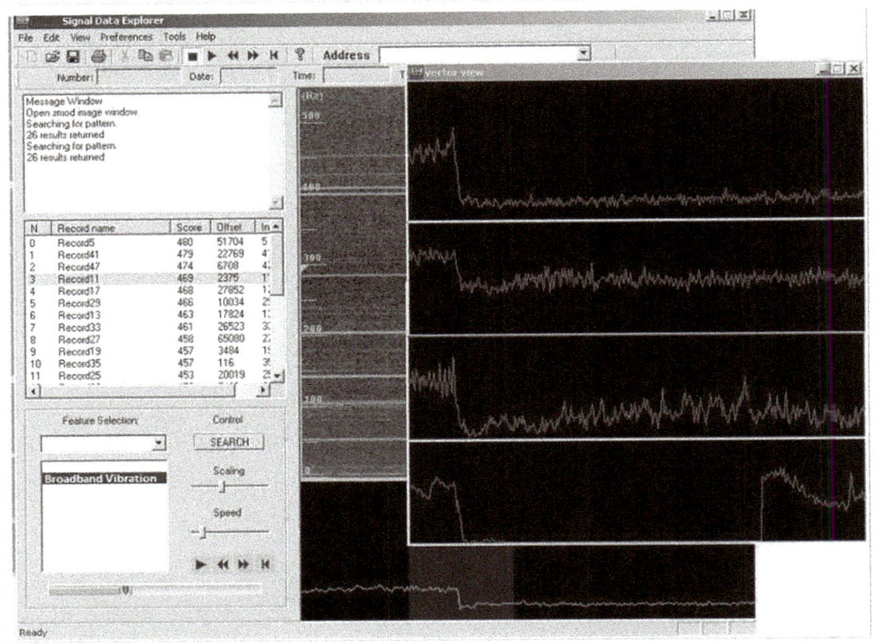

FIGURE B.24 Performance of neural network model of fuel cell system.

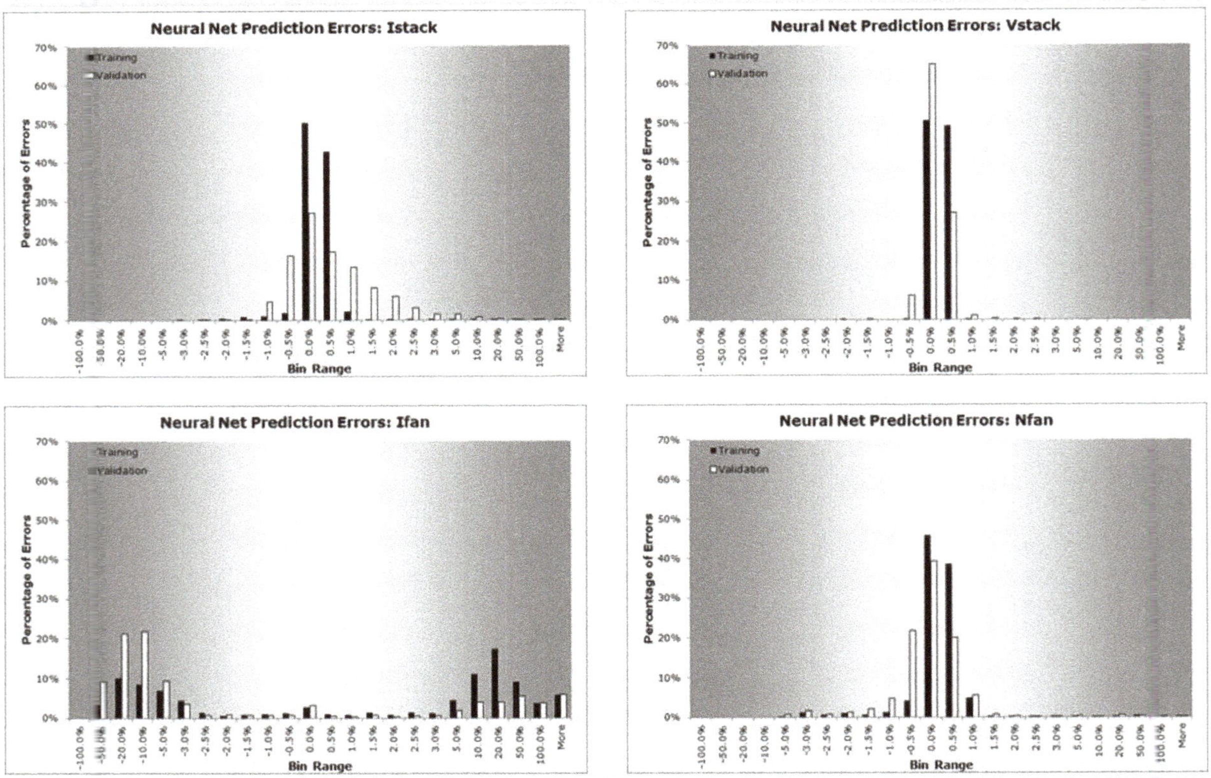

- Five hidden layer nodes.
- Each neural network modelled one of the following output parameters: fuel cell stack current (Istack), fuel cell stack voltage (Vstack), fuel cell system fan current (Ifan), and fuel cell system fan rpm (Nfan).
- The first six months of data from the fault-free fuel cell system was used as the training set.
- Each of the models was validated against the full annual set of data from the fault-free fuel cell system, using the mathematical model described earlier as the datum.

Figure B.24 shows that the neural network models for fuel cell stack current (Istack), fuel cell stack voltage (Vstack), and fuel cell system fan rpm (Nfan) performed well against the mathematical model, with most errors lying within a range of +2% to −1% for Istack, ±0.5% for Vstack, and ±1% for Nfan. However, the performance of the neural network model for fuel cell system fan current (Ifan) was very poor, with most errors lying outside the ±10% range. Clearly, neural networks have their place, but they are not a substitute for a good mathematical model if one is available.

Decision Trees

Decision trees are often used by operations personnel, who usually must respond quickly to what the data flowing from assets is telling them about the ability of those assets to carry out their assigned duties for customers. Using simulated data from one of the fictitious wind turbines making up the simulated power system described in this appendix that had been "seeded" with several faults, example plots have been produced of broadband vibration versus rotor rpm (Figure B.25) and gearbox temperature versus broadband vibration (Figure B.26), which shows distinctly different relationships between these parameters for the different fault cases listed in the plot legend. This "eyeball" cluster

FIGURE B.25 Broadband vibration vs rotor rpm for fictitious wind turbine.

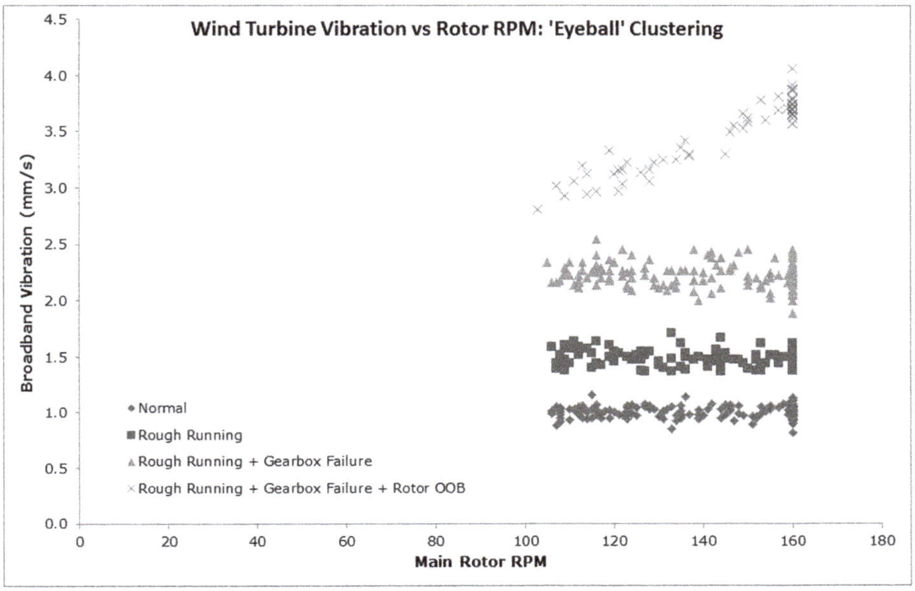

FIGURE B.26 "Eyeball" clustering of known faults in fictitious wind turb ne.

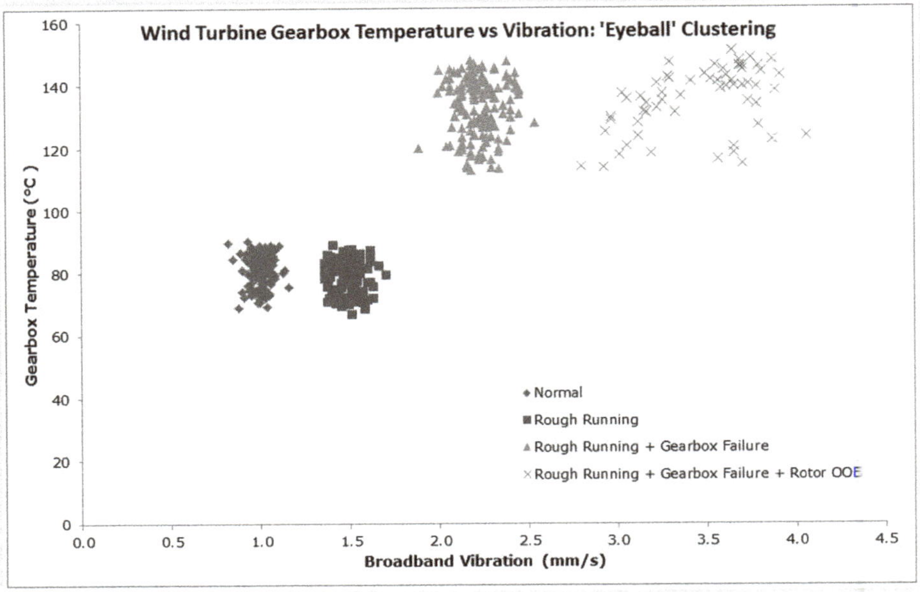

FIGURE B.27 Decision tree for fictitious wind turbine.

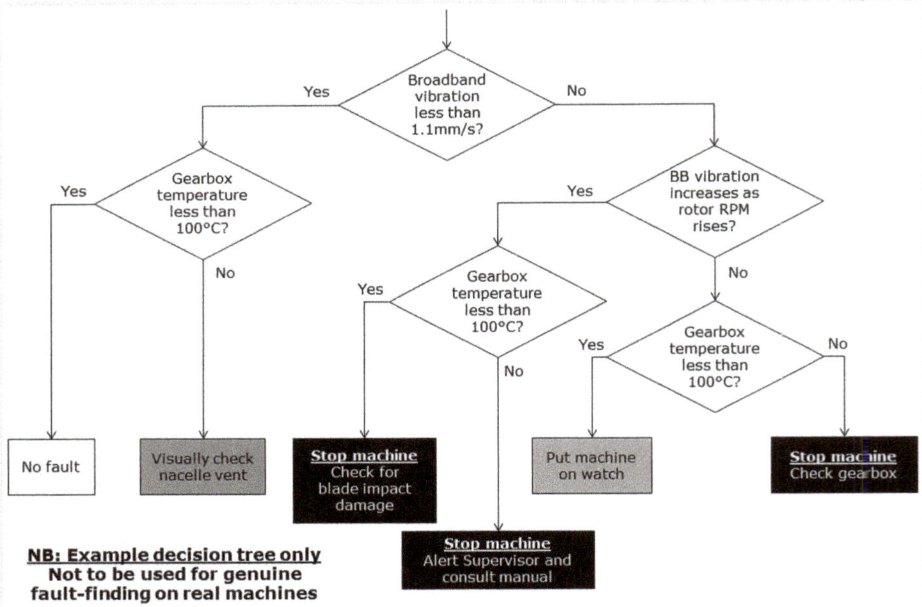

analysis, carried out with known fault data, can be used to create a decision tree such as the one in Figure B.27, which shows the actions that should be taken when broadband vibration and gearbox temperature fall within defined ranges in certain combinations.

However, it can be more useful to work with the same information shown as limits on time series graphs, such as those shown in Figure B.28.

Such displays can provide sufficient warning to allow operations staff to be proactive, rather than reactive, in their responses to issues as they arise.

FIGURE B.28 Gearbox temperature and broadband vibration limits for fictitious wind turbine.

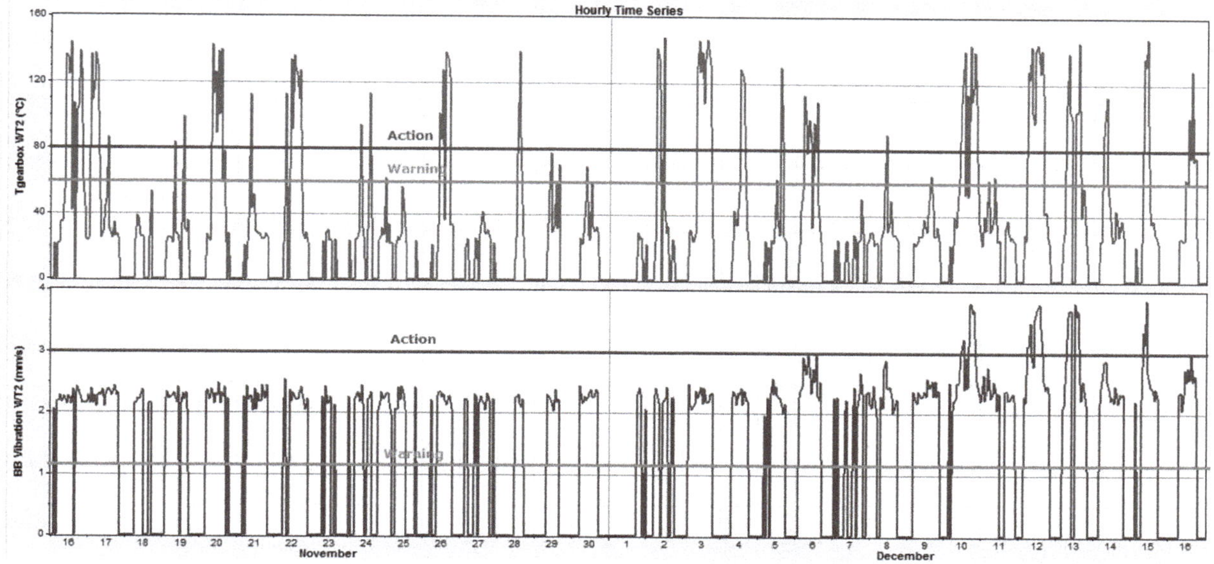

K Nearest Neighbors

One automated method that can be used to classify data snapshots as they arrive is known as K Nearest Neighbors (KNN), which is described in several of the references in Chapter 8. This method is based on determining the "distance" of each element in the new data snapshot from data snapshot elements in historical data that have already been analyzed and classified. "Distance" can be defined in several ways (*Battiti and Brunato 2017*):

- Manhattan distance: The sum of the absolute differences between each element in the new data snapshot and the corresponding element in a historical data snapshot. It gets its name from the analogy of the distance a driver would travel when negotiating a city laid out in a grid pattern.
- Euclidian distance: The square root of the sum of squares of the differences between each element in the new data snapshot and the corresponding element in a historical data snapshot. It is the everyday understanding of distance, based on Pythagoras' theorem.
- Mahalanobis distance: A modification of Euclidian distance, in that each of the differences is divided by a weighting factor, essentially allowing the user to give more importance to some measurements than others.

Having determined the total distance of all the elements in the new data snapshot from each of the corresponding elements in each of the data snapshots in the whole historical database, the new data snapshot is classified according to the median of the cluster of cases it is closest to, with the number of "close" cases K (hence K nearest neighbors…) being specified by the analyst beforehand.

Figure B.29 shows an example of the technique, using fictitious wind turbine data. It is seen that, in this case, the choice of distance measure is not critical. However, this method can be very inefficient if the historic dataset that the new data is compared with is very large.

FIGURE B.29 KNN analysis of fault in fictitious wind turbine (K=5).

Date	End Time	Rotor RPM	BB Vibration	Tgearbox	State	Distance Measures			Mahalanobis Weights		
			mm/s	°C		Manhattan	Euclidian	Mahalanobis	RPM	BB Vib	Tgbox
Test Data		102	0.98	83.3	Pitch Failure				0.5	1.0	2.0
16-Mar	09:00	102	0.97	84.8	Pitch Failure	1.5	1.5	0.8			
18-Mar	15:00	101	0.99	83.2	Pitch Failure	1.1	1.0	2.0			
17-Mar	16:00	101	1.03	84.5	Pitch Failure	2.3	1.6	2.1			
28-Mar	10:00	104	1.06	85.2	Pitch Failure	4.0	2.8	4.1			
21-Mar	06:00	100	0.94	85.4	Pitch Failure	4.1	2.9	4.1			
24-Mar	11:00	100	1.02	88.5	Pitch Failure	7.2	5.6	4.8			
27-Mar	07:00	99	1.03	82.9	Pitch Failure	3.4	3.0	6.0			
15-Apr	13:00	104	1.02	73.6	OK	11.7	9.9	6.3			
14-May	15:00	105	0.99	76.2	OK	10.1	7.7	7.0			
03-Jun	05:00	104	1.01	71.6	OK	13.7	11.9	7.1			
24-Mar	09:00	106	1.05	88.0	Pitch Failure	8.8	6.2	8.3			
18-Mar	14:00	98	1.02	88.5	Pitch Failure	9.2	6.6	8.4			
27-Feb	15:00	106	1.01	74.1	OK	13.2	10.0	9.2			
01-Mar	11:00	106	0.99	73.7	OK	13.6	10.4	9.3			
23-May	11:00	106	0.97	73.7	OK	13.6	10.4	9.3			
03-Aug	15:00	106	0.97	71.3	OK	16.0	12.6	10.0			
19-Mar	12:00	97	1.10	83.1	Pitch Failure	5.3	5.0	10.0			
30-Mar	11:00	107	0.98	86.8	Pitch Failure	8.5	6.1	10.2			

Unsupervised Cluster Analysis

The neural network, decision tree, and KNN demonstrations given above are examples of what is known as supervised learning; historical or training data is available that can be used to either set up the method or compare new data snapshots with directly. Two examples of unsupervised learning are shown in this section: one uses neural networks (which create what are known as Kohonen or Self-Organizing Maps (SOM); Figure B.30) while the other uses K means clustering (Figure B.31), where data is automatically grouped around a user-specified number of "centers of gravity" of the data.

FIGURE B.30 Neural net clustering of known faults in fictitious wind turbine.

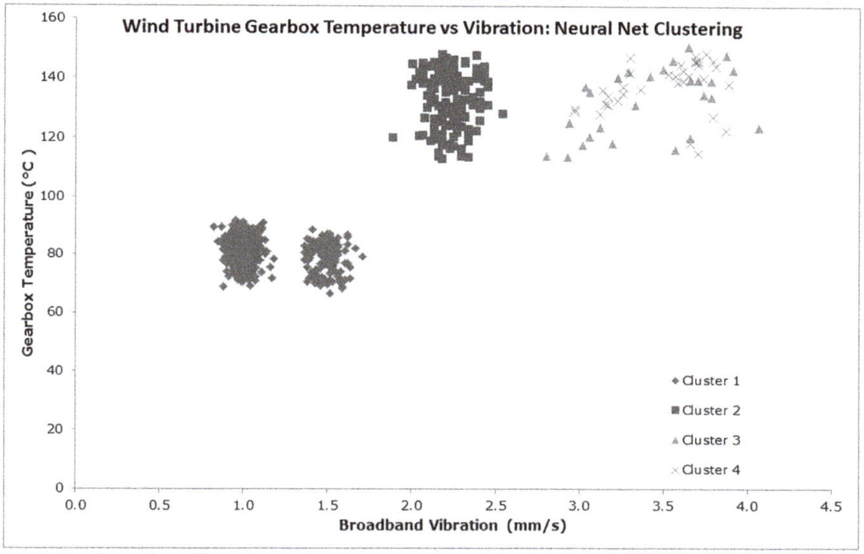

FIGURE B.31 K means clustering of known faults in fictitious wind turbine.

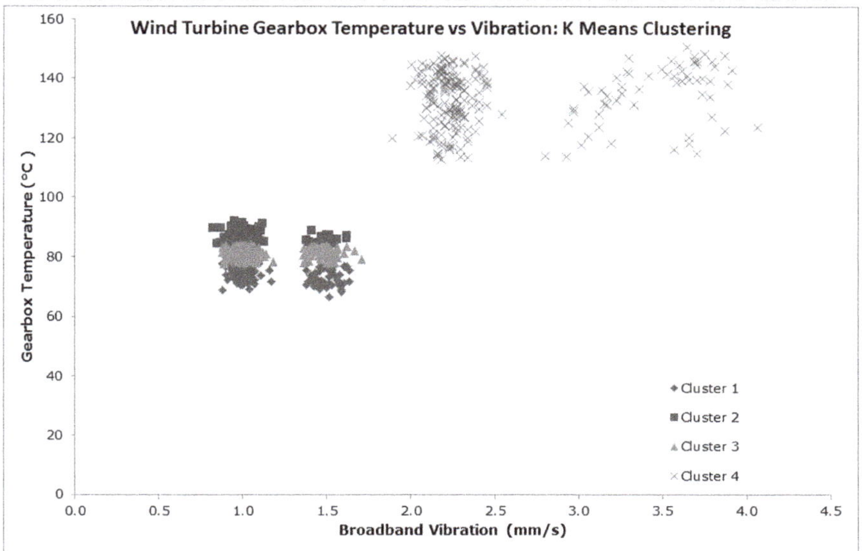

It is seen from inspection when comparing both these figures with Figure B.26 that the results from these methods do not accurately reflect the true distributions of how the faults modelled appear in the plotted parameters. Unsupervised learning methods need to be used with care to avoid drawing misleading conclusions from asset data.

Support Vector Machines

A final method of separating clusters of data, Support Vector Machines (SVM), is briefly discussed here by way of a pictorial example. Figure B.32 shows the principle behind the rather complex mathematics behind this technique in diagrammatic form. The objective is to find the best boundary between two clusters of data, in this case the clusters labelled "rough running" and "rough running + gearbox failure." This boundary is defined as the line that passes through the center of the widest gap that can be found between the two clusters, as shown in stages 1 to 6 of Figure B.32. This is equivalent, in physical terms, to finding how to place a pair of parallel rulers between the clusters in a way that maximizes the distance between them when they only just touch the data points on the boundaries of the clusters. Stage 6 shows where the (rather intimidating…) term for the technique comes from: dropping perpendicular lines from the boundary line to the two nearest points in one cluster and the nearest point in the other cluster produces the "support vectors" that give the method its name.

The equation of the line that passes halfway between the two rulers when placed in this position is shown in Figure B.33. Data snapshots with values of gearbox temperature that are larger than [(142 - 22.889 × broadband vibration(mm/s))°C] would be considered to lie in the "rough running + gearbox failure" cluster, while values below this would be considered to fall in the "rough running" cluster. Clearly this technique becomes more useful when more than two measurements are considered, and the boundary line becomes a multidimensional hyperplane. The reader is referred to the references in Chapter 8 for mathematical details.

A Final Thought

A good rule of thumb, attributed to Walter Shewhart and W. Edwards Deming and quoted in *Goldratt and Cox* (*2013*), is that trying to be more accurate than the noise in a system does not improve things: it actually makes them worse. Readers should bear this in mind when carrying out any sort of modelling, simulation, or analysis of data.

Readers should also remember that even the very best analysis techniques will be ineffective if the end users do not trust, understand, or have confidence in them.

FIGURE B.32 SVM explanation.

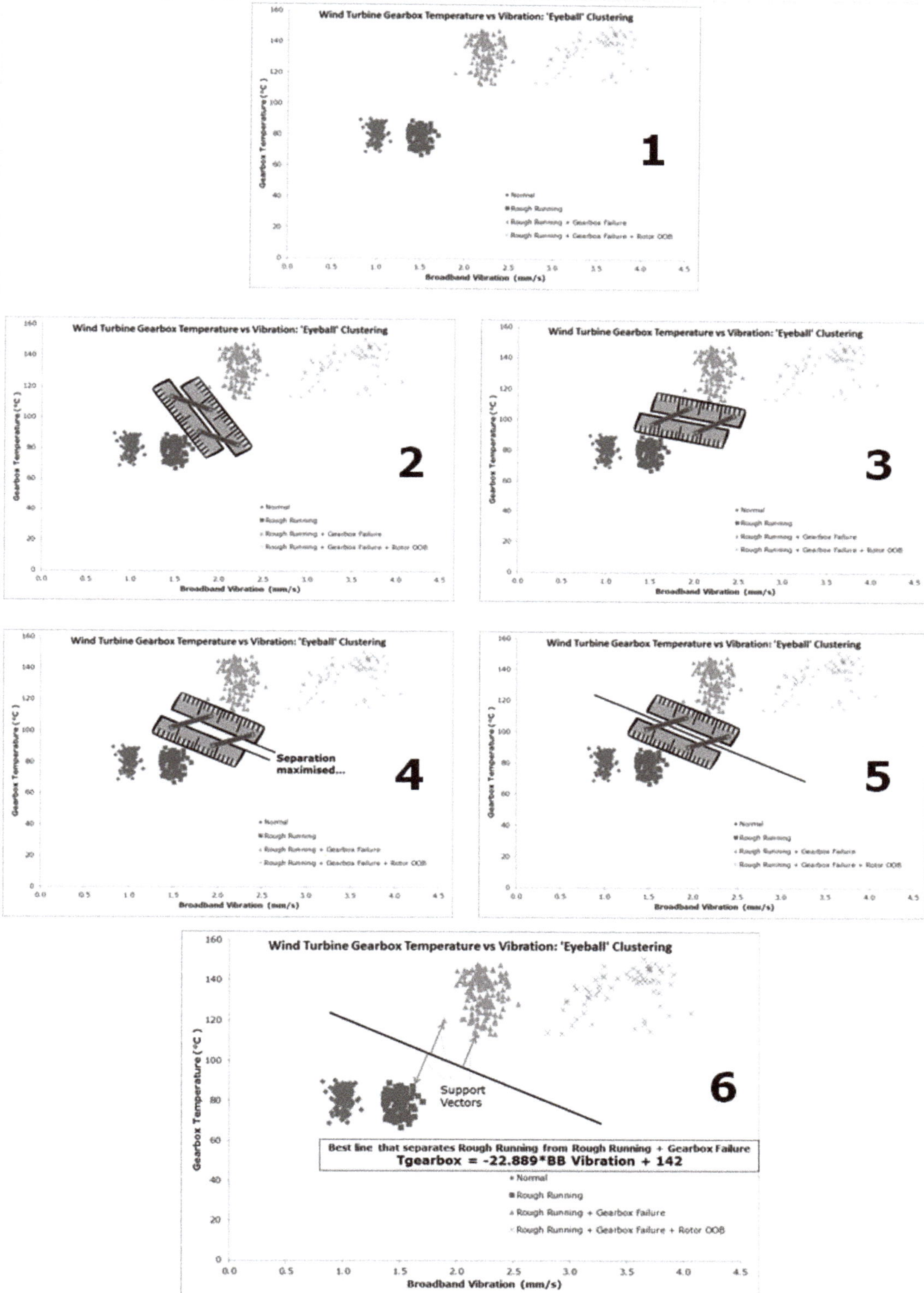

FIGURE B.33 "Eyeball" SVM for fictitious wind turbine.

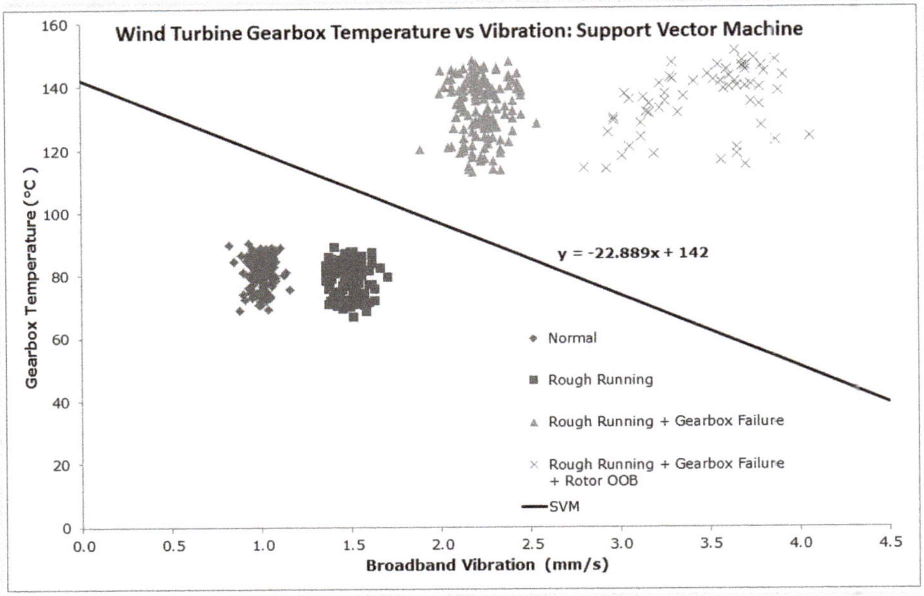

References

* denotes sources that the author found particularly helpful and/or inspiring.

Ampair, *Westwind 20kW Wind Turbine* (Milborne St. Andrew, UK: Ampair Energy Ltd, 2013). Pg. 257

* Battiti, R. and Brunato, M., *The LION Way: Machine Learning Plus Intelligent Optimization – Version 3.0* (Seattle WA: CreateSpace/Amazon.com Inc., 2017). Pg. 277

Gou, B., Na, W. and Diong, B., *Fuel Cells: Modeling, Control and Applications* (Boca Raton, FL: CRC Press/Taylor & Francis, 2010). Pg. 257

Lambert, T., Gilman, P. and Lilienthal, P., "Micropower System Modeling with HOMER," in Farret, F. and Simões, M., (eds.), *Integration of Alternative Sources of Energy* (Hoboken, NJ: Wiley-IEEE Press, 2006), 379-418. Pg. 255

Larminie, J. and Dicks, A., *Fuel Cell Systems Explained*, 2nd ed. (Chichester, UK: John Wiley & Sons Ltd, 2003). Pg. 257

Mathioudakis, K. and Sieverding, C. (eds.), *Gas Turbine Condition Monitoring & Fault Diagnosis* (Rhode Saint Genèse, Belgium: Von Karman Institute for Fluid Dynamics, 2003). Pg. 268

* Mills, S., *Vibration Monitoring and Analysis Handbook* (Northampton, UK: British Institute of Non-Destructive Testing, 2010). Pg. 258

* Provost, M., "The Use of Optimal Estimation Techniques in the Analysis of Gas Turbines," PhD thesis, Department of Propulsion and Power, School of Engineering, Cranfield University, Cranfield, UK, 1994. Pgs. 268, 268

Provost, M. and Nevell, D., "Data Processing System Using a Kalman Filter, US Patent 5,105,372. Filed January 30, 1991, granted April 14, 1992, 1992. Pg. 268

Provost, M. and Nevell, D., "Performance Data Processing System," European Patent 0 315 307 B1, Filed May 10, 1989, granted June 22, 1994, 1994. Pg. 268

Reeves, C., "*Vibration Monitoring Handbook* (Oxford, UK: Coxmoor, 1998). Pg. 258

Renewable Energy Development Division, *The Cook Islands Renewable Electricity Chart* (Avarua, Rarotonga, Cook Islands: Renewable Energy Development Division, Office of the Prime Minister, Government of the Cook Islands, 2011). Pg. 255

Westwind Wind Turbines, *Westwind 20kW Wind Turbine* (Crumlin, UK: JA Graham Renewable Energy Services, 2014). Pg. 258

Websites

HOMER Energy Pg. 255	www.homerenergy.com
Fuel Cell Stack Calculations, Michigan Technological University Pg. 258	www.chem.mtu.edu/~jmkeith/fuelcellcalculator/h2fuelcellpvcalculator.swf
Ampair Energy Ltd. Pg. 258	www.ampair.com
Westwind Wind Turbines Pg. 258	www.westwindturbines.co.uk
US National Renewable Energy Laboratory Pg. 264	www.nrel.gov

looking after planes, trains, clean energy, and human health: 23 important lessons learned the hard way...

This chapter was originally produced for the Society of Automotive Engineers (SAE, now SAE International) (*Jennions 2014*): it has also been presented as a paper at the Institution of Engineering and Technology/Institute of Asset Management (IET/IAM) 2014 Asset Management Conference in London, UK (*Provost 2014*). It is reproduced with permission from both SAE International and the IET.

1. **It's what an asset does, not what it is, that is important.**
   Train, bus, aircraft, and truck operators really want the ability to carry passengers and freight, not planes or vehicles. Utilities really want clean and efficient electricity generation, not turbines, boilers, etc. Manufacturers of electronic equipment really want components placed on a circuit board, not the machinery that does this. Bottling plants really want filled bottles, not bottle-filling machinery. Airlines really want thrust, not jet engines. The literature is full of examples of the seismic shift from customers wanting the asset itself (and negotiating the price down as far as possible to reduce capital expenditure) to them wanting the capability the asset provides (and being more willing to pay a predictable, de-risked operational cost that can more easily be passed on to their end customers). Providing an asset management capability and ensuring that this is delivered to your customers consistently and efficiently over the lifetime of the assets you make is a very lucrative, stable, and long-lasting business proposition which you need to fully embrace, before your competitors and other third parties do so.
2. **Asset management is a business issue.**
   It is not just another IT problem; it impacts the whole structure of your business and the way you think about and relate to your customers. Every aspect of a business, including strategy, development, management, partnerships, mergers and acquisitions, projects, marketing, sales, engineering, information technology (IT), communications, resource planning, configuration management, logistics, training, field support, and customer billing, is profoundly affected.
3. **Without senior management support, asset management goes nowhere.**
   The changes to your business that asset management demands are so huge that, without the full support of management at the very highest levels, it will never get "off the ground," never mind succeed. The vested interests, new and expanded mind-sets and understandings, changes in existing organizational power

structures, and general corporate inertia can only be overturned by committed leadership from the very top of your organization.

4. **Know your assets.**
A deep understanding of your assets; the business contexts and ambient environments they operate in; their failure modes, frequencies, and consequences; and how they need to be looked after is much more important than knowledge of IT, databases, and analysis methods. Your business has knowledge gained during the design, development, and production phases of the assets you make and sell that can be used to generate value for the business time and time again over the whole life of those assets, rather than being used once only during manufacture.

5. **Asset management needs to be incorporated at the start, not brought in later as an "add-on gimmick."**
Don't let the need for low first cost destroy the greater need to create lifetime value. There is a conflict to be resolved between the production demands for low first cost and high product and spares sales volumes and the asset management needs for data gathering hardware and more finely-tuned product and spares offerings that may be the best solution for your customers. Many production-focused organizations see asset management as a threat, both to sales of new products and to sales of spares; many asset management initiatives have been deliberately strangled at birth because of the perceived loss of spares revenues (or even new products) that could result. Mechanisms need to be found for the long-term rewards of asset management to be fed back to the production arm of the business, which, after all, provides the "entry ticket" for all future asset management value streams. Retrofitting sensors and remodelling organizations are very expensive and time-consuming processes.

6. **Break down the silos, open your mind, and look for what your customers really need to help them make money and serve their markets better.**
The benefits of asset management may not be where you first thought. Bringing all the measurements and analysis together into one "single source of truth" can produce many savings and unexpected value-creating synergies. Once the measurements are in one place, it becomes much easier for you to weed out unnecessary duplications, see the whole picture across both individual assets and asset fleets, and carry out analyses that go beyond the initial aims of monitoring asset health and operation and move toward answering the real questions that your customers may be asking about their businesses.

7. **Share to gain.**
Partnerships between all interested parties create the "win-win" situations that ensure success. Customers and end users have deep knowledge of asset performance in their own operations, but lack information about the same assets used by others elsewhere. Asset owners have a vested interest in "cradle to grave" asset performance and maintenance, but usually lack asset technical and operational skills. Subsystem suppliers have deep knowledge about the performance of components that they produce, but often fail to see their operational contexts. IT and communications vendors keep abreast of the latest developments in their fields of expertise, but lack asset technical and operational experience. Asset manufacturers and systems integrators have access to the asset design, modelling, and analysis tools and data, as well as the ability to bring all the above parties together; however, they will only succeed in offering asset management services to their customers if they offer the means for everyone involved to collaborate for mutual benefit.

8. **Never underestimate the persuading you will have to do, at all levels in your organization, or the power of the "heroes" who feel threatened by strange new ways of doing things.**
   No one ever got fired by doing what worked yesterday, whereas plenty of people have lost their jobs by deviating from the status quo. Many people, from upper management to those at the "sharp end" of your business, will feel that their jobs are being put at risk if their unique "head knowledge" and ways of working (which will have made them key organizational players and offered lucrative opportunities for enhancing their power and earnings) are made more generally accessible to others by asset management initiatives. They will fight the changes brought about by asset management with all the energies and internal politics that they can muster. If the person who needs to act on asset management information isn't convinced by it, you have wasted everybody's time and money.
9. **Beware of people who just ask for "data" and/or confuse data with information.**
   Many people don't know what they want to do with data. Help them to articulate what they really need, by talking about identifiable measurements and focused information requirements, not "data"; this approach rapidly improves the chances of success. People can rarely explain to you what they need on a "blank sheet," but are much more willing and able to provide constructive feedback when you show them examples of what can be done.
10. **Asset management is a chain, from sensor to business action.**
   The process of data gathering from sensors fitted to assets, data gathering and transmittal to a central location, data storage, data visualization, data analysis, and problem diagnosis, followed by assembling together the right resources (information, tools, spares, and qualified people) at the right time and in the right place to take the required actions to keep your customers happy is a complex and fragile chain. If any link in this chain breaks, no matter how trivial, the whole process collapses.
11. **Don't put the data cart before the business horse.**
   Many asset management initiatives start with specifying the data to be recorded on assets, without thinking about what is to be done with it. This usually results in the wrong data being recorded and/or transmitted at the wrong times and frequencies, and opportunities for gathering what data is really needed are lost. Alternatively, the request is to "measure everything, all the time," in the belief that transmission bandwidth is infinite, IT is zero-cost and the question of what is to be done with the data will be resolved eventually. This usually results in unmanageable data volumes in which the real signals are lost in a quagmire of digits in a database. Business needs drive analysis requirements, which in turn drive data gathering (both sensors and frequency of data gathering and transmittal). You can measure anything; you could measure everything; you should only measure what creates value.
12. **Keep it simple.**
   The asset management literature is full of analysis methods that are poorly explained, steeped in obscure mathematics, lack clarity or obvious engineering relevance, and seem to be aimed more at demonstrating the cleverness and academic credentials of the author(s) than enlightening humanity. The end result of any analysis has to be action, usually taken by someone with practical, rather than academic, intelligence. Techniques, such as applying thermal paint to a component that changes color when that component overheats, are considerably cheaper and more quickly and easily understood by those in the workshop (who may not have

much time to understand and fix problems) than more complex data gathering, transmittal, and remote analysis. The smartest analysis or visualization in the world is useless if nobody else understands and trusts it enough to act on it.

13. **A physics-based asset model is a very powerful business and technical tool.**
It builds the foundations for full understanding of asset and business dynamics. Such a model (or set of models) improves communication within and between all interested parties both inside and outside your business and provides consistent and traceable predictions and baselines of asset, project, and business performance. It also provides you with rapid assessments of how assets should behave in different environments and operational contexts and forms the basis of fast, consistent, and accurate assessment of asset performance in the field. Physics-based asset models can support the application of many advanced analysis techniques that would not otherwise be practical and enable the optimization of asset technical and business performance, profoundly improving the cost, speed, efficiency, and effectiveness of asset development and in-service support.

14. **If at all possible, compare all your asset measurements to a baseline, which ideally takes account of all known external drivers of the recorded values (e.g., load variation, ambient condition changes, and other quantified effects).**
There is always a baseline somewhere (from a number in someone's head to a full physics-based asset model) against which measurements can be compared; find it and make it visible, so everybody can easily see what is good and what is bad. Residuals (the differences between measurements and baselines) are much easier to understand and analyze than raw measurements and provide order-of-magnitude improvements in the "granularity" of your analyses that significantly increase the timeliness and effectiveness of asset health and operational assessments.

15. **A good measurement and/or analysis visualization, tailored to the person you are talking to, will make all the difference.**
Some visualizations (such as time series, X-Y plots, bar and column charts, dashboards, alerts, and interactive drill-down) will always be useful, while others (like mapping, statistical displays, reports, and system synoptics) may find more specialized niches in the organization or with customers. Use visualization to persuade and excite; people rarely know what they want to see and how they want to interact with the data but will provide enthusiastic feedback when you can show them examples of what can be done.

16. **The appropriateness of the analysis is more important than the "bigness" of the data.**
"Big data" is all the rage, with many commentators and IT consultants seeing the advent of massive unstructured databases, off-the-shelf analytics, and cheap "cloud" storage and processing as panaceas for most asset management issues. While such approaches can work well in the "softer" areas of retail, social science, and financial asset management, there are more appropriate tools and thought processes that you can and should use for analysis of the performance and operation of physical assets. The use of "smart analytics" to back-calculate what a good physics-based model of the asset could have told you gives a false sense of progress and potentially confuses failure signals with the noise of operational variation. "Black box analytics" also make it too easy to "overfit" data (producing an analysis that is not valid for new data when it arrives) and/or find spurious patterns or correlations in large datasets that don't make logical sense. Asset management must be based on sound technical and business logic; subcontracting the thinking to the latest IT hype can quickly lead you to expensive failure and loss of credibility.

17. **Cost is not value: keep reminding the cost-cutters of this.**
Many people confuse the cost of a component with the impact it has on your customers' operations, with the result that many items that are critical end up being ignored purely because they are technically undemanding, cheap, or generally "boring." The $1 component that generates costs of $1,000,000 when it goes wrong is worth monitoring and taking care of. Many of the best asset management programs owe their success to looking after the "boring but important" items in their asset inventories very well.
18. **An asset measurement without a timestamp (preferably GMT/UTC, which avoids time zone and daylight-saving time issues), unique asset identifier, and some measure of operating stress and environment is a random number from which useful information can only rarely be retrieved.**
A sensor reading from an asset means nothing if you can't place it in context or relate it to other readings from the same or related assets.
19. **Inadequate asset configuration knowledge and/or asset configuration control makes meaningful asset management impossible.**
There are significant differences between the "as designed," "as built," and "as maintained" state of all your customers' individual assets once they have been deployed in the field; these differences are critical and can be the source of many asset management failures, from not understanding data signals to delivering wrong spare parts to the maintainer in the field. Watch out for undocumented "temporary" fixes and modifications to assets and working practices that solve short-term issues but cause damage and play havoc later.
20. **Some people just don't "get it."**
Either reeducate them or remove them. Asset management demands such a huge change of organizational mind-set that it is inevitable that many people at all levels in your organization either can't or won't see what it's all about. At best, the unbelievers will sit on the sidelines and hope you will go away; at worst, they will actively sabotage the necessary business process reengineering. If necessary, spin off asset management into a separate organization to allow it to develop and grow and free it from malign influences.
21. **Don't assume anything.**
It's easy, given the massive complexities of asset management, for you to assume that the data, people, processes, and tools you will need are (or have been) thought about by others and will be made available for you to use. If you don't ask, you don't get.
22. **Know the limits of what you know and learn to appreciate the contributions everyone at all levels can make to the whole asset management process.**
No one person has all the answers and asset management insights can and do come from anywhere, both inside and outside the business. There will be many twists and turns in your asset management journey and changes in emphasis as you learn what is really important and what really will generate value. Data is not information. Information is not knowledge. Knowledge is not wisdom. Listen to anyone and everyone. Humility is a virtue; it opens you up to the knowledge and experience the people you have to work with can bring to the asset management enterprise.
23. **Push, but be patient.**
Success breeds more success, interest and enthusiasm will grow, the pace will quicken, and recognition and rewards will flow eventually (sometimes from the most unexpected directions...). It can be like a game of Snakes and Ladders; there are many ups and downs on the road to success.

Examples and Stories

Over the last forty years, the author has come across many examples and stories where success has depended on application of the lessons detailed above. They include:

In the 1970s, the chairman of a well-respected European airline, when hearing that an experimental aircraft engine condition monitoring program could have prevented a turnback of a wide-body airliner if its output had been heeded, demanded that it be put into fleet-wide use immediately. He didn't require formal justification; he knew that his airline's technical and financial performance and reputation would benefit if this was done.

Since the 1980s, many airlines have used engine monitoring to optimally dispatch aircraft, sending those with "hot" engines to cooler destinations and vice versa. This strategy extends engine on-wing lives and results in fewer engine over-temperature events, avoiding service disruptions.

Data collection doesn't have to be expensive and complex. In the 1980s, one major European airline equipped all their check-in desks worldwide with optical character readers, so that passenger service staff could feed engine and aircraft data to their main engineering base from cockpit printouts when they were not serving paying customers. A worldwide data gathering network, riding on the back of the ticketing system, was created for a few tens of thousands of dollars.

Another major European airline has amassed so much data on the performance of the aircraft, engines, and other subsystems that they operate that suppliers regularly use this "treasure trove" to initiate design changes to in-service aircraft. In one case, the hydraulic system of a wide-body aircraft was completely redesigned based on data from one takeoff during which an uncommanded pitch down was recorded.

In the 1980s, one somewhat skeptical power station manager in the United Kingdom shut down a large steam turbine on the basis of the output from an experimental vibration monitoring system. When the turbine was opened up and inspected, a crack was found in the main shaft that would have resulted in catastrophic failure and potential fatalities had the turbine run for another thirty minutes. He was convinced!

The aircraft gas turbine industry depends heavily on physical models, which have reached such a degree of accuracy and sophistication that they form the basis of operational and maintenance forecasts that can be produced for each customer covering the whole life of an engine fleet. Thanks to these models, engine development programs are now used to validate the engineering understanding the models have already produced, rather than generating that understanding "from scratch," resulting in huge savings of time and money. The models also create foundations for a great many sophisticated analytical approaches to condition monitoring.

One major gas turbine manufacturer found it necessary to create a separate company to develop condition monitoring and other aftermarket service capabilities in order to prevent the prevalent "manufacturing mind-set" killing off the ideas being developed before they had a chance to prove themselves.

There are many examples in the railway industry of sensors being fitted for one purpose generating more value when being used for something else. For example, air suspension pressures are used to produce estimates of passenger count, while electrical faults and wheel slip protection system activations observed across a fleet are mapped to indicate areas of the rail network that require maintenance action and data recorded for potential incident and accident investigations is used to find the causes of service delays and attribute penalty payments appropriately.

One major UK rail operator has eliminated the need for passenger door fault-finding activities at their engineering depots by relying entirely on the data such as opening and closing times and door actuator motor currents from millions of door operation cycles gathered from the in-service train fleet to accurately predict and schedule any necessary door maintenance activities.

Another UK train operator transmits a "mimic" of each driver's control panel to a central control room in real time, enabling support staff to give timely advice to drivers and other train crew.

One major UK truck manufacturer discovered that the sensors used to monitor diesel engines can be used to monitor driver behavior. They now offer a service that uses this data to progressively improve driving styles, producing significant reductions in trip delays, accidents, insurance premiums, and fuel consumption. The customers and drivers share the benefits, producing the necessary positive feedback to ensure success.

One major Formula 1 team uses condition monitoring data to model the performance of each and every car in a race in real time, using these models to predict race outcomes and run "what-if" analyses to optimize their refueling, choice of tires, and pit stops.

Many van and truck businesses now use real-time GPS and other vehicle data to track fleet performance, thus reducing costs and improving customer service. At least two major car manufacturers are extending this philosophy into the consumer arena, to offer comprehensive real-time advice and support to private motorists. This is felt to be particularly useful for battery-powered private vehicles, to overcome "range anxiety" and instill confidence in new technologies.

Almost all road vehicles are now fitted with comprehensive on-board diagnostics, reducing maintenance times and costs. Even owners with the right smartphone "app" can now access detailed real-time information on the performance of their vehicles.

Remote diagnostics are crucial for both maintenance and operational planning of a wide variety of critical, high-value, or difficult-to-reach plant, including petrochemical and other process plant; water, gas, and electricity networks; wind turbines (particularly those placed offshore); power stations; backup power units for mobile phone masts; etc.

Human health monitoring is becoming increasingly important, as populations grow older and healthcare resources are stretched. Very simple tools and visualizations can achieve huge improvements in personal health and well-being and possibly save lives without needing huge investments in new technologies.

These are just a small selection of the many success stories that are emerging as companies in many industrial sectors begin to appreciate the power of knowledge about how their assets they make and use are behaving in service and how this knowledge can be used to improve their businesses and satisfy their customers.

Conclusions

Figure L.1 summarizes graphically one of the most important points about achieving success with servitization and asset management: it is vital to escape from "financial flatland" by giving as much "weight" to the unquantified "intangible" returns, such as enhanced company reputation, improved customer satisfaction, better competitive

FIGURE L.1 Achieving success with asset management (source *Provost 2015*).

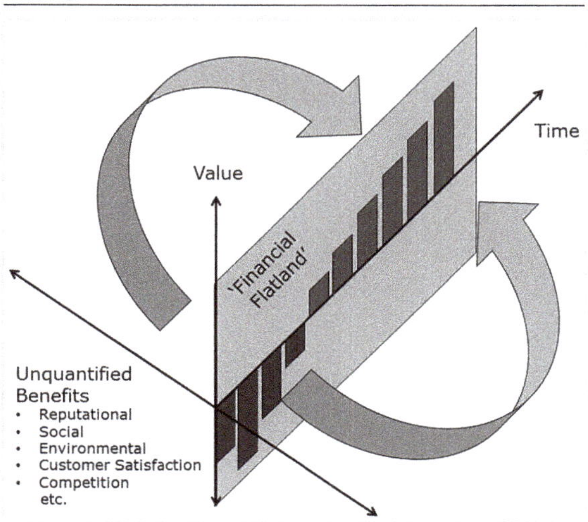

position, enhanced social and environmental performance, etc., as to the "tangible" economic impacts, which can mislead and impede progress if they are allowed to dominate the thinking of an organization. This point is well made by Armen Papazian in his papers about the economics of space exploration (*Papazian 2012, 2014*).

This chapter summarizes what the author has learned about servitization and asset management during the last forty years of his professional life. Crucially, asset management has to be business focused, because a business is affected by what an asset does, not what it is. It's also very beneficial to take a broad view, since the benefits of asset management may not lie where you first think; building on this, data integration gives synergies that create unexpected value and delight customers. Asset knowledge is critical; physics-based models build the foundations for full understanding of asset and business dynamics. Things must be kept simple and visible, if you want your efforts to be accepted and acted on. It must always be remembered that cost is not value; the cost-cutters need reminding of this frequently. An asset management business cannot be created without developing the three key ingredients: people, processes, and tools. Human factors must never be underestimated, because they will dominate your efforts. Finally, the author has found that perseverance always pays off, both personally and professionally.

References

* denotes source that the author found particularly helpful and/or inspiring.

* Jennions, I. Eds., *Integrated Vehicle Health Management: Implementation and Lessons Learned* (Warrendale, PA: SAE International, 2014). [Pg. 285]

Papazian, A., *Space Exploration and Money Mechanics: An Evolutionary Challenge* (Exeter, UK: Keipr Ltd., 2012). [Pg. 292]

Papazian, A., *Space Value of Money: The Spatial Dynamics of Monetary Value* (Exeter, UK: Keipr Ltd., 2014). [Pg. 292]

Provost, M., "Looking after Planes, Trains, Clean Energy and Human Health: 23 Lessons I've Learned the Hard Way," *IET/IAM 2014 Asset Management Conference*, Millennium Gloucester Hotel, London, UK, November 27-28, 2014 (Stevenage, UK: Institution of Engineering and Technology). [Pg. 285]

Provost, M., "Servitization? Asset Management? Why Should I Bother?," *Spring Servitization Conference 2015*, Aston Business School, Birmingham, UK, May 18-19, 2015 (Birmingham, UK: Aston Business School, Aston University. [Pg. 291]

about the author

Dr. Michael Provost has recently set up Michael Provost Consulting Ltd. to both help and inspire businesses to make the servitization journey and to teach others what Mike has learned during the nearly four decades involved in the modeling, simulation, analysis, condition monitoring, and management of physical assets and contributing to the many business transformations that servitization involves. He spent 27 years at Rolls-Royce plc working on the modeling and analysis of civil aeroengine performance, aeroengine condition monitoring (where he won the Chairman's Award for Technical Innovation in 2000 for his analytical contributions to Rolls-Royce's COMPASS™ Condition Monitoring system), corporate strategic planning, and advanced civil aeroengine and more electric aircraft systems design. He then spent two years at Data Systems & Solutions Ltd., a Rolls-Royce plc spin-off company, working on applications of aerospace asset management techniques in areas such as railways, before spending five years at Bombardier Transportation, where he led the team developing the engineering analysis and visualization methods used in Bombardier ORBITA™, a world-leading railway asset management system. He then spent five years at Intelligent Energy Ltd., a company focused on the development and commercialization of fuel cell systems, where he was responsible for guiding the development and roll-out of a range of asset management techniques across that company's automotive, consumer electronics and distributed power product lines. This book summarizes his experiences of asset management, monitoring, data visualization and analysis for planes, trains, and clean energy. Mike applies the principles he has developed for monitoring planes, trains, and clean energy assets to the management of his own health, with very beneficial results.

Mike graduated from Trinity College, Cambridge, in 1976 with an honors degree in Engineering, and gained a PhD in Thermal Power from Cranfield University in 1994. He is a Fellow of both the Institution of Mechanical Engineers and the Institution of Engineering and Technology and is also a Member of the Institute of Asset Management. Mike lives in Nottingham, England, and can be contacted on mike@michaelprovostconsulting.com or +44 (0)7811 944990. He can also be contacted via LinkedIn®.

index

www.ingramcontent.com/pod-product-compliance
Lightning Source LLC
LaVergne TN
LVHW070849200926
848482LV00016B/897